Powering Empire

Powering Empire

*How Coal Made the Middle East and
Sparked Global Carbonization*

On Barak

UNIVERSITY OF CALIFORNIA PRESS

University of California Press
Oakland, California

© 2020 by On Barak

Library of Congress Cataloging-in-Publication Data

Names: Barak, On, author.
Title: Powering empire : how coal made the Middle East
    and sparked global carbonization / On Barak.
Description: Oakland, California : University of
    California Press, [2020] | Includes bibliographical
    references and index.
Identifiers: LCCN 2019024337 | ISBN 9780520310728
    (cloth) | ISBN 9780520339675 (paperback) |
    ISBN 9780520973930 (ebook)
Subjects: LCSH: Coal trade—Middle East—History. |
    Coal trade—Political aspects.
Classification: LCC HD9558.M54 B38 2020 |
    DDC 382/.424—dc23
LC record available at https://lccn.loc.gov/2019024337

Manufactured in the United States of America

29 28 27 26 25 24 23 22 21 20
10 9 8 7 6 5 4 3 2 1

To my parents, Dvora and Moshe Barak

Contents

List of Illustrations

Note on Transliteration

In the interest of simplifying the reading, I have used a modified version of the transliteration system of the *International Journal of Middle East Studies,* excluding diacritical marks except for the ʿayn (ʿ) and the hamza (ʾ). For names and terms that have a common transliteration in English (e.g., Mecca), I have used that spelling.

Acknowledgments

Like former adherents of Islamic thought, the Ottomans were committed to the idea of *tabaqat*—the notion that an analogy could be made between cohorts of religious scholars and the layers of the Earth. This book tries to salvage such alternatives to the nature/culture divide and redeploy them as resources for addressing a present in which human activity is again conceived in geological terms, albeit in a less discerning manner. In this vein, I loosely employ such pious stratigraphy in what follows to simultaneously parcel up and parcel out collective authorship as an organizing principle of the multilayered gratitude I feel as I send this book to press.

I formally started the project at Princeton University in 2010, though its foundations were laid earlier at NYU, during the work on my previous project. Thus, I am indebted to teachers whose mentorship extends beyond the demarcation of diplomas. I thank Khaled Fahmy, Zachary Lockman, Timothy Mitchell, Michael Gilsenan, and Arvind Rajagopal for establishing this substratum of solid bedrock.

In Princeton during the early stages, this project benefited from the abundant intellectual and material resources available at the Society of Fellows and the History Department, as well as from a grant from the Committee on Research in the Humanities & Social Sciences for preliminary research, and from the treasures buried in the Firestone Library. The shiniest gems I pocketed there came from Lucia Allais, Eduardo Canedo, Mischa Gabowitsch, Dotan Leshem, and Daniel Stolz. Their help in times of need, incisive critiques of drafts of my manuscript over almost a decade, and their ability to distract me from it when I was in danger of becoming a one-trick pit pony have left me with precious deposits of our friendships long after leaving the place in which they were forged.

In my current academic home, at Tel Aviv University, with coconspirators Avner Wishnitzer and Liat Kozma (nearby, at the Hebrew University), with whom I have shared years of camaraderie and schemes of keyboard activism, I feel part of a troika, a small *tabaqa,* for the first time. Together we strive to consolidate this in real time within Israeli academia which is retroactively organized in generations. Avner and Liat were also the canaries that accompanied me into the pits of this book, sounding the alarm whenever the shaft threatened collapse. Avner miraculously might have read more drafts than I have written. In contrast to these canaries, Michal Shapira took a different bird's-eye view which helped me situate this project within a broader perspective; her friendship held me steady and kept me from losing my way.

The book proved to be the fossil fuel for several other academic careers. Between 2013 and 2018, grants from the Marie Skłodowska-Curie program and the Israel Science Foundation permitted me to pursue sources across Europe, the United States, India, Turkey, Egypt, and Israel. This generous funding also allowed me to work with young research assistants, graduate students, and advisees, thus helping to shape the next generation of scholars infected with the black lung of historical mining. I thank Nimrod Ben Zeev, Alaaddin Tok, Shira Pinhas, Itamar Toussia Cohen, Ismail Keskin, and Ran Levy, as well as Roy Bar Sadeh and Dotan Halevy, for their efforts and enthusiasm and for their increasing sophistication which kept challenging me to avoid ossification. During this time I published two articles that laid foundations for the book manuscript: "Three Watersheds in the History of Energy," *Comparative Studies of South Asia, Africa and the Middle East* (2014), and "Outsourcing: Energy and Empire in the Age of Coal, 1820–1911," *International Journal of Middle East Studies* (2015).

I finalized the manuscript during a sabbatical year at re:work, Berlin, a global labor history institute for advanced studies. Beyond ample intellectual stimulation and material support, the cohort that Andreas Eckert, Jürgen Kocka, and Felicitas Hentschke assembled there was a perfect example of group spirit and solidarity. Included in this were Alina-Sandra Cucu, Görkem Akgöz, David Mayer, Benedetta Rossi, Daniel Tödt, Hannah Ahlheim, Oisin Gilmore, Michael Hoffmann, Juliane Schiel,

Judy Fudge, and Deborah James. Ulrike Freitag, who read and helped improve several chapter drafts, deserves special thanks, as does Bridget Kenny, who helped me get the measure of Berlin on foot. Re:work was another intersection with the insight and aid generously offered by Gadi Algazy. Other Berliners—permanent or transitory—enriched this project in various ways. Sebastian Conrad invited me to present it and then read the entire manuscript; Yoav Di-Capua helped prepare a cooked version of the project with dishes ranging from Aden Yemeni soup to Port Said Koshari, washed down with India Pale Ale and Mocha coffee; Christopher Otter commented on chapters; Mischa Gabowitsch regularly came to my aid; Sarah Rifky discussed the film version with me; and Alia Mossallam, Nazan Maksudyan, Valeska Huber, Joseph Ben Prestel, and Sonja Hegasy provided good company and much wisdom.

Not for the last time, I will highlight what geology cannot capture, like the unfailing mentorship of Israel Gershoni, or my debt to the organizers of the various series, panels, and conferences where I tried out versions of this project. The latter include: Jennifer Derr, Fredrik Meiton, Brian Edwards, Matúš Mišík and Nada Kujundžić, Israel Gershoni and Yoram Meital, Anupama Rau, Roy Bar Sadeh and Lotte Houwink ten Cate, Chris Manjapra, Alina-Sandra Cucu, Sebastian Conrad, Charles-François Mathis and Geneviève Massard-Guilbaud, Shira Pinhas, Sahar Bazzaz, Anat Leibler, Toby C. Jones and Gabrielle Hecht, and Michael Gordin. Now imagine to how many participants I am indebted.

Niels Hooper, my publisher at the University of California Press, accompanied this project and from its inception contributed his intuition and timely interventions. Robin Manley smoothly steered the book on the road to production. My thanks are also due to friends and colleagues who shared materials or helped with the translation process: Uri Ben Ari, Selim Karlitekin, Pauline Lewis, Omri Eilat, Michael Christopher Low, Lucia Carminati, and particularly to Nimrod Ben Zeev and Daniel Stolz who accompanied the project through all its stages and put the rigidity of stratigraphy to shame.

Which brings me to family. My parents, Dvora and Moshe Barak, whose practical-poetical, responsible-mischievous outlook I find myself reproducing in everything I write; my sisters, Ayelet and Chen Barak, whose wit and unlimited support no family WhatsApp can capture or dampen; aunt Lesley Marks, who polished my English and focused my arguments, uncle Avishai Ehrlich, who fine-tuned my Marxism, and brother-in-law Gilad Seliktar, who threw my images into sharper relief.

My daughter, Tamuz Barak, alchemist, transformer of academic conferences into jazz concerts, expressed a healthy disinterest in bare coal,
forcing me to up the ante with stories of sea monsters and shipwreck
which, in the parental model of Tolkien or Rowling, were eventually
textualized here. Finally, my partner-in-crime, Galit Seliktar Barak,
minimized the environmental degradation that writing a book creates
beyond pulped trees or carbon footprint. Refusing to accept my difficulty with multitasking or the moodiness I tried to excuse as authorial,
her generosity, creativity, and optimism keep me human in more ways
than I can admit.

Introduction

Fossil fuels and Western imperialism are widely recognized as key elements that shaped the modern world. Today, they are also acknowledged as major forces that threaten future human existence. However, our historical understanding of these powers, and especially of their complex relationship to one another, is still vastly misinformed. For example, the Middle East, which is now mostly associated with oil extraction and American power, was in its history turned into a coherent region by British coal and imperial interventionism. This legacy provides an opportunity for a reappraisal of the entanglements of energy and empire, of classical- and neoimperialism, and of coal and oil. Unsettling the familiar geographies of extraction and combustion, coal's peculiar Middle Eastern career exposes both these processes and the connections between them to inquiry. In short, it could help us understand the complex process by which the hydrocarbon economy was created and globalized.

In relative terms, only a small portion of the coal mined in the British Isles was exported outside of Europe, and only a small amount of that was shipped to the Middle East. Yet the relative perspective of statistics is misleading, as it obscures the fact that this was enough to fuel a revolution of steamboat imperialism and eventually bring coal mining to life in the Ottoman Empire, as well as in India, China, and elsewhere. Depositing "black diamonds" in Ottoman territories en route to the British Indian Crown Jewel allowed the global fossil fuels economy to pick up steam and take shape during the long nineteenth century. The discovery of liquid "black gold" in this territory at the beginning of the twentieth century is one of the legacies of this passage. Indeed, under the mark of British fossil-fueled imperialism in these settings, coal (both imported and local) was transformed from a useless "black stone" into a valuable "treasure," rendering it into a resource that could be managed by Islamic and capitalist ethics, which sometimes competed and sometimes complemented one another. Thus, beside its familiar role in fueling industrialization in western Europe, the coal transported from Europe to today's oil-producing regions set in motion crucial yet overlooked circulations and calculations, which connected the world with durable carbon fibers. Essentially, these are the historical global underpinnings of our current global warming.

Alfred Thayer Mahan who coined, or at least widely popularized, the term *Middle East* in 1902, exemplifies Britain's historic carbon footprint. As an American naval strategist, he gained much of his fame and what he thought about the world from the maritime history of the British Empire.[1] Mahan's understanding of the sea as "a system of highways,"[2] and his consideration of one of these maritime corridors—a string of British coaling depots used as "bases of refit, of supply, and in case of disaster, of security"[3]—as "Middle East" drew on a British nineteenth-century perspective. From the 1830s, Britons tended to refer to the ocean, once seen as a barrier, as "the highway of the nations," and to the steamship as "the railway train *minus* the longitudinal pair of metal rails."[4] Coal thus served as the main building block for the Middle East, both physically and conceptually. Through this vital carbonized hyphen between Europe and Asia, fossil fuels could be unleashed at large.

The British engineers who pioneered thermodynamics and the new science of energy around the 1840s were among the first to make pronouncements like those later adopted by Mahan.[5] The solidification of once-liquid barriers into transmaritime connectors complemented the world these experts promoted conceptually, in which previously distinct physical realities could be made commensurable by means of the abstraction of energy. Histories of empire had neglected thermodynamics and its own epistemic imperialism—the fact that "energy" rapidly became a crucial organizing principle for scientific, and gradually also social and political action—just as current histories of thermodynamics have neglected the imperial aspects of this story.

However, beyond and behind abstraction and the British Isles, a great deal of concrete work and terraforming accompanied the solidification and globalization of the coal economy through avenues like the one between the port cities of Aden and Port Said—the corridor that runs through the heart of this book. The very geography of this region, the basic modes of provisioning water and food, and key forms of conviviality, sociality, and politics owe their existence—and often even their current shape—to mineral coal shipped from the British Isles. In key ways, we are still trapped in coal's amber.

Insisting on regarding coal exclusively as an energy source renders some of these important aspects—the very dimensions that make coal so detrimental—transparent. After all, it is not its use as an energy source, but rather factors like the emissions from combustion and land degradation resulting from mining and transporting coal, or the carcinogenicity of many coal-based dyestuffs used in synthetic chemistry[6] that should concern us most. Why, then, do we keep using the master's tools, fuels, and energies to try to dismantle the master's house? To decarbonize our world, we need to decolonize our terminology and our history and loosen energy's grip.

That the world we associate with oil in fact rests on the foundations of coal is indeed a symptom of a larger problem in our thinking about—and with—energy. After all, our own "age of oil," and by some accounts even "post-oil," is currently witnessing unprecedented coal burning. Rather than a story of "transitions" between different "energy regimes," this book reveals a great intensification of the existing forces that coal itself supposedly replaced. These included the power of water, human and animal muscles, as well as less tangible forces such as Islamic piety, competing against and converging with notions of risk management tied to finance capitalism, which were also on the rise. Tracing such historically specific entanglements, this book anchors the annals of the Middle East in the broader history of fossil fuels and what we call the Anthropocene, while at the same time asking what the history of this region, with its particular ethical dispositions, ideas about the body, about solidarity and community, and about nature, might offer in the face of our shared planetary conundrum. Any comprehensive scheme of decarbonization must begin by addressing the double historical nexus of how different energy sources are connected to one another and of the role of non-Western settings and actors in the global march of hydrocarbons.

COALONIALISM

Britain's industrialization and imperialism were not separate processes: both—not only the former—were predicated on coal. In the second quarter of the nineteenth century, a gradual reorientation of the British coal industry began, moving away from London household consumption and toward overseas export. Between 1816 and 1840, exports rose as a proportion of the output of England's northeast—itself continuously on the rise until the early twentieth century—from approximately 4 percent to nearly 13 percent. By 1900, British coal constituted about 85 percent of the entire international trade. England used coal exports to project its power, offshoring and outsourcing the Industrial Revolution by building an infrastructure that could support it overseas and connect it to other facets of the imperial project. This resulted in the development of "landscapes of intensification," which simultaneously stimulated an increase in production as well as new uses and demands for British coal, and eventually for coal mined overseas as well.[7]

From the 1830s onwards, a system of regularly spaced coaling depots sprang up in places like Gibraltar, Malta, Port Said, Mocha, Aden, and Bombay in a reverse domino effect. This development concurred with another one: the fact that Britain is an island, which was previously understood as a disadvantage, came to be seen during the Victorian era as a "wise dispensation of Providence," a means to access the wider world.[8] The new artificial archipelago soon began providing coal to various interiors, animating riverboats, irrigation pumps, railways, telegraphs, streetlights, and tramways in what would soon become the Middle East. This process offers a grounded substitute for the vague designation "modernity," which we implicitly ascribe to the political, social, spatial, and temporal effects of the aforementioned technologies. Time/space compression, integration into the global economy, the rise of the interventionist state, urbanization, and the emergence of cash-cropping were all energized by these carbon fibers. In the Ottoman Empire, the lion's share of the coal flowing through these ports was British. Coal depots provided perfect pretexts for military presence and securitization; they were thus also footholds for British and other European colonial officials, as well as for Ottoman, Egyptian, and other local powers.

Historicizing energy at these meeting points repoliticizes coal and casts new light on energy politics. If in the British Isles coal appeared to be "political" mainly in the sense of labor- and class politics, lending itself neatly to a framework of capitalism and accordingly pushing us to diagnose our present fossil-fueled planetary crisis as "Capitalocene," here a more complex carbon politics animated by interimperial rivalry was hard to miss. This book probes how these ostensibly separate political registers were in fact mutually constitutive.

Coal bunkers were indeed established throughout the British Empire for reasons other and often more pressing than fueling alone. The most obvious of these motivations was territorial expansion. Regarding coal today exclusively as an energy source misses this crucial part of the story, and makes us complicit with nineteenth-century imperialist excuses. It was not only European powers but also Cairo and Istanbul that played this game. However, the British usually came out on top. Ruling the waves in the Mediterranean and Indian Ocean and being the world's largest coal exporter during the nineteenth century, Britain regularly established coal depots in order to extend the British Isles. The neologism "coalonialism" thus seeks to capture the confluence of energy and empire and bind them analytically together. Simultaneously, it also seeks to estrange and puncture holes in the gravity of both concepts. As historians of empire have already done this repeatedly, let us begin with energy.

AGAINST ENERGY

This is a book about coal, the emblematic energy source. It is also a book against energy, the supposed transparent and ambient essence of all motive powers, which vaporizes materiality and specificity. In order to discuss coal as more than just a fuel—indeed, even to understand more fully how its use as fuel affects other domains—it is essential to set energy aside. What better way is there to recognize and resist energy's abstractions than by historicizing them?

Energy is a child of its time, the nineteenth century, despite feigning perpetuity. (According to the first law of thermodynamics, the total amount of energy in the universe remains constant, and energy cannot be created or destroyed.) The term took on its current meaning at the beginning of the century, indexing new assumptions about the convertibility of heat, motion, and work. Energy picked up steam from the 1840s on with the emerging science of thermodynamics. Over the next decades it inflated and absorbed more and more domains, from the life sciences to the social sciences. A child, but not a brainchild or otherwise immaculately conceived: like several other major physics abstractions, the science of energy did not spring out of pure theoretical contemplation. It was, rather, the messy challenges of handling steam engines, as well as less obvious activities like brewing beer or sailing in foreign waters, that gave rise to the general laws of thermodynamics. We know relatively little about these latter contexts of emergence and, erroneously, we tend to situate the former squarely within the British Isles.

Only by shedding specificities and disguising its birthmarks and foreign accent could energy become a universal force present in all matter, capable of converting itself into innumerable forms, yet inalterable and constant, a power perceived only in terms of its effects.[9]

However, it is worth insisting on the messy and worldly nature of epistemology. Like other nineteenth-century universals, energy depended on various imperial hierarchies and forms of praxis. This book rewrites them into the story, investigating the historical connection of energy and empire. It is a link that has ongoing implications: from our continuing fossil-fueled acceleration into geopolitical and environmental crises and their codependencies to the ongoing division into winners and losers from their mutual fueling.

Historians have catalogued the religious Presbyterian assumptions informing the science of energy and its debt to Romanticism. In this Presbyterian version, energy was fashioned against metropolitan, anti-Christian materialists and naturalists, attributing to God alone the power to regenerate a fallen man and a fallen nature.[10] They have revealed energy's class politics and its role in struggles between cotton capitalists and organized labor and between elite northern British men of science and "practical men." Most recently, the rise of work as the essence of energy, at the expense of heat and motion, has been anchored in the context of racial and antiproletarian maneuvers and in the beliefs of energy's Scottish Presbyterian promoters.[11] Perhaps most crucially, such studies demonstrate that energy and the laws of thermodynamics were not simply out there waiting to be discovered, as an older cohort of historians and philosophers believed. Rather, they were scientific constructs predicated on existing traditions of thought and action. The importance of prevailing forces in shaping the efficacy of coal is a key insight this book builds on and develops. However, historians have yet to attend to the importance of empire-making in the annals of thermodynamics, and vice versa—to that of coal in the annals of empire.[12]

Thermodynamics needed an imperial context for its emergence and growth. Consider the scientists crowned by Thomas Kuhn as having led the race to "discover" the first law of thermodynamics during the early years of the 1840s.[13] The first, Julius Robert Mayer, developed his insights about the relationship between heat and work in East Java as a ship's doctor aboard a Dutch vessel trading with the East Indies during the peak years of the coffee trade. Mayer noticed that venous blood he let from a European seaman was lighter in the tropics than in Europe, and this pushed him to connect heat, motion, and work (revealed in blood oxygenation) via a unifying "living force." A Christian theist abroad, Mayer also pushed this ontology of force as a counterweight to philosophical materialism.[14]

The second contender, James Joule, arrived at his own insights in his father's Manchester beer brewery. Joule was a member of a new generation of "scientific brewers" who aimed at achieving a quantitatively controlled standardized brewing process, and whose most valued skill was the accurate measurement of liquid temperature.[15] The industrialization of the brewing sector during the following decades had a less familiar imperial dimension: the rapidly mechanizing English beer industry relied on barley grown overseas, by Bedouins in the Negev desert, for example. This particular short-stapled barley, high in sugar and low in protein because of its arid habitat, was a winter crop taken by camel to Gaza in time to arrive by steamer at English breweries early in the summer. It helped promote a shift from heavy Porters to lighter pale ales brewed to be transported to the colonies, as well as the shift to year-round beer drinking.[16] Joule was by no means indifferent to oceanic steam navigation or to the ecological effects of replacing British coal with imported corn, wheat, and meat. And he was clearly interested in imperial politics in the remote settings from which barley arrived at England or beer left it.[17] Like Mayer, Thompson, and other pioneers of thermodynamics, Joule realized that temperature variations mattered on a scale ranging from a jug of ale to the vastness of the empire itself.

The coal business was also stimulated by the coffee trade in places like the port of Mocha, the first coal depot in the Red Sea. The global rise of coal and the attendant science of energy were also linked to the "tea races" to deliver to Europe Asian leaves whose price stood in direct proportion to their freshness, galvanizing the shift from sailing ships to steamers. Water desalination in Suez and Aden, Chinese tea, Mocha or Java coffee, and Gaza barley turned to India Pale Ale, not to mention European and even non-European blood and sweat—from the 1840s and on these and other loci and flows examined in the following pages were all frontiers where heat, cold, steam, and coal were tested, developed, and theorized.

These aspects and peripheries are largely absent from the familiar stories and histories of thermodynamics. They matter not only as lacunae, but mostly because in these peripheries thermodynamics stalled and often required other—sometimes conflicting—epistemologies in order to work and pick up steam. In this sense, the empire created the "constitutive outside" of thermodynamics, a world where not only the scientific but also the economic and political presuppositions that informed European history during the long nineteenth century relied on a set of mirror images and inflections. Rather than the familiar stories of gradual or stalled westernization (taking place in many domains during this time), a transregional and transmaritime history of energy reveals that the Europeanization of Europe depended on consolidating existing forces and often stirring up unforeseen ones in the colonies. Such a history reveals how democracy, individualism, liberalism, and secularism in Europe depended on making places like the Middle East more racialist and racialized, sectarian, authoritarian, and quite differently Islamic.

A TENTACULAR HISTORY OF GLOBAL CARBONIZATION

A more global history of energy reveals what is lost and added in its articulation in other tongues, times, and settings. It divulges energy's outlandish support systems, recounting but also calling out its artifices. This book takes up coal—which in a thermodynamic perspective is little more than energy's external shell—and examines its career over the long nineteenth century outside of the usual context of the industrializing British Isles. Yet coal could accomplish very little by itself. Rather than a history of coal's innate energy or intrinsic agency, this is a history of the multiple, nonlinear juxtapositions and chains of agency that spawned new objects, addictions, and consumption practices. In all these respects of history, materiality, and peripherality, I seek to provincialize energy.

Provincializing is not simply a geographic move away from the metropole, though such decentering helps defamiliarize energy in other ways. Figure 1, Charles Joseph Minard's 1864 depiction of British coal exports, described by contemporaries as "a giant octopus," suggests how. This French pioneer of "thematic cartography" (today we call it infographics) sought to capture the trajectory of the British coal economy, represented by the lines' direction, and its volume, represented by their width. Trying to depict the maritime nature of this flow, Minard was compelled to expand the Straits of Gibraltar almost beyond recognition. His fossil octopus thus exposes some of the choices, tensions, and contradictions between competing regimes of empiricism; here a statistical logic trumped that of cartography. This triumph is symptomatic of the rise of relative ratios and of actuarial models of the future,[18] which also underpins our fixation with notions like "energy transitions." Minard's map illustrates both how abstractions divert attention away

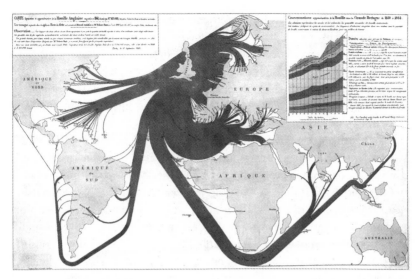

FIGURE 1. Charles Joseph Minard, *British Coal Export*, 1864.

from the material world and how they function as harbingers of very concrete terraforming that would follow suit in the coaling system. It thus simultaneously cautions us to resist the pull of these tentacles as analytical categories, while documenting their effect as categories of praxis and instruments of world-making.

The octopus is also a good metaphor for thinking about empire, another creature whose tentacles are often smarter than its actual brain. With wits distributed in ways that favor "boots on the ground" and proximity to junctions of tension, and with the ample acumen generated in these interfaces, both empires and octopuses have more neurons closer to the action.[19] This fact alone justifies attending to the sign language of the tentacles at least as much as to the declarations of the mouth.

Especially when dealing with corridors such as the one between Aden and Port Said or between other coaling stations or port cities, tentacles, those elongated muscles studded with suckers, joining line and node, help illustrate the active and agentive nature of connectors. Such a tentacular focus helps extend insights like those in Kenneth Pomeranz's seminal study, which also addresses energy and empire yet connects them in their separateness. Comparing the importance of coal mining for the "great divergence" of England vis-à-vis China, Pomeranz discusses the spaces he compares as linked by British imperialism, while ignoring the actual bridges—territories like the Middle East—that physically bind them to one another.[20] Such tentacles, through which much of the British Asia traffic flowed from the nineteenth century, were narrow channels composed of coal depots, artificial canals, and deep-water ports, as well as elaborate new political, legal, religious, and economic arrangements. Rather than functioning as passive thorough-fares, they actively promoted steam navigation and both the fuel and the politics necessary for its progress.

Finally, the octopus offers a useful metaphor for portraying this book's structure. Each of the chapters independently addresses one or more key aspects of coal that the framework of an "energy source" would occlude or impoverish. If from the 1840s energy became coal's essence, its pulsing heart, taking note of coal's materiality, weight, and nonthermodynamic exchangeability with other motive forces and examining the geopolitics and ethics that were informed by and in turn animated it outside Europe are other ways of provincializing energy. Each chapter charts a transfor-mation that begins with the adoption of coal and ends with the arrival of oil in the early twentieth century. *Powering Empire* may then be divided into two clusters. The book's first three chapters disrupt the prevalent notion of "energy regimes," which organizes our thinking about the times before and after the industrial adoption of coal. Why is it that even though we have yet to see an "energy transition"—in the sense of actually leaving an old power source behind—we are convinced that steam power competed with, and then replaced reliance on, biomass or waterpower, when in fact we only see more of the latter? Where others see "replace-ment," I see intensification and an enrichment of fuel baskets; where oth-ers talk of "transition," I discuss a reordering or repurposing. If we care more about coal's environmental impacts than about its energy potency, as we should, we have to acknowledge that alongside the undeniable ruptures and transformations it wrought, coal is also responsible for sig-nificant prolongations with detrimental effects. The book's following three chapters are each devoted to an environment shaped by coal, be it material, financial or spiritual, maritime or subterranean.

In a cephalopodic manner, the introductory chapter holds the tenta-cles and clusters together, thus showing how cohesive this creature is and marking its trajectory as well as what drives it. This exposition also seeks to flesh out the benefits of the proposed thought experiment of sidelining energy, thereby clearing the ground—as well as the water-ways and underground—for a closer consideration of coal. Rather than seeking alternatives to energy or making a case for removing it alto-gether, I ask what other frames of reference reveal about things that usually fall within energy's domain and identify foreclosed conversations that may be reopened if energy is muzzled.

What follows is a survey of the book's main themes according to an organizing principle of growing complexification, one akin to an evolutionary tale: from water and marine biota, to fish and water birds. Next, it moves to larger mammals, and then to humans, first attending to their bodies and physicality, thereafter to their manual labor arrangements, forms of togetherness, and finally to abstractions, including cerebral and emotional forms of calculation and devotion and other kinds of intellectual, affective, and spiritual labor. While the composition of a drop of water might be infinitely more complex than an ideology, there are several advantages to pursuing this scheme. Energy was an imperialist project epistemologically as well as literally: as part of the ascent of physics in the nineteenth century, energy impinged on biology, physiology, labor science, and neoclassical economics. Through Darwinism, Spencerism, and Marxism, it traveled to the social sciences. Retracing and reverse-engineering carbon's footprints inevitably goes through this intellectual trajectory. Pursuing an evolutionary course into a book about coal also helps to historicize and connect the "species thinking" at the base of modern anthropocentrism with fossil fuels, to link biopolitics and thermopolitics,[21] as well as nonenergy aspects of coal.

WATER: THE SUBSTANCE OF HEAT AND COLD

British imperial extension often happened in the most literal sense of the word when raw chunks of material from beneath the British Isles were shipped and distributed around the world. To appraise this terraforming in nonthermodynamic ways, chapter 1 pays attention to water, which furnished the conveyer belt along which much of this substance moved. Such an aqueous account goes back to the roots of the development of the steam engine and coal-fueled industrialization in the British Isles rather than other global coal regions. It was the proximity of English coal mines to waterways that could carry this material with relative ease, and the related problem of mine flooding, that resulted in the development of fire engines that were connected to mine pumps. Coal moved across the empire in very much the same fashion, via artificial and natural canals, rivers, and other waterways. In addition, it was similarly used to pump water, rather than simply evaporate it into steam to produce mechanical energy. As Minard's straits-expanding map also demonstrates, there was nothing metaphoric about the fact that coal flowed.

Alongside this horizontal flow, the nexus of coal and water also reveals the importance of verticality. Initially, coal was transported to the Middle East in sailing ships. From around the middle of the century, as more and more ships began to burn carbonized fuel during steam navigation, a ballasting problem arose. Steamers were rising in the water—often more than an inch a day—as they consumed their coal supplies. Coal-as-fuel detracted from coal-as-makeweight, and this required various kinds of solutions, such as water-ballasting systems: coal-fired steam pumps that drew seawater to compensate for the loss in coal—a loss to which they contributed by their own operation.

Early thermodynamics focused on a single dynamic inside steam engines, the heating of liquid water into steam, pressure from which turned wheels and produced movement. Yet coal's transformation into a global energy source was also tied to how it was carried via waterways and replaced with liquid seawater. Therefore, I insist on retaining water in the analytical picture of the age of coal, as historians of oil have done.[22]

Another transparent yet highly significant dimension of water in the age of steam was salt—particularly so with seawater, of which salt is the main mineral constituent. From a limited steam-engineering perspective, salty and thus dense seawater was detrimental to engines and boilers. It required more fuel to heat and resulted in steam that was less elastic and therefore inferior to freshwater steam. In the early 1850s, this challenge animated the development of desalination technology on board steamers. Solutions to these problems of steam navigation, such as coal-fired desalting condensers, eventually led to an increase in the populations of coaling stations in arid environments. Thus, growing numbers of European merchants and troops were able to survive in places like Aden and Port Said, where fresh water was very limited. To this day, this is one of the main sources of political power in the Arabian Peninsula and the Gulf: Saudi and Emirati oil-based desalination as well as politics rest not only on the abundance of oil, but also on the shoulders of British coal and imperialism.

The salt ejected in the desalination process, and more so through the system of land reclamation, usually by solar evaporation of seawater—the vast process of dehydration whereby many depots were established—enabled the emergence of economic sectors predicated on conserving fish and meat. These, in turn, supported the growth of human populations in these ports. Like other parts of the Middle East associated with oil, several of the chief coal depots were, in the words Abdel Rahman Munif used to describe Arabia's oil boomtowns, "cities of salt."

MULTISPECIES BOOMTOWNS

Stressing coal's synergies with water over the entire spectrum of its functions, states of matter, and chemical composition also yields insight because water is the medium for multiple life forms and processes that steam evaporates from sight. Both fresh- and seawater were brimming with life that engineers often chose to ignore until they could no longer do so. Consider water ballast and the related need to replace liquid make-weights, solid cargos, and fossil fuels in different ports of call. Transporting earth-matter and then water across the world as makeweight involved other circulations, intended and not. Exotic flora sprouting on ballast hills in the British Isles clearly revealed this about terrestrial domains.[23] The dramas underwater were in the beginning less visible, but in retrospect no less significant. Fossil fuels are associated nowadays with reducing biodiversity and with "the sixth mass extinction"—the vast annihilation of wildlife in recent decades. But during their initial spread they actually put in motion new circulations of species and spurred ecosystems in productive, rather than only destructive, ways.

Water ballasting triggered what biologists call "propagule pressure" and a mass movement of marine biota inside steamers' water tanks. Ballast water is nowadays the largest vector for the invasions of nonnative aquatic species. Since the late 1860s, these flows blended with what marine biologists term "Lessepsian migration" (after Ferdinand de Lesseps, founder of the Suez Canal Company), another unintended mass movement of species along the canal between the Indian Ocean and the Mediterranean. It was triggered not only by the artificial connection of these bodies of water, but also by the standardization of their salinity as they came into contact with one another.

For better or worse, predation does not exclude collaboration. In the context of coal-fired transformations underwater—but also on land and even in the air—the axis of both intra- and interspecies collaboration and predation characterized the flows of creatures along steamer, rail, and telegraph lines. Coal depots were often multispecies boomtowns. Coaling stations were shaped by the multiplicity and heterogeneity and no less by the unprecedented rapidity and regularity of the arrival of numerous species. Although in the past many of these places had been visited on occasion or seasonally by foreign ships, most migratory life forms find it difficult to reproduce in low-density populations. Colonization could begin in earnest only with regular and rapid all-season communication.

Colonizing species were often not the benign ones but rather the most resilient, adaptable, and guileful. Anna Tsing researched species attracted to industrial environments and especially to those that thrive in postapocalyptic habitats.[24] European railroads similarly attracted multiple life forms;[25] so did the Hijaz Railway, a gigantic pollinator that standardized the flora of this Ottoman overland counterpoint to the maritime British network described here.[26]

As chapters 2 and 3 show, respectively, if coal did not simply replace water and hydropower, the same is true of animal and human muscle power. At the peripheries of empire, the entanglements of fossil fuel and other driving forces are thrown into sharp relief. Because colonial coal depots often created new environments, it is easier to recognize in them the invigorated inflow of such forces that coal supposedly obviated.

Both Aden and Port Said offer examples of the complex synergy between the aforementioned variables: steam navigation, submarine and surface migrations, the coexistence of hydraulic, muscular and fossil powers, and salting, culminating in the creation of new multispecies imperial geographies. Unpacking this synergy here is also an opportunity to get familiar with the geography covered in this book. Only 150 people lived in Port Said in 1859. In 1869, when the canal was opened for navigation, the town's population numbered 10,000, and by the 1882 British occupation, it had reached 17,580. Port Said's physical development involved massive land reclamation, mostly in the form of the extension of wharves into the sea and solar evaporation of seawater. Lessepsian migration, the arrival of steamships and their water-ballast tanks at its docks, and the pulsating life onshore increased the number of fish and produced a vibrant fishery. Simultaneously, salt, made available by the above-mentioned processes of seawater evaporation and land reclamation, allowed a fish-salting sector to emerge. Together with the installation of coal-burning water-desalination condensers and other engines that solidified this desalted liquid into artificial ice, salted and frozen Port Said fish were increasingly traded all around Egypt and its coastal neighbors. The town's expanding fisheries and fish-salting industry attracted, in turn, a growing number of water birds that could also be hunted, salted or frozen, and similarly shipped across the region.

As chapter 2 documents, this was not necessarily a rosy picture of convivencia (like figure 2); such processes soon created a host of problems resulting, for example, from differential fishing and bird hunting permits (foreigners were allowed to shoot from boats whereas local Egyptians could only use nets). Property rights on land reclaimed from

FIGURE 2. Squids, pelicans, and children together. Paul Reymond, *Le Port de Port-Saïd*.

the sea became another bone of contention and were articulated as tensions between "natives" and "foreigners." This canal town, which was initially a tabula rasa with no one then "native" to it, grew into a dual city split into Arab and European quarters marked by intercommunal tensions.

Similar multispecies tensions (and their communal modes of reduction) animated life in Aden, where human and other populations were also exploding. Aden's population soared from 600 in the year 1839 to 20,738 in 1856. By 1891, it numbered 40,926 inhabitants. Other life forms followed suit, and proximity between species quickly became an object of fascination and experimentation in this coal depot. For example, European steamer travelers developed a habit of standing on deck and throwing coins for the Arab and Somali boys to retrieve from the shark-infested bay (see figure 3). The racialized economy of lives that this practice revealed prompted disapproving comments by Islamic pilgrims in their travelogues.

Moving British troops from India to Aden entailed making environmental changes that were simultaneously planned yet had unintended consequences. For instance, during the 1840s, the British started importing cacti from the Deccan Plateau to plant living barriers (or precursors to barbed wire)[27] that were supposed to ward off attacks by nearby tribes.[28] When these did not work, they coopted tribal leaders by buying their animals for consumption. To provide increasing quantities of meat to the growing troop and steam-passenger traffic in Aden, the British eventually took over Somalia, which became known as "Aden's butcher shop."[29] Indeed, meat consumption was another outcome and engine of coalonialism, a result of the transformation of livestock from workforce into fuel, and from a relationship of collaboration with humans to one of predation. Animal biomass and related greenhouse gas emissions, soil erosion, and water depletion have mounted during and since the

FIGURE 3. Boys diving for money at Aden. *The Graphic*, November 1875, and *Illustrated London News*, October 17, 1891. Credit: Mary Evans Picture Library; Getty Images.

nineteenth century to the extent that today ecologists claim giving up beef would help curtail global warming more than giving up cars.

CARBON AUTOCRACY

To simultaneously tease out this process and flex our "provincializing thermodynamics" muscle once more, let us examine ice machines (which used coal power to produce cold by solidifying water rather than heating and vaporizing it, as with steam engines). These condensers made places like Aden cooler and homelier for Europeans. Standardizing the temperatures of empire was one of the most exciting potentials of artificial cold production. As explained in the 1860s by an anonymous European, "If beverages can be cooled by means of ice; if meat and other articles of food can be preserved in good condition for some time by its agency . . . if these things be so, then some, at least, of the miseries that press upon the white man in a hot climate might be alleviated, and we might then really see what northern muscles can effect in southern regions."[30] Yet again, coal power emerges as an enabler rather than a replacement for muscle power.

As chapter 3 reveals, ice eventually cooled human bodies, and not only those endowed with northern muscles. Moreover, it was not only a product but also an enabler of steam power. Below the decks of steamers commanded by Europeans, the engine rooms were manned by dark-skinned stokers or firemen—among the hardest, most dangerous, and most undesirable jobs in the age of steam. On steamships, these positions were usually left for non-Europeans and nonwhites, who were considered racially suited to the furnace-like temperatures of the engine room, which could reach seventy degrees Celsius in passages through the Red Sea (see figure 4). Due to their dark skin, thousands of Arabs and Somalis were hired annually at Aden. These racial assumptions were combined with the coal-fired technology of artificial ice production into a protocol whereby stokers were locked in the engine room to work until they fainted from heat stroke, at which point they were resuscitated with an ice bath on deck.

While Europeans referred to the stokehole as "inferno" and "Hades," Egyptian and Ottoman steamer captains punished unruly passengers by sending them to load coal in the engine room. By the same token, Indian Islamic reformers partitioned steamers into upper deck "heavens" and lower deck "hells."[31] These harsh employment practices, racial theories, and mythological netherworlds illustrate "the constitutive outside" of thermodynamics, being at once external and indispensable for enabling imperial steam navigation. During the second half of the nineteenth century, the model of the thermodynamic body, equating human physique to a fuel-burning motor, became the general rule in western Europe and North America. Yet differential racial theories about the connection between dark skin and heat resistance allowed brutal labor arrangements to be sustained on Europe's peripheries.

The science of energy did not replace racial thinking: it actually fed off it, flaming it in turn. Steam power too did not replace hydro- or animal muscle power but rather intensified them. Steam engines, frequently understood as labor-saving devices, likewise did not replace or reduce human labor force. But as chapter 3 reveals, they certainly changed it. Diversions from the thermo- and biopolitical standards of Europe went hand in hand with labor arrangements that were similarly hard to reconcile with the ideal of free-waged work that supposedly accompanied labor-saving devices using coal instead of muscle power.

Port Said and Aden were frontiers for experimenting with work schemes and contracts that were not quite free. Examining an interconnected energy labor market composed of coal miners in Wales and also along the Ottoman coal coast, stokers in Aden and coal heavers in Port Said, chapter 3 demonstrates that what Timothy Mitchell called "carbon democracy" in the British Isles depended on limiting participatory politics, workers' rights, and ideals of freedom in the Middle East.

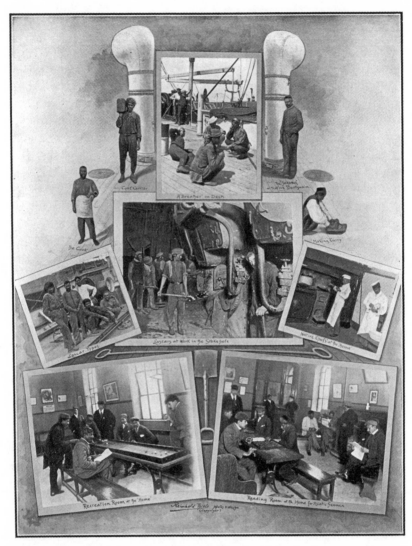

FIGURE 4. "Human Salamanders," *Illustrated London News*, October 27, 1906.
Credit: Mary Evans Picture Library.

W.E.B. Du Bois saw as early as 1925 that with the global division of labor along a color line and "with a democratic face at home modern imperialism turns a visage of stern and unyielding autocracy toward its darker colonies."[32] If energy animated a liberal political and economic dynamics in Europe, this was based on a carbon autocracy and racism elsewhere.

GOING NATIVE

Together with Indian troops in the service of the East India Company, one of the first non-Arab communities to strike roots in Aden was a group of Parsis who arrived with the occupying army from Bombay in the late 1830s or early 1840s. These Zoroastrians became a global trading diaspora with the Canton opium trade, in which they, and not the more famous British and American actors, played a leading role. They brought opium to Aden too, as well as other habit-forming substances like salted meat and ice, which enhanced the consumption of meat, milk, and eggs on board steamships. Such menus set the table for our current "fossil fuel addiction."

As Aden demonstrates, the artificial archipelago of coal depots that connected London to Bombay—and which is the subject of chapter 4—was "Indianized" on multiple levels. It ranged from cacti and landscape to the introduction of the Mughal institution of *Durbar,* allowing British officials to include Aden's "natives" in longstanding British policies of nonintervention in Indian cultural affairs.[33]

This multispecies dynamic left its mark on human classifications. Deriving social categories from natural ones was all the rage in the second half of the nineteenth century. This was, after all, the great moment of "species talk." Charles Darwin's *On the Origins of Species* was published in 1859. Darwin was especially interested in islands and archipelagos as terrains that allowed for mapping the spread of nonindigenous species and their naturalization.[34] Darwinism symptomatized a broader transformation. For Michel Foucault, beginning with Georges Cuvier's late-eighteenth-century comparisons of living and fossilized creatures (which I discuss in the last chapter) and accelerating with Darwin, life—both human and nonhuman—came to be seen as subject to temporal change, making the framing of humanity as a species conceivable.[35] As a species, humans became exposed to "biopolitics," the politics of biological existence.[36] This book probes how biopolitics depended on one's place in the imperial food chain.[37]

Steam navigation also set human populations in motion, this too on an unprecedented scale. Seasonal work, gunboat abolition, coal-fired military conflicts, pilgrimages, economic bonanzas and shortages pushed, pulled, and reshuffled large chunks of humanity. During the nineteenth century, many people left what would become the Middle East, and many more arrived at its shores. These itinerants developed, or were compelled to adhere to, various procedures of domestication and of becoming native, a process which generated new categories of belonging and transformed existing ones. Words old and new, including *native, Muslim, Arab, India(n), Britain/British*, among others, acquired new meanings when stretched over the massive scale and rapidity of the arrivals.

While such transformations were diverse, it is safe to generalize that many of these designations acquired a new biological fixity as denoting human subspecies. Bloodline, genealogy, race, or divine choice did not disappear in the face of the new species-based frameworks; earlier forms were refolded and reinscribed into the new biopolitical paradigm. As with supposedly competing energy regimes, schemes like man-as-species did not replace man-as-body and earlier ideas of race. Chapter 4 recounts the fruitful coexistence of these epistemes[38] and how convergence often infused them with violence. It examines how the umbrella of a single species fanned a narcissism of minor differences.

RISK AND *RIZQ*

The mirroring and mutually flaming effect of individualism and communalism, liberalism and racism, and carbon democracy and authoritarianism is an important theme in this book, evident in multiple points along fossil-fueled colonial infrastructures. As Sven Beckert has shown, it is important to think of war capitalism, industrial capitalism, and I would add finance or speculative capitalism, together.[39] The 1870s saw the rise of new ideas about imperial defense or "security," especially of the merchant fleet, which revolved around the fortification and securitization of coal depots. These were meant to allow the Royal Navy to extend and demonstrate its security and deterrence force worldwide. The coaling process and its depots were now at the center of imperial defense and imperial trade, making them the key settings for exchanging military and economic rationales.[40] Chapter 5 shows that this new way of thinking about fossil fuels, which today we call "energy security" (another agenda wrongly attributed to oil), merged with another set of calculations, namely "risk management."

Risk sharing was a longstanding engine of imperialism, as exemplified by the East India Company and similar joint-stock companies from the seventeenth century. But colonizing corporations were only one type of several risk-managing bodies that intertwined with the political sphere.[41] The fruitful slippage between security and risk management was especially evident in the Red Sea. This maritime space was transformed during the second half of the century from a place of danger (both environmental in the form of gales and reefs, and human in the form of pirates and fanatic Muslim inhabitants or cholera-ridden pilgrims) to a space where risk management and security converged.[42]

A key risk-management body that helped shape the age of steam was the insurance company. It is important to note that risk management was not about risk reduction—taking chances was the lifeblood of economic growth. Rather, it was about diverting risks away from investors in ways that constantly expanded the insured infrastructures. Insurance was one means of encouraging the experimental spirit that animated many of the above-mentioned developments. Specifically, maritime insurance firms were key in translating the accidents and daily failures of steamers and coaling depos—from boiler explosion and shipwreck to food poisoning and heat stroke—into effective pressure toward government investment and intervention, technological innovation and the development of political and economic arrangements to handle hardware, cargos, crews, and passengers more efficiently and lucratively.

At the same time, the fossil-fueled march of our global risk society was animated by, and in turn stimulated, hidden forces such as existing forms of piety and attendant ecologies of submission (whether to God or through debt-bondage to Muslim steamship owners). Passengers were among the most lucrative naval consignments, especially for early steamers that were limited in space (passengers could be carried on deck and they conveniently loaded and off-loaded themselves). The Indian Ocean hajj was especially important for the development of this sector, responsible in some periods for about 5 percent of the entire global steamer traffic. With the pilgrimage abiding by the lunar calendar and therefore sometimes occurring during the monsoon period, Islamic piety and the ability to compress nonwhite passengers in tin-can conditions became a vital condition for the rise of the all-season steamer and the hydrocarbon economy.

If insurance managed peril and uncertainty through statistical calculations of risk and security, Islamic thinkers and even uneducated pilgrims rejected such a probabilistic attitude to the future, stressing instead the importance of *rizq*, or divine reward. In the Islamic Indian Ocean and Red Sea, a communal rather than individual subject, stress on responsibility rather than unimpeded license, and intergenerational solidarity rather than abandonment of future generations, were often the norm. Human compression on board steamers and the resulting deprival of bodily autonomy further bolstered this modus operandi of thinking and acting collectively.

ETHICS AT GREATER DEPTH

Chapter 6 grounds these and similar Islamic alternatives to the amoral dispositions of insurance and risk and to Presbyterian growth-centered carbon ethics in an Egyptian and Ottoman canon of translations of European geological books. It probes different articulations and potentials of "fossil" and the underground, rather than reducing coal to "fuel" alone and to the propulsion and emissions resulting from combustion. Sidelining energy one last time, the chapter examines what we call the Anthropocene from the perspective of Ottoman geology.

It is in such residues of discontinued attitudes and horizons, and in the new and unintended consequences of thermodynamics, that we may find renewed ways of approaching our common predicament. Amplifying unacknowledged or vilified attitudes is significant exactly because the global coaling system did not chart a direct course to the Anthropocene. There is nothing inherent in fossil fuels that flames intercommunal tension, securitization, abandonment, or the financialization of the future. Hydrocarbons alone do not even cause global warming. Rather, this nexus was the result of the combination of material features, religious and imperial trends, capitalism, and racism. The very contradictions inherent in the processes of bringing them together also opened up possibilities for unexpected inflections, coalitions, and alternative ethical horizons found in translation. All around the beaten track of thermodynamic universals, there existed other dynamics of community, bodily comportment, and cross-generational accountability. Existing in the shadows or discontinued for a century, these do not offer readily available counterweights to the managerial, securitized, and economic attitudes that accompany energy. Some of these non-Western alternatives are every bit as bad. What they offer are new/old potential beginnings. They force us to measure our best analytics and politics of liberation against nonliberal epistemologies of subjugation to higher forces, to discarded ideas of unity between humanity and the world, and to political modes of sacrifice (ones that are not synonymous with austerity), passivity, and avoidance, ways of doing less that contrast with prevailing notions of activism and agency predicated on the coal-fired metaphysics of growth.

This history at once divulges hidden foundations for the present world and helps undermine both our fixation with energy regimes and energy's own authoritarian regime. Much in contemporary energy politics consists of products of the secularist, capitalist, and imperialist frameworks that coal created as it spread globally in the long nineteenth century. Provincializing thermodynamics divulges forgotten attitudes toward fossils, toward the underground and underwater, and toward the bodies of people and animals. It conjures up spiritual, affective, and corporeal nonsecular frameworks that enhance the scope of our thinking and responses to the combined fossil-fueled climatic and geopolitical crisis. As we continue to extract more coal and oil, this book unearths some of the hidden counter-energies fossil fuels unleashed in the colonies. These are still raw resources, in need of distilling, adapting, and renewing. But they are forces which exist in an intellectual repertoire that, like our fuel basket, is richer and older than we admit. Before seeking and tapping new forces, let us take stock and make use of what we already have.

CHAPTER I

Water

DOWN WITH ENERGY REGIMES!

A ghost is haunting the field of energy studies and energy history; it is the specter of "energy" itself. The abstraction of different natural powers into energy directed attention away from their very materiality and from other features that made them specific. This scheme of commensurability reduced diverse features into constituents of "energy regimes" projected across space and back in time. "Every society," writes John McNeill, one of the pioneers of this conceptualization in history writing, "has its 'energy regime,' the collection of arrangements whereby energy is harvested from the sun (or uranium atoms), directed, stored, bought, sold, used for work or wasted, and ultimately dissipated."[1]

The notion of an energy regime, together with similar heuristic devices, such as periodization into "the age of . . . , " were not without their advantages for narrating historical transformations in broad strokes and for identifying characteristic features of different systems. But there was a price: energy regimes are implicitly thought to be predicated on a hegemonic, modeling energy source, both technologically and geographically. The nineteenth-century adoption of fossil fuels, for example, seemed to separate England—the first place to take up coal industrially—from the rest of the world, a "great divergence" as it were.[2] The divergence perspective mapped neatly onto divides between industry and agriculture, modernity and tradition, the artificial (polluting, abnormal) and natural (renewable, sustainable), and the (energetic) West and (lethargic) East, obfuscating important connections among these worlds as well as other kinds of entanglements among machines, humans, animals, and other forces. Contrasting urbanizing, industrial, coal-rich Europe with the non- or deindustrialized, non- or underdeveloped agricultural peripheries that fed its working classes and factories missed two key facts: not only were these peripheries themselves dependent on coal steam power from about the same time they were fully adopted in England, but English industrialization itself depended on these remote settings as markets and laboratories for coal and coal-burning technologies.

Another aspect that "energy" tends to evaporate is context-specificity. Steam power and fossil fuels generally are seen as portable and independent, disconnected from their immediate environments. Steamships could move with only small crews against the current or the wind. The portability of their coal allowed them to compete with, and eventually replace, sail ships and rowboats and allowed railways to successfully compete with coach lines. But there are two things to remember about this apparent ease of motion. First, that it was hard won and effortfully maintained: it took many decades to create a world that supports steam navigation technically, politically, and economically. Second, that independence from surroundings was always partial: steam engines were born out of a nexus of coal, water, and muscle, and in order to work they had to reproduce certain conditions wherever they went. To generate steam, coal and freshwater supplies were needed in large quantities and at evenly spaced intervals. Like the railway and steam engineers then, historians today might concern themselves with where coal and freshwater could be obtained and on whose back they could be carried to their destination.

From an "energy regimes" perspective, industrialization and the coal that animated it stand apart from agrarian society and its hydraulic and animal forces. McNeill's formula, taken from *Something New under the Sun,* sounds eerily familiar: "The windmill gives you society with the feudal lord; the steam mill, society with the industrial capitalist."[3] (Energy regimes indeed seem like a rearticulation of the Marxist "modes of production" theory in environmentalist terms.) This chapter examines whether such rifts—between motive powers and geographical regions—were indeed so clear-cut. What would happen if we attended more to continuities and connections? After briefly examining steam power's aquatic and faunal traces in England, we explore coal's introduction into the region between Aden and Port Said; first, via the Red

Sea into Ottoman Egypt along similarly liquid and brawny sinews. Coal's second trajectory, into Egyptian-controlled Arabia, also began at depots along the Red Sea. It was informed not by existing waterways, but by the lack of drinking water: coal's advance into Arabia was enabled by animal forces, and by the fact that it could transform seawater into freshwater and, in the process, into a significant political force.

MAINSTREAMING FOSSIL FUELS

Coalfields, as well as the principles and early artifacts of basic steam engines, could be found in many places outside of England for millennia. It was the geographical proximity of English coal to water transportation networks, as well as the related problem of mine flooding, that informed the development of coal uses and the steam engine in the eighteenth and nineteenth centuries. This particular configuration of coal mines and water explains the development of Newcomen's atmospheric steam engine at the beginning of the eighteenth century and its function to pump water out of Newcastle mine shafts instead of (though frequently, alongside) ponies, mules, and other work animals.[4] During that century, and until the early 1830s, steam power penetrated several important industrial sectors in England. Nevertheless, it still presented little competition to the stronger and more developed waterwheels. England's decisive shift to steam power finally happened during the 1820s and 1830s in a niche market:[5] the cotton industry, which was subsequently transformed into the main engine of the industrial revolution.

The shift occurred in the context of the prevailing wave of labor unrest among weavers struggling for the Ten Hour Work Day Act. As Andreas Malm demonstrates, tensions resulted from the fluctuation and irregularities of waterpower: when the river was high, workers were forced to stay at waterwheel-powered mills long beyond the mandatory twelve-hour work day. (E.P. Thompson has shown how especially in cotton mills, where clock discipline was most rigorously imposed, the contest over time became most intense. It was there that the blue-collar protagonists of his classic article about clock time and work discipline adopted their employers' equation of time and money to better fight them at their own game.)[6] Steam engines offered capitalists a way out of this conundrum: by 1830 these machines were mobile enough to compete with waterpower, which had bound production to riverbanks. Mobile generators and portable fuels allowed mills to be moved to urban centers, where workers were abundant and thus replaceable and more manageable. Against the ebbs and flows of water and work time, the regularity of coal and steam power afforded better exploitation of human work.[7]

However, in order to produce this first constitutive rupture, the mobility of coal and the steam engine was dependent on the power of water in several important respects. The Newcomen fire engine had an unexpected accomplice that allowed it to break out of the mine shaft, leave the coalfield, and gradually attain the mobility that would eventually help its progenies defeat the wheel: it was the waterwheel itself. In order to mitigate their dependence on seasonally variable water levels, during the eighteenth century English water-powered ironworks and cotton mills began installing steam engines to pump back and recycle the otherwise wasted water that had gone through the wheel. Coal-burning fire engines generated heat that vaporized water, creating a vacuum inside sealed chambers used to pump and animate an "aqueous driving belt" that allowed the Newcomen engine, incapable of rotary movement, to feed the water back to the wheel.[8] Toward the century's end, this hybrid was replaced in the cotton industry by Watt's low-pressure engine that included a separate condenser, using steam rather than the atmosphere to depress the pistons, and producing the rotary motion of the waterwheel, which it could now gradually replace.

Other interfaces between waterwheels and steam engines provided the impetus for the development of thermodynamics and the notion of energy itself. One such context was long-distance steam navigation, dependent on coal-burning engines to turn wheels in the water to generate motion. In 1844, engineer James Thompson—older brother and hidden partner to one of the claimed inventors of the first law of thermodynamics—asked his brother, mathematician William Thompson, for an account of "the motive power of heat in terms of the mechanical effect of the 'fall' of a quantity of heat from a state of intensity (high temperature as in a steam-engine boiler) to a state of diffusion (low temperature as in the condenser), analogous to the fall of a quantity of water from a high to a low level in the case of waterwheels."[9] The ensuing history of thermodynamics is rife with other instances in which water was the main medium for solving practical and conceptual problems of steam power.[10] Yet its story was never simply one of acknowledged and egalitarian collaboration between these two powers, or between the two brothers Thompson for that matter. Rather, thermodynamics emerged out of denying the practical basis and historicity of these interfaces, and hence also James's role.

Innovations in steam technology informed by existing platforms of emergence were not unique to water. Steam engines also developed through interfaces with other powers, replacing—while also continually extending and imitating—generators of wind and muscle power. For the steam engine to become a model for the human body, muscle power, both human and animal, had first to be inserted into the engine. However, a simple dichotomy of mechanics and anthropomorphism was less the starting point of the process than its result. As different actors gradually left the early hybrid machines behind, they also imprinted many of their own unique traces on and in these machines.

> In the Newcomen steam engine, the piston followed the condensing steam, pushed by atmospheric pressure, that was thus made to lend its strength to the pump that extracted the water, that flooded the coal mine, that made the pit useless. . . . When it reached the end of the cylinder, a new flow of steam had to be injected through a valve opened by a worker who then closed it again when the piston reached the top of its stroke. But why leave the opening and closing of the valve to a weary, underpaid and unreliable worker, when the piston moves up and down and could be *made to tell* the valve when to open and when to close? The mechanic who linked the piston with a cam to the valve transformed the piston into its own inspector—the story is that he was a tired lazy boy.[11]

Bruno Latour uses this account to theorize how machines acquired human properties such as supervision and observation. Later we will see that they internalized animalistic properties as well, and vice versa: under steam, people and animals could more easily be equated to machines. But the myth about Newcomen's boy, who connected the piston and cam so that he could go fishing, also betrays a similarly apocryphal belief in the labor-saving and liberating effects of steam engines. In fact, it was quite the reverse: such technology reorganized and streamlined labor, intensifying it along regular, machine-like lines.[12] Chapter 3 is devoted to this process.

The realization that machines acquired some of the traits of their animate parts (human and animal) was not wasted on contemporaries. Observers in England and France (and a little later also in Istanbul and Cairo) noticed this affinity. The French engineer Bernard Forest de Bélidor (1697–1761), for example, stated, "There is not a single other [apparatus] of which the mechanism has so much resemblance to that of animals. Heat is the cause of its motion, a circulation takes place in its different tubes like that of blood in the veins; it has valves that open and close at the proper moment; it feeds itself, it rejects what it has used at regular intervals, it draws from its own work everything that is required for its support."[13] James Watt, who coined the term *horse-power* in order to promote the steam engine that eventually replaced Newcomen's, can be seen in the same light. Commensurability between different kinds of power generators depended on both citation and departure: steam engines could be made comparable to waterwheels only when they were put to the same task, and once there were independent measures—in this case units taken from the world of animal power (a mediating sphere bracketed from actual animation of the machinery)—to quantify their outputs. Energy depended on writing under erasure the circumstances of its production.

Notably, however, mine animals—especially mules—continued to be an important part of coal production well into the twentieth century, often competing successfully with steam engines, which by this point were quite developed.[14] While by 1900 humans and animals (though mainly the latter) produced only 25 percent of the total power in Britain and steam power constituted 70 percent, the former represented a steady and unwavering increase from 1760 onwards.[15] Rather than doing away with animal power, in total terms, steam engines in fact stimulated it.

It was no coincidence that steam power vied with muscle- and hydro-power in the cotton sector. Cotton mills saw the replacement of animals by steam engines, and by the more developed and powerful technology of the waterwheel, at about the same time, during the 1780s.[16] The newly introduced generators were used to power newly developed spinning machines, key among which was Samuel Crompton's spinning mule, so called because it was a hybrid of Richard Arkwright's "water frame" and James Hargreaves's "spinning jenny," just as a mule is the product of crossbreeding a horse with a donkey. The spinning mule, invented in the late 1770s, was powered by animals until 1790, when it was first powered by water.[17] It was replaced by another breakthrough, Richard Roberts's 1825 steam-powered "self-acting mule," which had the advantage of rendering factory owners "independent of the working spinners whose combinations and stoppages of work have often been extremely annoying to the masters."[18] Addressing labor unrest was not only a priority for and feature of the mobile steam engines, used, as shown above, to subdue workers' militancy, but also of other machinery these engines powered in the cotton industry.

Over the following decades, engines became smaller, required less coal, and thus were less dependent on being situated near mines. However, the proximity of coalfields to river transportation networks ensured a steady supply of coal to these proliferating machines. The steam engine's departure from the coal mine was again aided by water. Once they were portable enough, such engines were assembled on riverboats and on railways running alongside rivers or narrower watercourses. These water sources were as important as a supply of coal for generating steam and cooling boilers. The first railroads in the British Isles were extended from collieries and later industrial centers rife with coal along such waterways. This was the case with Richard Trevithick's 1804 Merthyr Tydfil Railway near Cardiff, John Blenkinsop's 1811 Middleton Railway, George Stephenson's 1825 Stockton and Darlington Railway, and the 1830 Liverpool and Manchester Railway, which were the four pioneering British railways.

These local and, by the 1830s, transmaritime waterways connected and synchronized the cotton industry to global fields and markets. The trend toward clock time was reinforced by steam power in the cotton sector, and it coincided with a rapid synchronization in the overseas transport sector, also greatly aided by the introduction of steam navigation. Lancaster cotton mills were connected to American plantations and increasingly also to Indian and Egyptian cotton fields, serving as producers of raw material and as markets for finished goods. As early as the 1840s, travel time between India and England was calculated *by the minute*.[19] Such transportation routes were themselves becoming a lucrative industry whose profits depended on regularity and synchronization. Synchronization was aided by the rapidly expanding global system of coal depots. The 1830s thus also saw the launching of English and Welsh coal's overseas career, a result of changed patent regimes and tax structures, as well as ever-more mobile steam engines that were put on ships and exported to the colonies. If the adoption of coal was the main driving force of industrialization, this was the watershed decade; and if we have already identified coal energy's tendency to wipe out its aqueous and animalistic infrastructures, it might be useful to follow the waterways and animal tracks that led to the global proliferation of coal and examine the support systems for steam's globalization.

The cotton sector was not only the most mechanized and regularized, it was also the most global.[20] In fact, these features were intertwined with one another: the growing synchronization of the overseas transport sector from the second quarter of the century onwards matched the regularization of labor and production in the cotton factories. Cotton was the main raw material, and textiles were the main finished goods of the modern global economy, a fact accounting for cotton's centrality in the history of globalization. But as we have seen, defeating impediments in large-scale textile production required predicating weaving, transport, and eventually also cotton cultivation on coal. Consideration of these carbon fibers therefore allows us to provincialize the history of the steam engine.

INTO EGYPT

In the late eighteenth century, Ottoman Egypt started adopting a new approach to the management of the Nile. Partly in order to adapt to a series of environmental calamities including epizootics that decimated livestock and animal-dependent modes of irrigation, Egyptian rulers opted for the centralization of existing resources and for their rapid intensification.[21] Especially after the turn of the nineteenth century, the country saw ever-grander schemes for larger artificial waterways whose construction, maintenance, and eventual failure led to a doubling down and to the investment of even more funds, tools, men, and animals to dig, dredge, and maintain broad canals and irrigation works. New waterworks were designed to promote the adoption of cash-cropping long-staple cotton, which was sold to Lancashire's mills to finance a new army created for regional domination.

When British coal first arrived in this dynamic setting, it was to serve completely different agendas and actors. During the 1830s, consignments of coal started landing regularly in dumps and bunkers built for steamers servicing a new British overland route to India via Egypt, which began replacing the sea voyage around Africa, expediting and regularizing England's communication with its Crown Jewel. But as was usually the case with British imperialism, this coalonialism soon encountered local powers and processes, entangling with existing water, animal, and human power. British coal and Egyptian water and animal power were mutually constitutive, connecting perennial Egyptian irrigation and perennial British all-weather steam navigation.

Egypt's expansionist visions similarly relied on existing resources, whether British, Ottoman, Bedouin, hydraulic, faunal, or mechanical. During the 1830s, Egypt had taken control of the entire east coast of the Red Sea, from Suez and Aqaba to the first coal depot of Mocha. It also stretched to the north, towards Greater Syria. The conquests were designed by Mehmet Ali, Egypt's energetic Ottoman governor, who was seeking to augment diminishing wood supplies and declining animal power, as well as find effective ways to protect himself from similar centralization policies in Istanbul, which was then also beginning to adopt coal power. British coal facilitated these advances until the end of the decade, when it became sufficiently entrenched to curtail them in both the south and north simultaneously.

To begin explicating coal's entry into Egypt, let us attend to the reports of Colonel Patrick Campbell, the British consul general and East India Company (EIC) agent "in Egypt and her dependencies"—those newly acquired territories. Campbell not only related some important facts about the process, but also divulged attitudes that still dominate our own perspective about the hierarchy of coal and other prime movers. In 1838, Campbell reported a major consignment of fifteen hundred tons of British coal to be sent by sail on the long voyage via the Cape of Good Hope to Mocha, and two thousand tons of coal sent to Alexandria to be transported along the Nile to Cairo, from whence it was to be sent by camel to Suez. The monsoon prevented the coal from arriving at Alexandria between mid-July and mid-September, during which period "the *djerms* [local Nile sailing boats] with the coal for Cairo could easily have passed the bar at Rosetta to enter the Nile [but] did not arrive in Alexandria until 15th November, when the Nile was low and the water over the bar shallow. Moreover, the winter had then set in and for many successive days no vessel could leave Alexandria to Rosetta."[22] Other complications included the lack of shovels, weighing machines, and coal sacks by which the coal could have been sent by the Mahmudiyyah Canal linking Alexandria and the Nile. Yet another obstacle—an outcome of previous delays—was the difficulty of procuring camels to transport the coal from Cairo to Suez. "This has arisen from two causes, first, the arrival of Osman Pacha, sent by the sultan, with the sacred coverings for the tomb of the prophet at Medina. . . . Secondly, this is now the time when the pilgrims also flock from all parts of the Turkish empire . . . [passing through] Alexandria on their pilgrimage to Mekka; and as they give a very large price for camels, our means of obtaining them are necessarily limited and circumscribed."[23]

In the following decades, the flow of coal via Egypt to key British fueling stations would be greatly improved by absorbing more and more sectors of the Egyptian economy into the coal-burning global system. The development of this global fuel economy was inseparable from Egypt's adoption of fossil fuel. But if existing motive powers appeared to Campbell as impediments to the steady flow of coal, coal's dependence on wind, water, and animals was also apparent. The account thus simultaneously reveals and subverts the origins of our current "energy regime" biases.

The infrastructure for the introduction of coal included the sailing ships that stocked the coal stores by sea and river and the camel caravans that did the same by land. Moreover, Campbell describes as an obstacle to the steady flow of coal the *mahmal* or ritualistic procession annually leaving Cairo with the *kiswah*—a set of ornamented curtains used for draping the *ka'bah*. But if such religious rituals seem like superfluous obstructions to the march of progress, it is important to note that two decades earlier, facilitating and expediting the pilgrimage to Mecca was cited as an important impetus to digging the Mahmudiyyah Canal along which the coal was now passing.[24] Moreover, around the time the report was written, the same reason was given for building a railway on the very route Campbell described.[25] When it materialized, this railway depended on the revenue generated by religious pilgrims (to Mecca as well as to numerous local festivals in Egypt). The steady and cheap flow of coal, in other words, was tied to existing forces and agendas and, in turn, also animated them.

As was the case with animal power, water power, and human labor in England, entanglements with existing local forces in Egypt often resulted in their intensification. If ruptures occurred, as they frequently did with carbon-based fuels, they occurred much later. Each of the elements mentioned in Campbell's report—sail ships, camels, rivers, and man-made canals—illustrates this point. Let us take them one by one. Campbell begins with riverboats and sail ships, important in the 1830s but also in the following decades, well into the so-called "age of steam." In fact, the great exaggeration in reports about the supposed demise of the sailing ship demonstrates that the road to rupture could last for decades. Naval historians have corrected the prevalent assumption that the invention of the steamer or the 1869 opening of the Suez Canal dealt a death blow to sail ships, and have in fact proven that the opposite was the case.[26] In the Red Sea and the Arabian Sea, the arrival of the steamer actually stimulated the use of dhows and the redistribution of maritime traffic along the lines of trunk and feeder routes, where steamers sail in straight lines (hence their name, liners) between deep-water ports furnished by coal depots, and native sail ships move goods and passengers between these ports and smaller harbors (*banders*), sometimes illegally. Very often, passengers would use steamships to defy the monsoon and then continue in a dhow to subvert quarantine or antislavery regulations.[27]

The British and Ottoman archives are full of accounts of literal and metaphoric collision between these kinds of vessels, which were gradually cast as modern and traditional, licit and illicit, in a positive feedback-loop dynamic: the more dhows defied the orders imposed by steamers, the more steamers were deployed to patrol maritime spaces and fight contraband, piracy, and various kinds of trafficking.[28] What emerges, then, is not only the diversification of transportation and energy expenditure options, but also the way part of this rich spectrum was rendered invisible.

The same pattern characterized other places across the Middle East. Even if in 1900 sail ships accounted for only 5 percent of the ships that called on the port of Istanbul (compared to the 1860s, when their numbers were four times that of steamers), in total this represented more sail ships than in any preceding year, not to mention the sharp increase in rowboats servicing both these types of vessel. Similar figures applied in Jeddah, to which we will soon turn, and elsewhere in the Red Sea during these years.[29]

Animals, like the camels mentioned by Campbell, also demonstrate the reorganization into a liner-and-feeder pattern. Coal's rapid adoption in the Middle East was facilitated by the fact that between the 1780s and 1810s Egypt and other parts of the Mediterranean had experienced a series of epizootics that decimated much of their animal population. During these decades, camels, oxen, and water buffaloes—whose muscle power was key to Egyptian agriculture—were replaced by human labor, and then by fossil fuels. As Alan Mikhail puts it, following McNeill, there was "a fundamental transformation in the energy regime of Ottoman Egypt":

> Of course, animals were still used when humans could not match their strength or stamina—to pull heavy materials or waste, to transport items to market, and so forth. But even these remaining economic functions would soon come to an end with the introduction of rail and coal in Egypt and with other advances in mechanization and machine technology in the first half of the nineteenth century. The result of all this was the decreasing economic importance of animals, and therefore an overall reduction in the centrality of domesticated animals in Egypt.[30]

Mikhail ventures to suggest that "if the American example is any indication, it seems safe to assume that rail initially actually increased the use of horses in Egypt—both in the original construction of rail lines and in the movement of people and goods to, from, and between trains"[31]—and this was indeed the case, as settings closer to home also suggest. Throughout the Ottoman Empire, in the absence of rail feeders (smaller lines leading to a larger main line), the number of animals bringing goods to trunk lines grew enormously during the late decades of the nineteenth century and early decades of the twentieth. While caravan lines parallel to the main railways eventually went out of business, there are various reports of stations where thousands and sometimes tens of thousands of camels waited to unload the trains. Animals were also indispensable for carrying coal inside and away from Ottoman coal mines to depots and train stations.[32]

This was true also in Egypt, which had ample feeder lines. Egyptian trains could not move without camels, donkeys, and horses. These animals, especially camels, carried railway tracks and telegraph poles for the communications network built alongside the railway (inter alia to prevent collisions between trains and camels or water buffaloes); they transported water from the nearby Nile to train stations, and they moved leftover merchandise and passengers to destinations that were beyond the reach of the railway or its capacity to transport. During its first years, the Egyptian railway generated a steady increase in transportation by camel caravan (from fifty camels to more than twenty-five hundred in less than three years in the early 1850s).[33]

Eventually, trains did away with camel caravans that ran parallel to certain lines, and often this prompted sabotage against the railway. In other cases, they downgraded camels from "desert ships" to "desert boats," as it were, increasingly using them for shorter distances between rail and tram lines.[34] Nineteenth-century observers, tolerant only of a single model of progress, were constantly befuddled by "the slow and primeval camel carrying his burden at competitive cost by the side of the fast and last improved locomotive, of the locomotive competing with a camel . . . Verily there is some comfort in knowing that the locomotive must beat in time."[35] It was this perspective, assuming inherent competition between the primeval camel and modern train, which obfuscated the synergies between these actors.

The main sphere in which steam engines were employed for tasks previously performed by animals was irrigation. This was a principal channel for coal's diffusion into the Egyptian interior. As with the cotton mills emblematizing the English Industrial Revolution, the industrialization of Egyptian agriculture that provided these mills with raw material started before the adoption of coal as the key motive force. Beginning in the second decade of the nineteenth century, it entailed a shift from subsistence agriculture based on flood irrigation to cash crops, mainly long-staple cotton, sugar, and tobacco, which required perennial irrigation. And as in the English textile sector, the introduction of steam power in Egypt in the 1840s significantly reinforced (agricultural) industrialization.

Consider the Mahmudiyyah Canal, completed in 1820 but thereafter soon proving too shallow for passage, which is why Campbell's ships could not sail it when the Nile was low, and why new measures, now aided by steam power, were adopted in the 1840s to deepen it. This was one of Mehmet Ali's major engineering projects, meant to connect Alexandria to the Nile, provide summer irrigation to nearby cotton fields, and facilitate the passage of pilgrims to Arabia.[36] The canal provided Alexandria with freshwater and cotton cargos, allowing it to more than quadruple its size within a few years and become Egypt's key port.[37] The new waterway offered a connection to other modes of transport, again along the trunk-and-branches line. After 1820, several Bedouin tribes were settled on its banks, as part of Mehmet Ali's cooptation of local forces, using their animals to transport produce from the canal overland. Later, their Shaykhs secured licenses for coal-enabled pumping stations.[38]

The adjacent *khazzan*, a new artificial reservoir feeding the canal, proved wasteful in terms of evaporation and ground absorption, and in 1849 two pumping stations were constructed to feed the canal directly from the river. Until the 1890s they were periodically enlarged and furnished with newer steam pumps, combining a system of pumps and waterwheels connected to "inverted marine steam engines" of the type developed and used in steamships; this nearly doubled their output every decade.[39] By the 1890s, portable steam engines were also the norm for watering fields. Usually an eight-horsepower engine connected to an eight-inch pump was placed on a plot of level ground, and the pump was supported on a wooden trestle or fixed inside a masonry well.[40] In 1899, there were more than thirty-three hundred portable engines with a total output of about thirty thousand horsepower in the country, mainly burning British coal.[41]

One of the main advantages claimed for pumping clean water from the river to irrigation canals was the fact that it made dredging the canals less urgent.[42] Simultaneously, steam power itself became increasingly important also for canal maintenance and dredging (see figure 5). The two main dredging companies, the Behera Company and the Egyptian Dredging Company, employed various kinds of mechanics, such as the Priestman dredger (from the engineering firm that later built the Priestman oil engine, a prototype for the internal combustion engine), and mainly twelve-inch centrifugal sand and mud pumps which required thirty-five- to forty-horsepower engines.[43]

During the decade-and-a-half before the 1869 inauguration of the Suez Canal, Egypt had seen the largest concentration of mechanical

FIGURE 5. Caravan passing a dredging barge, circa 1880.
Museum für Kunst und Gewerbe, Hamburg.

energy in the world, mainly in the form of dredging machines that dug the canal in tandem with human workers. The extent of technological trickling-down from this ambitious project to irrigation ventures is unclear. What is obvious is the steady increase of the use of dredging machines in this field and the declining proportion of corvée workers—from one-fourth of the general male population under Mehmet Ali to one-eighth by 1881, a year before the British occupation. During the 1880s, actual numbers of corvée workers were falling in similar proportions.[44] This was a nonlinear process, rife with experimentation and failure: for example, in 1885, when the government contracted out the dredging of the Mahmudiyyah and Ismailiya canals, where a technologically intensive approach was applied relying largely on Priestman dredgers, the project was seen as a failure. In 1887, a significant improvement was achieved by using sand pumps, and by the end of the decade the dredging corvée was eliminated.[45]

River steamboats capitalized on the fact that pumping stations and their coal dumps were stationed along the navigable canals and the Nile (for this network see figure 6). Yet with the expansion of rail transport from 1852 on and alongside the same waterways (according to William Willcocks, the chief British irrigation engineer in Egypt, canals and railways were designed together),[46] river-bound steam navigation found itself in direct competition with a government monopoly—the Egyptian State Railways (ESR)—therefore facing heavy tolls and other govern-

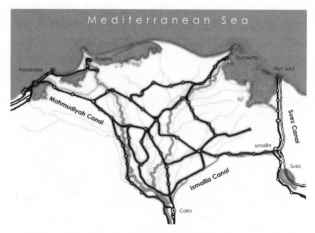

FIGURE 6. Late-nineteenth-century Egyptian waterways.

mental obstructions meant to encourage the railway.[47] The ESR, whose main branch stretched along the Nile and depended on the river as a source of water for steam generation and boiler cooling, and on animals to carry this water, was a key consumer of British coal.

The Suez Canal was partly modeled on the Mahmudiyyah. The connection of Alexandria to the Nile via the sweet-water Mahmudiyyah Canal enabled the dramatic expansion of the port city. This was also the case with the Suez Canal town of Port Said, the emergence of which as one of the major global coaling stations after the canal's 1869 inauguration depended on regular water provision and on an effective aqueous driving belt for transport. In the town's early years, Nile water was carried to it by boat and on camelback. In the late 1850s, three thousand camels and donkeys were sent regularly from Alexandria to Port Said, and about eight hundred animals carried water from Qalyubiyyah. This was compounded in the summer of 1859 by two steam condensers (with the addition of a third in 1861), each capable of desalinating five thousand liters daily (see figure 7). This method of desalination required one kilogram of coal per liter of water and was three times costlier than getting barreled water. Water was distributed to houses on the backs of men and animals, attracting to the port city farmers who found their livelihood in this profession. These *saqa'un* quickly monopolized water distribution in the city.[48]

During the early 1860s, Port Said was connected to Ismailia via pipes through which drinking water was pumped by steam power, and then by

FIGURE 7. Water desalination in Port Said, from Paul Reymond, Le Port de Port-Saïd.

the sweet-water Ismailia Canal, providing Suez and Port Said with drinking water and facilitating transportation of materials for the construction of the Suez Canal, food for the workers, and coal for the dredgers. In turn, the canal required regular mechanical coal-fueled dredging. From the mid-1860s till the 1890s the town's water supply developed further with pipes, canals, and distribution facilities.[49] Water provision was a constant bone of contention in Port Said, leading to regular water theft and clashes between residents, as well as attempts at regulating the quantity and price of water, which had to be distributed between private residences, the docking warships, telegraph and government offices, the animal quarantine, etc. Generally, foreigners enjoyed larger quantities of water day and night, compared to residents of the Arab quarters, who received smaller portions only during the day.[50] Carbon politics animated hydro politics and prepared the ground for the emerging social hierarchies of the city.

The British agenda of perennial transportation, the Egyptian agenda of perennial irrigation and cash cropping, and the imperial aspirations of both flowed into one another via waterways that were gradually centralized and becoming coal-dependent. These sweet-water irrigation and transportation canals entangled the industrialization of textile production in England with the industrialization of cotton cultivation in Egypt, stamping both with the sign of coal. Coal's disembarkation onshore and much of its career thereafter was dependent on animals, waterways, and other movers. Coal-fired vehicles gradually became mainstream, pushing preexisting powers to the sidelines where, rather than disappearing, they proliferated under the radar. During this period, the nexus of coal, animal, and water power that increasingly informed transformations in Egypt as a whole and animated the development of entrepôts like Port Said and Alexandria soon also became instrumental in reform in Cairo.[51]

CONDENSING EMPIRE

Key among the existing Egyptian powers that allowed coal's advance was the Egyptian ruler's enthusiasm for steam. Mehmet Ali, Campbell reiterates, "gives to the steam communication every and the most cordial aid and assistance in his power, indeed without this aid our difficulties from the circumstances which I have already mentioned would be great, and almost insuperable." Campbell referred to technical help in the iron foundries of Cairo, tasked with repairing broken steamer engine parts; to governmental license to establish coal depots in Cairo, Suez, Alexandria, and the Red Sea Islands of Kamaran and Cosseir (built, in fact, by the Egyptians for the use of the EIC); and to camels and boats to transport coal to these depots. Moreover, the coal depots of Jeddah and Mocha were under the auspices of the EIC in Egypt.[52] To this should be added Egypt's expansionism as the British at once benefited from it and wanted to appropriate more directly the thrust it propelled.

Egypt's influence in the Red Sea and Arabia was significant not only for the shape of the coaling complex in and before 1838, but also for its future development, and especially for the conquest of Aden a year later and Aden's emergence as the key midway coaling depot instead of Mocha. As Campbell recommended, "the possession of [Aden] would be most advantageous to our interests; and so, as from the present war carried on by Mahomet Ali against the Wahabees and the Assyr Arabs, it is more than likely that the commerce in Mocha coffee would pass through Aden; and besides, the possession of Aden would completely put an end to any ulterior designs of conquest which Mahomet Ali might have towards that quarter."[53] The Egyptian ruler, "who evidently has some distrust of having the British as such near neighbours to Mocha," proposed that if the English only wanted a coaling station, he'd operate one for them in Aden, and pressed on Campbell the advantage of a coal depot without any expense of garrison or administration on the part of the Indian Government. Prime Minister Palmerston instructed Campbell to squarely reject this offer.[54] A political imperialist logic trumped economic prudence. Soon thereafter, British steam power indeed put a stop to Egyptian expansion along the eastern shores of the Red Sea, as it did a year later in Syria.

In several respects, the introduction of coal into the Arabian Peninsula and the subsequent discovery of oil were the continuation of a process pioneered by and in Egypt. Egyptian canals were themselves extensions of the waterways that facilitated the emergence of steam power in the British Isles, pushing it from flooded coal mines to industrial centers, from whence it moved along large navigable rivers and the railways that paralleled them, to sea routes and steamer lines, and then to large rivers like the Nile, the Egyptian railways, and the country's canal system. The same can be said of Arabia, connected to Egypt by aquatic ties long before the arrival of coal: in 1820, when the Mahmudiyyah was inaugurated in Egypt and named after Mahmud II, this Ottoman sultan ordered Mehmet Ali, then in control of Arabia, to repair Mecca's waterworks.[55] Egypt continued to exert influence, at once extending and undermining Ottoman sovereignty in this imperial periphery via its hajj caravans, railways, port cities and steam engines, all of which framed the entry of coal onto this barren scene.

In Egypt steam power mostly helped to pump available freshwater, maintaining and extending existing trajectories for political power. A significant exception to this rule was to be found in Port Said's steam condensers, installed there in 1859 and 1861. But what was exceptional in Port Said was the rule in Arabia. Around the last third of the century, and especially in arid environments, steam power was used regularly to desalinate seawater, thereby allowing European navies to capture and maintain new footholds in the region. Alongside hydraulic or mechanical power, desalination represented political power and sovereignty as well. Until well into the nineteenth century, sovereignty was not conceived or gained through control of flattened spaces on a map, but rather as control over narrow corridors. At first, these were natural arteries, but increasingly included artificial ones too.[56] Water being the smoothest and fastest conveyer of warships, government officials, agents of colonizing corporations, and irrigation inspectors, coal blended the British thrust to rule the waves with the centralizing efforts of the Egyptians. Its career in Arabia demonstrates another aspect of the pact between thermo- and hydro-politics.

To begin probing the encounter of water and coal supply with sovereignty in Arabia, we may start with Aden, the coaling station that after 1839 replaced Mocha as the key British fueling hub on the Arabian coast, and one of the first coal depots to convert to desalinated water. These features soon made it one of the largest coaling stations in the world. Between the 1839 occupation and the mid-1850s, British occupiers used Aden's existing water infrastructure and even restored the city's medieval reservoirs.[57] But with the expansion of the fueling station, existing water sources were no longer sufficient (even after shifting to mechanical steam pumps in the late 1840s),[58] and after repeated security concerns about the ability to defend them,[59] in the 1860s they were augmented with a new aqueduct and, when this was no longer sufficient, with steam condensers capable of desalinating seawater.

Although separating salt from seawater by evaporation had been practiced in antiquity, the history of desalination began in earnest with the European "age of exploration" and colonial expansion during the late sixteenth and early seventeenth centuries, when stove-top desalting devices were installed as emergency measures in oceangoing sail ships.[60] In the nineteenth century, steam navigation allowed the ships' engines and condensers to be used to supply freshwater and clear cargo space. The first steamers, designed for river transportation, could replenish their boilers with the water around them. When seagoing steam vessels were built, they initially had to carry freshwater supplies or use seawater which, being denser and saltier than river water, resulted in saline deposits on the flues or at the bottom of the boilers. Consequently, boilers had to be "blown off\out" periodically to release heavily salted water. Nonetheless, salt water reduced the life of iron boilers, increased the danger of boiler explosion, required more fuel to heat, and resulted in less elastic steam that was inferior to freshwater steam in generating motion. These problems were solved beginning in the 1830s with the development of surface condensers, and further during the early 1850s with the inclusion of seawater evaporators in machines like the Pirsson Condenser, which could transform seawater into freshwater to be consumed by the boilers and crew.[61]

Such condensers were regularly fitted to seagoing steamers only by the end of the 1860s. The delay was partly due to the fact that the new condensers required temperature-stable lubricants, and those were made available only after the 1856 breakthrough introduction of mineral oil (i.e., petroleum) lubricants and greases.[62] The increased friction, heat, and wear and tear inside the steam engine—a result of the need to handle water of various densities, chemical compositions, and temperatures—

was important for the emergence of "petrolubes" and, at mid-century, a new and important impetus for the budding oil industry.

Newer condensers led the way to higher steam pressure and eventually to the development of compound engines. Paradoxically, the increased speed of ships furnished with these high-pressure engines and turbines decreased their time at sea, making it more practical on some voyages to store freshwater in tanks than to produce distilled water at sea. For a while, this deterred the impetus for improved maritime desalination; however, other voyages and missions in arid territories maintained this impetus.[63] It is therefore not a coincidence that the standard textbook *Fundamentals of Water Desalination* cites the 1882 British naval campaign to occupy Egypt as a watershed moment in the transition from stove-top to the already fairly developed steam-boiler evaporation. (The centrality of this campaign, during which Aden's condensers worked twenty-four hours a day at full capacity and were bolstered by new ones, is also clearly reflected in the archival record.)[64] Indeed, the first company to develop and apply multi-effect distillation to desalination applications did so after installing condensers in Suez and Aden, as early as the 1850s.[65] Another such watershed moment was associated with petroleum exploration in the Arabian Gulf at the beginning of the twentieth century.[66] The history of desalination reminds us of a simple fact: it was *having left* western Europe and North America, sometimes for the furthest reaches of the Ottoman Empire in alien lands and waters, that the steam engine was perfected.

On the heels of developments in Port Said, beginning in the 1860s, condensers were installed in the Red Sea and Persian Gulf to provide distilled water in ports where the local water supply was limited. By World War I, a British empire of condensers was hugging the Arabian Peninsula from all sides including in Kuwait, Bahrain, Qatar,[67] Hengam Island ("the Perim of the Persian Gulf"),[68] the original Island of Perim,[69] Kamaran, Suakin, Aden, Port Sudan, Bushehr, Muscat, and elsewhere.[70] One of the first places to see this innovation, in the 1860s, was Aden, where several condensers were deployed by the British-India government and private companies, such as the Peninsular and Oriental Co. (P&O), and Luke Thomas and Company, and especially by Parsi entrepreneurs who had aided the EIC's advance in India and China. The companies mainly served their own and other ships calling at the port. While during normal times all these condensers, save those belonging to P&O, sold desalinated water to the general public, they were deployed so that in an emergency water shortage they would be able to sustain the foreign population, which by the mid-1870s was a little over nine thousand people, each calculated to consume five gallons per day.[71]

While condensers allowed increasing numbers of foreigners to settle in such arid colonial posts, existing cisterns and wells supported local populations, who were said to prefer local well water anyway because of its "bite."[72] Several reports indicated that Arabs, Somalis, and animals had a distaste for condenser water in Aden, and that animals would even refuse to drink it altogether unless some seawater was added.[73] This local response was similar to the resistance by residents of Cairo to desalinated Nile water, which they regarded as "dead water." British representatives elsewhere in the region reported a similar disinclination expressed by Kuwaitis and Persians, who refused to drink "dead" condenser water.[74]

In Aden and nearby, desalination clung to the language of emergency that had characterized it from the beginning of imperial expansion. Condensers and the coal bunkers feeding them were indeed part of a securitized logic, and they were unabashedly seen by contemporaries as tools of empire. From the 1860s, British commercial steamer companies provided condenser-desalination support for troops on punitive expeditions in the region.[75] Similarly, they were effective cards in diplomatic interimperial rivalries. In 1872, for instance, they helped ward off an Ottoman proposition to build a quarantine station at Aden. This was an attempted Ottoman assertion of influence couched in the language of hygiene and the need to protect the town's local population from cholera outbreaks ascribed to the British steamer traffic from India, where the disease was endemic. In response, British sanitation officials claimed that Aden's condensers ensured the town's safety from disease and made quarantine pointless.[76]

The British themselves (along with other European imperial powers)[77] were making claims about water quality in Arabia to simultaneously stave off accusations about their own responsibility for cholera outbreaks, and to further interfere in Ottoman affairs. In 1881 they were voicing vicious complaints about the quality of Mecca's Zamzam water, just as the Ottomans were engaged in an ambitious overhaul of the region's water infrastructure. This major refurbishment was meant to demonstrate the competence of Sultan-Caliph Abdulhamit II (1876–1909) in providing potable water for the region's residents and pilgrims.[78] By the last third of the nineteenth century, such technical aptitude was the foremost benchmark of political legitimacy, both inside the empire and in the interimperial arena.[79] This was especially the case with the pilgrimage to Mecca. Like the British, Dutch, Russians, Egyptians, and French, the Ottomans too were engaged in industrializing a hajj that was increasingly flowing along rail lines (from Russia, Egypt, and India) linked to steamer lines all leading to the port of Jeddah and requiring new sanitation, logistical, financial, security, and religious arrangements. Ottoman hydraulic renovations were part of a wider "infrastructural turn." In the Hijaz in the following decades, this included telegraph and rail lines meant to compete with land under British naval dominance,[80] but it also characterized the Egyptian and Ottoman scramble for Africa.[81]

Because British medical and sanitation authorities tended not to differentiate between the condition of water in Mecca and Jeddah, in the mid-1880s Osman Nuri Pasha, the Ottoman governor of the Hijaz, included the port city in the overhaul scheme of Mecca's waterworks. As in the Holy City, Jeddah was supposed to receive its water freely by pipes from a nearby well. In both cases, there was also the intention to undermine the power of local elites of water tank owners who leased their reservoirs to water carriers who sold water for a handsome profit. And in both cases, the renovations encountered fierce resistance, mainly in the form of Bedouins sabotaging the pipes.[82] After repeated emergency measures, by the end of the following decade there was no choice but to resort to desalination, and a small condenser was brought from Geneva in 1899. Another machine was installed in 1907, but it soon broke down.[83] In 1911, condensers were brought from England, assembled in Jeddah, and operated in a festive ceremony with local notables and steamboat agents.[84] By 1912, a large and a small condenser were operating at Jeddah.[85] The coalonial bonds were further tightened with the British suggestion that these machines required Cardiff coal rather than Ottoman coal from Zonguldak, as the former burned more cleanly and provided more energy.[86]

As Michael Christopher Low has shown, struggles around water provision in the Ottoman Hijaz are a key yet unacknowledged prehistory of today's Saudi petro-state. The Saudi regime, using about 15 percent of the country's oil to desalinate water, thereby sustaining large-scale wheat agriculture and water and bread subsidies upon which its legitimacy and stability depend,[87] should be seen as a hydro-state resting on the pillars of earlier Ottoman and then British experiments with water supply. Yet the Saudi petro/hydro-state was in fact brought about by the arrival of coal and steam power into the region, and as such it bears the imprint of coal's relationship with water and animal power. To examine what this reframing contributes to our understanding of politics in the region, let us revisit Jeddah's fraught transition from water pipes to condensers.

Existing scholarship, following the reports of turn-of-the-century sanitation officials, ascribes sabotage of the Ottoman water pipes mainly to a competition between elites: local tank owners, representatives of the central Ottoman government, and the British. Dr. Frank G. Clemow, the British representative to the Constantinople Superior Health Council and the Crown's delegate to the 1911 International Sanitary Conference, suspected that "certain citizens of Jeddah, [who] own the majority of the water tanks, were the ones actually behind the Bedouin attacks. . . . These interests had damaged the water pipes in an attempt to retain their monopoly over the cistern water and the Ayn Bariman springs."[88] Ottoman officials echoed these concerns.[89] They also repeated the received wisdom about the distaste of local Arab populations for distilled water. "The owners of tanks will do all they can to prevent the sale of distilled water, if they do not actually attack the machines. The people themselves, strongly conservative by nature, will take time to get accustomed to distilled water in place of the other waters they have been used to drinking."[90] On the other hand, being located in British naval security zones, condensers were easier to protect and more difficult to sabotage.[91] This reading holds water, yet it is incomplete. Anchoring the condensers in the context of existing powers, which not only produced water but also carried it to consumers, highlights other aspects of the story.

First, there was coal and the steamships it fueled; and since 1860 there was more of both with every passing decade. In that year, for example, 472 tons of steamer coal were stored in Jeddah, fueling a few dozen steamers.[92] During the following decade, after the opening of the Suez Canal, British traders were regularly supplying over five times more merchant and warships with coal stored in this port city.[93] Even more than Jaffa, Alexandria, Port Sudan, or Aden, Jeddah expanded rapidly during this period as a result of the hajj shifting from land to sea routes, with the adoption of all-weather steamships undeterred by the monsoon.

Steamships created new hazards and solutions. Shortening the contagion vector of waterborne disease, they quickly spread cholera from India to Arabia. Similarly, the massive new volume of pilgrim traffic put unprecedented stress on the city's water supply. Yet steam engines also offered a solution, in the form of condensers and desalination devices, as we have seen in Aden and elsewhere.

Unlike piped fountain or well water, desalinated water did not reach consumers directly on its own: it required a "feeder." As we have seen in other contexts, here too coal, water, and animal power animated one another along liner-and-feeder lines. The same process that had caused

a reorientation of the hajj from overland to marine routes affected caravan traffic significantly, concentrating many a camel owner in the port
city, from where they would take pilgrims on a night-and-a-half journey
to Mecca, usually with two pilgrims per beast, providing them with
water along the way.[94] For pilgrims traveling between Jeddah and
Mecca, camel and water fares were the largest single item of expenditure, and this gave rise to a vast informal Bedouin economy of camel
owners, water sellers, and travel guides.[95] When a plan for constructing
a railway on this route as an extension of the Hijaz Railway was entertained in Ottoman circles at the turn of the twentieth century, several
Bedouin tribes objected vocally, violently, and eventually successfully.
The plan not only jeopardized their share in the pilgrim traffic from the
coast, but also threatened the regular seaborne barley imports that were
key for feeding the hajj-caravan camels across the Hijaz.[96]

Resistance to the water pipes plan should also be seen in this context.
In fact, these were two linked parts of a single scheme: between 1896
and 1902 the Grand Sharif of Mecca was collaborating with the Ottoman governor of the Hijaz in requisitioning iron water pipes from Istanbul to furnish Jeddah with spring water. In 1902, this was part of a
larger initiative to begin building a railway from Jeddah to Mecca and
then to Medina.[97] Bedouin resistance to the railway cannot be separated
from their sabotage of the pipe infrastructure, and vice versa. This was
also true with regard to other sections of the Hijaz Railway, which
tracked the available water supply and were accordingly resisted by
Bedouins. Undermining Jeddah's water supply was a means to undermine the railway.[98] That the Hijaz Railway was nicknamed "the sultan's
mule" captures how animal power framed this competition.

Transportation alone was not the sole cause of Bedouin antipathy to
the new water system that excluded animals. If in Egypt prior to epizootics and steam engines animals were crucial for rural and urban water
supply, canal dredging, and irrigation, this was even more the case in
the Arabian Peninsula, where animal husbandry and water supply
depended on one another intimately. For this reason, any significant
reform of the region's water system had to take account of this connection. Attempts that did not—such as the Ottoman piped water scheme—
failed. Attempts that accommodated animals and their owners—
and desalination was one such arrangement—proved more successful.
Beyond their significance in explicating the enduring material conditions for effective sovereignty in Arabia, examining the continued
importance of animals also offers another powerful case against the

"energy regime" paradigm that sets coal against preexisting powers and, in turn, against oil. By contrast, only through these existing powers could coal and oil take root.

Unlike water pipes, condensers in Jeddah and other Arabian port cities did not undermine the importance of animals and animal owners in water distribution. On the contrary, they provided more water sources for more tribesmen and their mules and camels to carry. The steam engines that operated them were not attached to railway locomotives but rather to the steamers that brought barley from overseas to feed the camels and the pilgrims who rode them. Given that Bedouins were evidently the pipe saboteurs, accounting for their agenda as animal owners reveals why condensers were more successful than pipelines. By improving the well-being of animal owners while lessening the grip of any single water supplier, condensers benefited from the mutually reinforcing effects of coal, water, and animal power. This formula allowed for a stable sovereignty in Arabia.

The adoption of condensers was also a way in which the Ottomans moved with the herd, and perhaps an indication of the continued Egyptian influence on imperial affairs. On the western shore of the Red Sea, directly across the water from Jeddah, Port Sudan had just abandoned a water pipe scheme and adopted condensers. The evolution of this port—the fruit of the thrust of Egyptian-British empire-building in the Sudan—required increased water supply to sustain rapid urban development led by a growing community of foreigners. In 1905, British officials contemplated extending the water pipes to a hill fountain in Khor Arbaat, especially as the town's Europeans would not drink water transported to the city by donkey from closer wells and consumed by the Sudanese and poorer Greeks. Eventually, however, they chose the condenser option. Such devices, brought from the southern port city of Suakin and put under the responsibility of the Sudan Government Railways, supplied freshwater distributed by animals for the next two decades. Coopting the city's animal-owning water distributers and boosting their numbers, these devices managed to support an extraordinary pace of urban development resulting in a quintupling of water consumption during the 1920s (from less than three hundred to more than fifteen hundred tons a day). In the end, this led to the resumption of the Khor Arbaat plan and the eventual replacement of the condensers with spring water.[99]

Comparing this case and that of Jeddah reveals that a shift to piped water depended on both control and securitization of production and cooptation of distribution and transportation. In 1926, Port Sudan's animal owners were used to transport iron pipes that arrived at the port and connected it to Khor Arbaat. Whereas in the age of condensers steamships in Port Sudan had to pay the prohibitive price of $2.50 per ton of desalinated seawater, the new arrangement decreased the price to slightly over 50 cents per ton.[100]

A 1926 *Popular Mechanics* article titled "Steel Pipes Conquer Desert Wastes" celebrated in its subtitle how, over the previous decade, "Oil and Water Lines, Carried across the Sands on Camel Back, Introduce Modern Transportation to Persia and the Sudan." In the article, the connection of Port Sudan to freshwater springs in the hills of Khor Arbaat earlier that year was wedded to the completion of the 150-mile-long double pipeline from what came to be called *midan-i-naftun* ("the place of oil") to the port of Abadan.[101] Previously oil had been transported in barrels by camel; now the new pipeline would allow it to flow directly to tankers waiting at Abadan. As the article suggested, oil too was connected to water. What follows suggests that steam power framed this connection.

The case of Iranian oil—the first instance of this resource being discovered and produced in the Middle East—is paradigmatic to current understandings of the coal-to-oil transition. Access to this oil allowed the British Royal Navy to abandon coal and eventually also the steam engine.[102] For our purposes, what is noteworthy is the fact that in the Iranian case too, oil, water, and animals have been connected in a context framed by the power of steam. When Rear Admiral Edmond Slade was sent by First Lord of the Admiralty Winston Churchill to Iran in 1913, with orders to investigate purchasing 51 percent of the Anglo-Persian oil company, he and his committee were taken to a water well in Lengah, on the coast of the Persian Gulf. The water drawn in their presence showed distinct visual and olfactory traces of oil. Such surface oil shows, as well as gas seeps, would guide oil exploration efforts during the first half of the century.[103] The visit pleased the committee members and they eventually recommended approving the transaction, which resulted in shifting the Royal Navy to oil.[104]

Anglo-Persian had in fact been digging artesian wells in search of both oil and water across the region. On another occasion, the company was approached by British officials involved in improving the water supply in Kuwait, previously dependent mainly on condensers, to sink a trial bore for ground water. As was often the case, here too the option of deploying another coal-fired condenser was eventually chosen, and Anglo-Persian was tasked with erecting it.[105] When an oil refinery was built in Abadan, the company employees were sustained by two such condensers and a filtration plant built in the 1920s.[106] In general, water was an important component in various aspects of oil drilling, requiring the construction of a water supply infrastructure: it was useful for clearing debris and for cooling drillers, and injecting it into the ground helped produce pressure that forced the oil out more quickly.

In Abadan, Persian crude filled steam-propelled tankers, a transportation method that continued until after World War II and developed out of the challenges of carrying and handling large amounts of water inside steamers. The pipelines taking the oil to Abadan (the first completed in 1912, the second in 1926) offer another example of how coal steam power framed the context of oil's emergence and then disappeared. Crude oil, being thicker and heavier than water (Persian crude, even compared to others, was exceptionally heavy), required strong steam-pump engines to make it flow along these new pipelines.

Other existing motive powers were also crucial: steam boilers for the pipeline's pumping station had to be dropped at sea near Abadan (before a harbor was built there), towed ashore by rowing boats, and carried 150 miles by camel to the oil field. Pipes were also carried using animal power, which helped orchestrate its own obsolesce: "When the oil was turned into those pipes, one of the most travelled camel-caravans in the world was abandoned for this purpose," stated *Popular Mechanics*.[107]

As with water pipes in Jeddah or Mecca, the stage was set for oil-pipe sabotage by those very camel drivers now left without any alternative means of subsistence. Charles Ritchie (who later became the manager of British Petroleum) was responsible in 1926 for supervising the three thousand tribesmen who erected the pipeline. He recounted how immediately after completing the project, workers unfastened the bolts holding the pipes together at night, so that they would be hired to refasten them the next day.[108]

These small-scale subversions should be understood in a broader context. Anglo-Persian was building its pipelines with an eye on Russian pipelines such as the one operated by the Nobel brothers since the 1870s to connect oil fields to distilleries in Baku. This pipeline had, in turn, been copied from the prototype assembled in the mid-1860s in Pennsylvania in order to remove the constraint of the teamsters who transported oil in barrels to a nearby railway in their horse-drawn wagons. These short trips were sometimes as or more expensive than the long-distance rail fare, and they represented a strong claim on the part of workers to a share in the profits of their employers. Ludwig Nobel—

who a decade later copied the idea and technical specifications of the Pennsylvania pipes, their metal joints, and the steam engines that pumped the oil through them—wanted to break a similar monopoly of Tatars who transported the oil from oil field to refinery in horse- or mule-drawn carts. This too was an expensive means of transport, interrupted by "heat, storms, and Moslem religious observance."[109]

As in Iran, in both Pennsylvania and Baku cart drivers sabotaged the new infrastructures that threatened to replace them. Yet the violent hostility between competing modes of transport and "energy regimes," and between members of the working and capitalist classes, obfuscates the fact that in all cases many of these camel or mule drivers eventually found employment transporting pipes and other heavy equipment, extending the new systems ever further.[110] The impetus of the fossil fuel infrastructure to expand, in other words, was partly animated by the need to redirect and co-opt potentially destructive existing forces. Although in the grand scheme of things animal owners may indeed have participated actively in bringing about their own obsolescence, the process took decades.

The very limited usefulness of entertaining an insulated "age of coal" and the impossibility of separating coal from water and animal power during most of the nineteenth century also disrupts the "energy regimes" perspective when moving forward or climbing the ladder of energy sources. Like coal, Iranian or Saudi oil—and by extension the "age of oil" in the Middle East—often seems to spring ex nihilo and out of context from under the desert sand. Yet oil too was connected to existing powers and resources. In the early twentieth century, these powers included an infrastructure created for transporting and burning coal. Recounting oil's connection to coal via animal and water power helps get a firmer and more concrete grasp of a slippery substance that tends to be understood today exclusively with abstractions like money and energy. Teasing out these longer legacies might inflect conventional understandings of current issues like water scarcity in Arabia and the Gulf. Anchoring desalination in the history of imported European coal, rather than as a legacy of oil and an offshoot of that cheap and easily accessible resource, throws into sharp relief the artificiality and unsustainability of current life in oil-producing states and their intimate relationship to Western imperialism.[111]

Instead of an "energy regimes" perspective, we have opted for a historical framework that does not a priori analytically and artificially separate fossil fuels from sustainable sources, coal from oil, steam engines and horses or camels, or even the British Empire, from the Bedouins who carried its desalinated water or oil pipes across Arabia. This is not to say that ruptures and shifts did not occur in these systems. But the rise of an alternative motive force was a necessary, though certainly not a sufficient, condition for removing existing ones from the picture. Coal is still used in most industrialized countries even in our so-called age of oil, and in fact its use is on the rise in India and China and the rest of Southeast Asia; and in places like post-Ottoman Egypt and Turkey, farm animals successfully compete with gasoline-burning tractors in agriculture.[112] Animals, coal, and water constantly carried one another as fuels and prime movers, frequently shifting roles, imitating each other, collaborating and competing. Their imbroglios were the heterogeneous infrastructures that simplifications such as "energy" targeted and sought to reduce to more manageable and docile categories. Historians should follow and recount these attempts at reduction but avoid reproducing them. We too have followed the interconnected sites of England, Egypt, and Arabia during consecutive historical moments: the emerging sector of British industrial textile spinning and weaving, the carbon fibers it extended overseas to the coal depots of Mocha and Aden, to the Egyptian industrialized agricultural periphery, which replaced subsistence agriculture with cash-cropping, and then the expansion of these fibers to the urban centers of Cairo and Alexandria, to Jeddah, Port Sudan, and Abadan. Rather than "transitions," we witnessed a diversification of fuels and prime movers. Global warming is a result of this accumulation: of emissions from various fossil fuels, animals, and the impact of hydroelectricity; economic growth too is a result of them piling up on top of one another, and of the need to co-opt existing forces which would otherwise disrupt the entry of a new motive power, a new force predicated on old infrastructures.

The various cases examined above allow us to generate insight about the impetus of such intensifications, about the fixation of coal and oil systems on growth—both material and economic. In several instances we have seen how the fear of sabotage was a powerful engine towards expansion: as steam engines and related rail or pipeline infrastructures inserted themselves into existing systems, they generated anxieties and resistance which, in turn, they redirected toward further expansion. The fragility of the new system, not only its power, was a key factor in its expansion. In such systems, the expectation of neat ruptures and clear transitions seems to be part of the problem rather than the solution.

CHAPTER 2

Animals

FROM ENGINE INTO FUEL

We have seen how water was transformed during the second third of the nineteenth century from a power source to a conveyer belt and, a few decades later, to a consumable product trickling out of coal-fired desalination machines. The current chapter attends to the fate of animal muscle power over the course of the nineteenth century, examining its internalization and incorporation within a human body now imagined as a thermodynamic engine. Whereas the previous chapter took up steam and desalting, this one will complement the spectrum of water's chemistry and states of mater and examine salt and ice. Chapter 1 looked at animals as engines; here we will look at how they became food.

The peak of the age of horsepower was much more recent than one might think: in the early twentieth century, horse-drawn vehicles became increasingly necessary for short-distance transportation in Europe's metropoles as trains and steamers brought goods and passengers from afar. The peak in horsepower in agriculture came even later.[1] In what follows we continue examining the simultaneous quantitative increase and qualitative sidelining of animal power, on a trajectory that began as a trunk evolving into a branch, and finally to edible fruit. In the same way, the previous chapter showed that steam engines produced new sources of potable water in arid settings. Changes in forms of transportation were indeed quite meaningful: in the past, a significant advantage of eating animals was the latter's ability to transport itself, as food, from pasture to butcher on its own four legs. The growing scale of meat consumption and the emergence of new transportation infrastructures such as rail and tramways that extended from ports to inner cities made such advantages redundant and even transformed them into an impediment, since relying on animal muscles stiffened their flesh and made it less palatable; and cattle-droving also meant that animals lost weight along the way.

Abandoning "energy regimes" does not mean forsaking change altogether. As this chapter documents, in relative terms animal muscle was indeed utilized less and less as an external driving force as the nineteenth century progressed. Yet adopting the energy regimes perspective by categorizing animals as externalities prevents us from acknowledging their persistence in roles other than driving forces. It also prevents us from taking stock of the environmental damage they wreak in these new roles. Finally, ignoring such transformations means ignoring changes in our own human nature and thus key resources for agency and resistance.

Animal power and waterpower were connected to, and animated by, the fossil fuels that supposedly took their place. They were also directly and deeply connected to one another both before and quite apart from the mediation of coal and then oil. This mediation was arguably the first step in their eventual sidelining. Unless one deploys an abstract organizing principle like "energy," it is hard to disentangle motive powers from one another materially. Especially in a setting like the Arabian Peninsula, taking as an example the overland connection between Aden and Port Said—the geography this chapter revisits—water and animals in particular were inextricably linked.

The water source is a good place from which to appreciate this deep connection. Wells were built to accommodate the anatomy and behavior of animals: well water was raised either by a contrivance of drum and buckets, the shaft of which was turned by a mule or drawn by a camel pulling a rope;[2] water then flowed to fill containers for men and animals, the latter lending their names to this process in the regional dialects. For example, *manhal* (pl. *manahil*), "a watering-place with an abundant flow of good water," is derived from *nahal*, "thirsty camels being watered at the well."[3]

The opposite was also the case, when hydraulic topography was modeled on animal anatomy, as with the word *thimayil*, "bellies" (e.g., of camels), but also denoting "small depressions in a rocky riverbed into which the rain water runs under the gravel," and "water-holes hidden under the ground, often in the soft loam under the gravel of a dry torrent channel."[4] Such connections defined the nexus that linked Bedouins, animals, and water. Thus, from the word *banah,* "a camel hair rope forming a belly girth,"[5] one could identify "those who pull the belly girth" (*min yimiss al-ban*), that is, the Bedouin riders, and thereby derive *bain,* "broad and deep valley without any channel," or "water-course."[6] Smaller and portable water containers also retained faunal connotations. The word *riwi,* "to quench one's thirst, to drink," was associated with *rawyah,* "water-skin made of camel leather," or "large receptacle made of camel skin used to hold and carry water on a camel's back."[7] Similarly, *girbah* was a "water-skin made of the hide of a goat, holding about 40 liters, smaller than a *rawyah.*"[8]

In short, water was drawn by animals from wells that mimicked dromedary bowels to nourish people and their quadrupeds, and it was carried by and consumed from receptacles made from animal organs. Drinking was an act of becoming one with the beast. Bedouin control of water sources in the desert also gave them control over animals and defined the patterns of shepherding. Animals were shackled in irons for work in the morning, and released to graze freely until high noon, when they would return on their own to drink.[9] These profound connections were significantly altered during the nineteenth century.

Whereas waterpower retained and even increased its significance in the age of fossil fuels, animals lost much of their importance as key engines in most economic sectors in the West.[10] An "energy regimes" perspective would therefore have us remove them from the picture of the age of carbonized motive powers. But a different perspective, one focused less on production and work-output and more on actual fauna, less on energy transition and more on enrichment, less on rupture and more on reconfiguration, would show that almost everywhere in the world animal biomass rose considerably during the nineteenth century and ever since (as seen repeatedly in chapter 1 for the nineteenth century; in 1900 there were roughly 1.3 billion large animals in the world, a century later, about 4.3 billion).[11]

If domestic animals were growing in numbers but were no longer needed for field or industrial work, how were they being used? Under steam, animals were transformed from workers to food as a result of their key economic and social roles. This is not merely a claim about animals' functionality, but a description of their physiology, and our own. As Christopher Otter shows, from the late eighteenth to the late nineteenth century, the ratio of muscle to fat in the bodies of livestock changed dramatically in favor of the latter, in what amounts to an evolutionary leap.[12] Of course, function and physiology were interconnected (in the bodies of both animals and humans who, as regular meat eaters, also changed in height and longevity). As muscle mass was transformed from an advantage, allowing an animal to work harder, into a disadvantage, as its flesh became stiffer, animals did less and less agricultural work and were moved away from grazing on weeds to passive feeding instead.

Again, this both was enabled by and, in turn, animated the combined power of coal and water. We can therefore continue provincializing the steam engine by examining the role of the Middle East and other British and Ottoman imperial peripheries in the emergence of a fossil-fueled global meat industry. And vice versa, as this chapter also asks how the democratization of frozen meat transformed the Middle East. The trend of carnivorous eating moved from Aden, Port Said, and other port cities into the region's interiors and large urban centers. In this process, crystal ice cubes paved the way for the progress of meat along a trajectory of black coal.

The previous chapter showed that we cannot think of steam engines and water without also thinking of animal power. This chapter suggests that we cannot think of animal power in the age of steam without considering water; and vice versa: coal-fired ice production and refrigeration were offshoots of the steam engine–animal nexus. The steam engine's initial function was to produce heat (hence the names "heat engine" or "fire engine" for early prototypes) which transformed water into vapor to use as steam pressure to generate motion. But during the second half of the nineteenth century, it became clear that steam engines could also gainfully do the opposite: transform water from liquid into solid form or produce cold temperatures that would reduce or stop motion altogether. The mechanical production of cold was inherently the fruit of the coal-animal-water complex: before it started chilling fruit, eggs, and milk,[13] the main object of refrigeration technology was freezing or chilling animal meat (specifically those parts of it composed of water) as well as the flesh of (European) humans in hot (colonial) climates. In the new complex, meat was usually slaughtered at the port, close to transportation infrastructures, and frozen on board. The global meat industry was therefore a significant setting for the technological developments of artificial ice production and refrigeration, but also specifically for the development of steam technology.[14]

SINEWS OF EMPIRE

From around 1500, forestalling the putrefaction of meat was the initial incentive for a European thrust into the Indian Ocean, and for a process of empire-building laced with trade in pepper, cardamom, and nutmeg. Pepper and nutmeg masked the taste of meat cured by salt, a process which prevented it from rotting. This was achieved by treating meat with salt (or using other techniques like smoking) to remove moisture by means of osmosis, making the product less hospitable to microbe growth and decreasing its rate of spoilage.

Whilst establishing an imperial dietary regime in the East, the British were establishing a "pemmican empire" in America, where other "Indians" and frontiersmen were hunting and selling bison meat and fat to British and Canadian companies. Starting from the 1780s, but gathering speed after the 1820s (until the collapse of buffalo herds by the 1880s), this food fueled the wide circuits of "York boats" that established a British colonial presence from Hudson Bay to the Arctic Ocean, as well as significant expansion inland.[15]

In the last third of the nineteenth century, the British Empire—and specifically the import of mutton and cattle from Australia and New Zealand to England—was the main engine driving the development of refrigeration technology.[16] The nineteenth century saw the "democratization of meat" and its transformation into a working-class staple food in western Europe, in tandem with medical justifications couched in the new language of "nutrition" and analogies between the human body and a steam engine requiring fuel.[17] The 1880s saw parallel experiments in live transports versus refrigeration. The former failed repeatedly, for instance when bad weather transformed a ship's deck into a gory mess of broken animal bodies attracting sharks, or led to fires such as that on board the SS *Egypt* in 1890, where hundreds of beasts were roasted alive. Such failures bolstered the latter option of refrigeration.[18]

The first successful consignment of frozen Australian meat arrived in London in 1880, "in such good condition that neither by its appearance in the butchers' shops, nor by any peculiarity of flavor when cooked for the table, could it be distinguished from freshly killed English meat."[19] Such descriptions were part of a press campaign that made frozen meat palatable to a public that had already enjoyed several decades of positive experience with Australian canned and salted meat. In early February, a few days after the arrival of the first consignment was celebrated with servings of cooked frozen meat, the press reported that "a splendid joint of beef, cooked that afternoon, was put on the table and pronounced to be fully equal to the best English-bred beef. The juice, flavour, and colour had not in any way been affected by the freezing process."[20] Another correspondent commented that "the mutton and lamb cutlets were as succulent and sweet as if the animals from which they were taken had roamed the Sussex Downs within the past fortnight."[21] During the 1880s, British newspaper readers were complaining about the unhygienic habit of butchers wrapping meat and fish in newspaper, bemoaning that "nothing could have been more unappetizing to look at than the steak with the daily news almost printed on it."[22] Nevertheless, they were conditioned to consume frozen meat with these material and discursive additions. The campaign was a huge success: whereas in 1880 Britain imported 400 sheep, in 1893 it imported 3.4 million.[23]

The nexus of steam, print, and frozen meat informed refrigeration from its earliest days. Anecdotally, the "grandfather of the refrigerator," Scottish printer James Harrison, discovered the principle of refrigeration in Australia while cleaning movable type with ether. He noticed that the evaporating fluid left the metal colder. A few years after building a faulty prototype in 1855, he built an operational ice maker that consumed one ton of coal daily and produced between four and five tons of ice. The device's main function was seen to be cooling animal meat, and "it is in hot climates, however, that the full value of the invention will be felt. Ice within the tropics will soon be looked upon as a necessary of life, as much so at least as fuel is a necessary in the winter of temperate regions."[24]

Some of the pioneers of thermodynamics were involved with experimentation in transoceanic meat preservation, lending the companies involved their experience, fame, and contacts. For example, Sir William Thompson, later Lord Kelvin, had introduced the brothers Henry and James Bell to J. J. Coleman, with whom they established the Bell-Coleman Mechanical Refrigeration Co.[25] The first successful shipment of frozen meat landed at Le Havre from Buenos Aires in 1877 on the SS *Paraguay*, which had been fitted with an ammonia compression machine. The overzealous engineers were so determined to succeed that during the entire passage they kept the temperature at about -17°F (for comparison, contemporary meat is shipped at approximately -1°F). The ship's cargo consisted of 5,500 carcasses of mutton, which arrived in "tip-top condition." Learning of this success, in 1880 the *Strathleven* left Australia equipped with an advanced Bell-Coleman refrigerator, loaded with forty tons of frozen beef and mutton, and landed in London on February 2, 1880.[26] Realizing the potential of New Zealand's enormous livestock population and of the frozen export trade evident in the first attempts from Australia, the New Zealand and Australian Land Company was established. In preparation for a pilot voyage to England, a makeshift slaughterhouse was erected by the dock, first-rate butchers procured, and the most sought-after brands of sheep selected for a trial consignment to be frozen on board.[27]

Another major shipment arrived in 1881, at a time when London was cut off from overland supplies owing to heavy snowstorms. Its success convinced Australian meat companies to establish a regular steamer line, "the Orient Line," to the British capital.[28] By the end of the decade, Londoners were so used to getting their meat by sea that when this flow was interrupted by the London dock strike in 1889, there was good reason to settle the dispute.[29]

Midway stations engaged this process in multiple and meaningful ways. They also witnessed the production of ice and the transition from salted to frozen meat under the auspices of coal. Consider Aden in the decades leading up to the 1880s: despite expectations that it would turn into a commercial entrepôt for the Red Sea by wrestling the coffee trade from the Egyptian-controlled port of Mocha,[30] in the early years of the settlement, hostilities from the hinterland sultanates severely reduced Aden's inland trade, and the port turned to maritime and coastal trade instead. Virtually the entire Red Sea region had some share in supplying the rapidly expanding British coaling station. Rice came from Calcutta; sorghum from Persia, Zanzibar, the ports of Yemen and India; wheat was supplied from Bombay, but also from Kutch and Persia; pulses came from a variety of ports in the Red Sea, Persian Gulf, and Arabian Sea; potatoes and onions were shipped in from Bombay and Egypt. Most supplies of fruit were local, but large quantities of oranges and pineapples came from Zanzibar.[31] However, the single most important foodstuff attainable in Aden was Somali meat, without which the European population and the British garrison at the depot—accustomed to the new meat-eating habits—would not be able to survive. Somalia was soon transformed into "Aden's butcher shop."[32]

Meat was consumed in Aden by both the wealthy and working classes. Europeans ate meat daily, as did the wealthier Arabs. Poorer Arabs subsisted on a leaner diet and rarely ate meat. Somalis consumed more dairy and meat, which they preferred part-baked with boiled rice.[33] Oxen for work and cows, goats, and sheep for milking were procured from the hinterland, whereas animals for food were imported from Africa. Fat-tailed black-faced Berbera sheep were the main meat import.[34] Animals for slaughter were shipped into the colony, where they were handled by local butchers who were mostly Muslim, but also included some Jews.[35] The annual demand for sheep and cattle for food was increasing annually. During the year 1875–76, some 63,262 sheep and goats and 1,104 cattle were imported from Africa by sea.[36] In 1899, 2,000 cattle and 82,700 sheep and goats were imported, of which 1,000 cattle and 79,400 sheep and goats came from the Somali coast; the rest came from Arabia.[37]

Initially, British troops were supplied with salted meat (which they ate with pickles, twice a week). This was delivered from Bombay, where a salt works based on solar-brine evaporation had existed for centuries.[38] Soon, however, the occupiers began to use contracts for fresh meat as a political inducement. As Mahalal ha'Adani, a nineteenth-century Jewish chronicler and ice factory owner in Aden put it, "when the Bedouins saw the Christian army they were taken by religious fervor . . . and began to gather and conspire. But the British opened their sacks and boxes full of silver and gold, and the coins were new and shiny. They began buying a lamb for a sovereign, a jug of milk for a half-sovereign, an egg for a silver coin; the locals sold everything for ten times its value and received pure silver in return."[39] Such co-optation was indeed a professed British policy, predicated on the assumption that, "if the Arab merchants of Aden . . . enter into any contract with the authorities here . . . they will fulfil their bond, with profit to themselves and their connections inland will ensure an ample supply—indeed not only the Sultan but the numerous Chiefs of the different divisions of the Abdali tribe, begin to feel the presence of the British necessary to them."[40]

In the mid-1840s, 'Ali Abu-Bakr, the Sultan of Lahej's former governor in Aden, and Menahem Moshe, a wealthy Jewish merchant who became the leader of Aden's Jewish community (and who was connected by marriage to Mahalal Ha'adani), were bidding against one another to provide fresh meat to the troops. In the following decades, a system of private contracts for meat—brought by steamers (during the monsoon) and dhows (at most other times) increasingly from Somalia— was deemed better than the alternative whereby the British themselves, or some of their dependents, would keep a herd of sheep at Aden. This was ruled out especially because watering the animals was too expensive.[41] It is also likely that the elaborate system of engagements with animal owners—which we saw was key to water provision in the area (and in Aden itself, 'Ali Abu Baker was also providing water in the port)[42]—played a role in appeasing powerful local forces. Indeed, from the 1860s the British made sure to recruit and pay a well-balanced number of representatives from the main tribes to accompany the Indian troops and provide information about the interior.[43]

Debates about different schemes for fresh meat provision during the two decades following the mid-1840s reveal how political questions blended with concerns about the costs of water and steamer transportation, and steamers' propensity for technical malfunction.[44] Grass for fodder—for which Aden depended on the Arabian hinterland—proved equally consequential: besides concerns with meat supplies, ample grass was required by the Engineer's Department for its camels and bullocks.[45]

An imperial dynamic is revealed in this process: like desalted water, salted meat was crucial for the military presence in settings cut off from regular supply lines. This was a product requiring especially large amounts of salt. As a British military doctor observed in 1858, "the salt meat for soldiers in the field has always been highly salted in order to keep for two years or more in every climate." In addition, treating this meat also required large quantities of potable water to remove the salt before cooking, a procedure that was wasteful in places like Aden where this resource was scarce and also reduced the nutritional content of the meat.[46] Thus, in Aden, concerns for the troops' nutrition and health (salted meat was correctly seen as a cause of disease) and water scarcity were bound up with the political benefits of procuring fresh meat from locals.[47] The smooth and economic operation of transmaritime coaling relied on ready-made local support systems and on optimizing their synergy.

Salted meat was useful for the setup phase of the Aden coaling station, but once fresh meat became the preferred option, other imperial logics emerged. Whereas inside Aden, animal purchases from local chiefs took the form of co-optation and soft power, the entrenchment and expansion of the station activated another, more aggressive modus operandi, manifest in the projection of British military force in Somalia. With the regularization of Aden's steamer connection with the African coast, and with the technical improvement of the high rate of malfunctioning that had made it an unreliable route for food provision in the past, the Somali option became increasingly attractive.

Unlike other imperial holdings, Somalia was not incorporated into Britain's emerging world system of cash-crop agriculture: Somaliland was devoid of mineral wealth, and its shallow soil, poor drainage, high alkalinity, and rockiness rendered the gypsum and limestone rangelands of the eastern Horn largely unusable for agriculture.[48] Its value lay solely in the production of livestock and their by-products. British suzerainty was extended over the northern Somali coast ostensibly, as former British viceroy of India Lord Curzon reflected in 1907, "to safeguard the food-supply of Aden, just as the Roman Protectorate was extended over Egypt to safeguard the corn-supply of Rome."[49] Curzon, who during the 1880s had objected to the use of famine-relief funds for the extension of the Indian railways, had no qualms about yoking Aden's food supply to sustaining and extending the British steamer infrastructure.[50]

To gain access to Somaliland's internally generated surplus livestock and to assure its perpetuation for export, Somali traders and livestock brokers mediated between European capitalists and local Somali breeders. Commodity and most livestock shipments were controlled by a few European companies, but mainly by Adenese Parsi family firms (e.g., Cowasji Dinshaw, Premji Brothers, K. Pitamber) that had come from Bombay with the British occupiers. Starting in the 1840s, these Indian merchants began to replace the customary Somali barter system by extending credit to caravan leaders who traveled inland and returned the following year with their purchases.[51] By 1852, Aden merchants were buying up practically everything that was sold at the Berbera fair.[52]

The gradually expanding British influence in Somalia, motivated mainly by the desire to secure food for the vital coal depot of Aden, demonstrates the multifaceted connection between steam power, empire building, and meat consumption. Meat was first used for co-opting local strongmen and nourishing foreign troops; then it encouraged further occupation and troop deployment. Colonial and local armies marched on their stomachs during this period in Aden, Somalia, and as we will see, also in Egypt. This was part of a broader global process of the weaponization of meat. The use of animal fat to grease rifle magazines and later the use of glycerin to make explosives suggest that meat could physically be transformed into a weapon. Yet more than its capacity to explode, it was meat's ability to fuel that became increasingly important. In the nineteenth century, British military power was increasingly seen as connected to servicemen's regular consumption of meat.[53] As Aden and Somalia reveal, "food security" and "energy security" were tied to one another in the colonies, putting flesh on the bones of what I call "coalonialism."

MEAT ON ICE

The Parsi families who sold Bombay salted meat to the British troops stationed at Aden and monopolized the Somali fresh-meat trade at this coal depot were from the mid-1850s also the main players in the port's mechanized water desalination, artificial ice-making, and then local salt production. The British-Indian hybrid form of resource and political governance in Aden is captured by the local word coined for the ice water sold there—*baneis,* combining the Hindi word for water, *pani,* and the English word, *ice.*[54] Before the arrival of onboard refrigeration in the 1880s, artificial ice loaded from machines stationed at the port onto steamers, as well as fresh meat, eggs, and milk, was crucial for nourishing the crews and passengers of such vessels.

It is hardly surprising that in a place like Aden, one of the frontiers of the global fossil-fueled transportation sector, coal, ice, water, salt, animals for work and food, and steam power were bound together in various synergic relations. Water was desalinated by coal-fueled steam condensers and distributed around Aden with mules and camels. Steam technology was also used to produce ice from the desalinated water, which was distributed in the same way. Both items were sold mainly to the European population and on board steamers calling at the harbor for refueling with coal loaded from floating wharves similarly control-led by the Parsi firms. Ice was also used to cool the ample meat and other animal products which the Parsi merchants sold for consumption on board steamers that stocked up at Aden before crossing the scorching Red Sea. (In 1870, for example, the three principal Parsi firms at Aden supplied 20,019 pounds of beef, 14,957 pounds of mutton, 1,022 sheep, 69 bullocks, 261 dozen fowls, 2,866 eggs and 13,192 pounds of vegetables to the 88 steamers and 18 sailing vessels carrying passengers through Aden harbor.)[55]

By the 1870s, consumers on land and sea seem to have grown accustomed to ice, and when supply failed in the hot season, as it did occasionally, "the deprivation is much felt by all who can afford to indulge in this luxury."[56] According to Rebecca Woods, standardizing the temperatures of the British Empire was seen as one of the most exciting possibilities of artificial cold production. This endeavor forged the link between animal and human flesh. We saw in the introduction how Europeans operated in such a way that equated ice-cooled beverages, frozen or chilled animal meat, and human "northern muscles" supported by ice in southern and eastern regions.[57] During the hot season, the Red Sea crossing between Aden and Port Said was the most sweltering part of the passage from Australia to England. On board iron ships, artificial ice was used not only to prevent meat (or milk and eggs) from spoiling too quickly, or to alleviate the hardships of indulgent European passengers and crewmen, but more importantly also to resuscitate Somali and Arab engine-room stokers who constantly suffered from heatstroke in the boiler room and who were revived in baths of ice loaded at Aden and Port Said.[58]

Although dark-skinned men were deemed racially suitable for the conditions of the stokehold, steamer sick lists reveal that they were actually the population most susceptible to illness on board. This was the result of heatstroke as well as cholera and other infectious disease that seemed to target the vessels' most exhausted crew members first. Such was the case of the *Samannud* in 1865. The adventures of this ship, under Egyptian command, and its 1,865 North African pilgrim passengers reveal that when stokers succumbed to cholera they were coercively replaced by passengers whose diet was reinforced with chicken, to make them fit for the task. After being diagnosed with the malady, one North African was even forced to eat chicken daily and pay for it himself, despite vociferously protesting that he didn't want the chicken and that the money was for his children.[59] Such accounts suggest that the shift to meat consumption was not always voluntary. It also reveals that Egyptians like Sulayman Qabudan Halawa, the *Samannud*'s commanding officer, shared many of the racial and medical theories of the Europeans of his day (including seeing his "uncouth" passengers in a way similar to the colonial construction of "natives").[60] The racial perceptions that underpinned such labor arrangements (explained in chapter 3) were part of the way nonwhite bodies were anchored in a universal thermodynamic physiology.

The pattern of steam navigation between coal depots and the prevalence of regular meat supply and ice machines in these settings made meat a staple item for steam passengers. Meat and ice were combined to standardize temperatures and make steamers into little portions of Europe, and their passengers and crew into human motors. In 1888, as young Mohandas Gandhi sailed from Bombay to London calling midway at Aden, Port Said, and Malta, he described in his diary how fellow travelers repeatedly warned him that after leaving Aden he would not be able to maintain his vegetarian diet. Such appeals were surely less violent compared to the tales of the *Samannud,* but they seem equally persistent. When Gandhi stood his ground, he was given the same advice about the Red Sea. As Europe approached, meat consumption was seen as crucial for survival in its cold climate.[61] This was no mere lip service: as I have already indicated, a year later, the inability to offload Australian meat and the interrupted flow of coal at the London docks led to a resolution of the milestone 1889 London Dock Strike, which Gandhi followed and supported.[62] Gandhi's vegetarianism, labor politics, and contrarian anti-imperial stance developed in tandem. The latter included a critique of the British taxation on salt, which Gandhi first articulated in an 1891 article in *The Vegetarian*.[63] As the journal editors said of their young Indian correspondent, vegetarianism was not a challenge in India; it became such only in, or en route to, England.[64] Indeed, in an interview with the journal that year Gandhi made clear that "my aversion to meat was not as strong as it is now. [In India] I was even betrayed into taking meat about six or seven times at the period when I allowed my friends to think for me. But in the steamer my ideas began to change . . . The fellow passengers in the steamer began to advise us (the friend who was with me and myself), to try it [meat]."[65] As this case demonstrates, the standard thermodynamic carnivorous body was inflected and politicized not only from below, but also above deck, and between deck and dock. Notably, the politics of vegetarianism was not of the activist kind, but rather a politics of passive resistance, of noncooperation, of avoiding rather than doing.

Similar to Aden during the last third of the century, the canal town of Port Said became one of Egypt's main gateways for salted, fresh, and then frozen meat. Local salt works facilitated meat preservation. Thereafter, ice machines and refrigeration facilities offered cold storage for beef and mutton (consumed by the town's foreign population) which was transported on board passing steamers, and by rail to other urban centers for consumption by increasing numbers of Egyptians. Migrants from southern Europe slaughtered large animals at Port Said roughly from the time the town was established in 1859. These animals—mostly cattle, favored by Europeans—came from Malta, Cyprus, Greater Syria, and elsewhere. In 1867, the Suez Canal Company and the Egyptian government established a slaughterhouse on the beach and began enforcing medical inspections and animal quarantine.[66] In the following decades, Port Said became the gateway for the entry of Australian frozen meat into Egypt. Refrigeration facilities established at Port Said in 1892 allowed increasing quantities of Australian meat to be consumed in Egypt and by steam passengers crossing the Suez Canal.[67]

Whereas many of these developments followed familiar patterns of urbanization in the region and abroad, and entailed the removal of animal processing from inner cities, other developments in Port Said's protein intake followed trajectories that were particular to coaling stations. As early as the 1860s, residents of Port Said began tapping the salt crust (revealed during the canal construction) in Lake Manzala and in El Mallaha near Port Said for salt export, mainly by foreigners, and subsequently for increasing use by local fish salters. Using salt unsuitable for export allowed fishermen to distribute their catch to the interior and around the Mediterranean coast even before widespread refrigeration.

At the same time, Port Said's expanding fisheries attracted water birds to the vicinity of the ships, stimulating duck, stork, and other kinds of bird hunting which provided a significant portion of the city's growing demand for meat. This new ecology of interspecies predation translated into intraspecies predation as well, not only between large and small fish, large and small birds, and water birds and fish, but among humans too: fishing and hunting regulations prevented Egyptians from using firearms to shoot birds from their boats, whereas non-Egyptians were exempt from this restriction.[68]

In tandem with the rise of a salted meat sector, and together with the endeavor to separate salt from water to make it potable, a parallel large-scale commercial process of separating water from salt by means of basin solar evaporation characterized coaling stations like Colombo, Mauritius, Aden, Perim, Port Said, and others. The upsurge of this sector was the direct development of the land reclamation efforts that helped expand these stations (discussed in chapter 4), as well as the "salt law of ballast"—the ability of colliers from Europe or India to take on salt and thus avoid returning home empty or in ballast.[69] This, too, will be discussed in chapter 4. For our purposes here, it is significant to point out that the rise of the Red Sea salt sector in the 1880s, targeting mainly the Indian market and competing with salt producers in the subcontinent, was a key context for the famous Salt March led in 1930 by the then much more experienced Gandhi. Contesting British salt taxes was personalized and popularized by the fact that salt was used everywhere in India to replace minerals lost in profuse sweating in the tropical climate. The barefooted Mahatma framed his march to the Arabian coast as a "pilgrimage," imbuing salt with both universalist and particular Hindu mythologies. The emotional and religious mechanisms that turned it into such an effective politics were again inseparable from the biological, economic, and material infrastructures of empire.

FROM VEGETARIANS TO CARNIVORES

Following the lead of port cities during the last third of the nineteenth century, ice machines were introduced to most large urban centers in the Ottoman Empire. They had been familiar in Cairo as early as the 1870s.[70] In 1886, Sultan Abdulhamit II issued a concession to a partnership between an Ottoman official, Salim Agha, and a British company: a twenty-five-year monopoly on the production and sale of artificial ice in Istanbul and several other locations across the empire (the capital had used natural ice and snow for centuries, and had imported ice from Norway). The city's butchers, fishmongers, and steamship owners were among the main intended clients of the new product.[71] Izmir, Salonika, and Baghdad followed suit. Additionally, in the subsequent decade in Greater Syria ice machines could be found in Beirut, Aleppo, Damascus, Jaffa, Haifa, and Jerusalem, among other places.[72] Before the century's end, artificial ice was sold (and given freely to the poor) even in Mecca and Medina.[73] In all these places the ice was used to refrigerate meat and helped promote widespread meat consumption. While Ottomans from most walks of life were far from strangers to the taste of meat,[74] with its price now lowered and its freshness prolonged by artificial ice and later refrigeration, meat in the twentieth century became a key import item.

Let us examine the process in Egypt, where meat consumption began before the age of coal-powered refrigeration but was significantly invigorated by it. By the 1810s, disease, extreme weather conditions, warfare, and diminished pastureland led to a major reduction in the number of domesticated animals in Egypt, and subsequently to a shortage in meat supplies.[75] Although meat was a luxury for the majority of rural cultivators, most members of Egypt's governmental elite ate it regularly. Each of the three main policies associated with the regime founded by Mehmet Ali and continued by his heirs had a direct effect on meat production and consumption.

State centralization: The government aimed to control the supply of animals for both work and meat by seizing them, centralizing their use in new governmental institutions such as a central meat depot, and trying to prevent others from producing meat by setting up roadblocks to prevent smuggling and administering animal audits and confiscations, all aimed at increasing governmental control to maximize state access to animal protein.[76] In 1833, the Pasha established *Maslahat al-Mawashi al-Sudaniyya* (the Department of Sudanese Cattle), tasked with the import and distribution of Sudanese livestock.[77]

Supplies of animals for agriculture, warfare, and consumption usually came from the desert: Bedouins were the traditional livestock breeders of Egypt.[78] The Bedouin increased or decreased the size of their herds according to market fluctuations, or to meet army demand before military campaigns, especially to the Hijaz. Mehmet Ali also imported sheep from Europe, incorporating the Bedouin—who continued to raise livestock and sell animal produce—into his monopoly: sheikhs mediated between the state and their tribe and were required to provide the government with animals and animal products according to government needs.[79]

Militarism/defensive developmentalism: I claimed that in Aden stomachs were militarized, and this was also the case in Egypt. Throughout the Ottoman Empire, meat eaters belonged to the ʿAskary or martial social rank. In Egypt, mandatory conscription of Reʿaya (the other main Ottoman social category of "subjects") came with the expansion of meat consumption down the social ranks. The Pasha's unprecedented decision to recruit Egyptian farmers to his new army meant that the soldiers' eating habits spread among the wider public.

In the early modern period, Egyptian peasants seldom, if ever, killed their own work animals for meat. Military service extended meat consumption to include nonelite members of society. Besides properly baked bread, rice, lentils, and beans, soldiers were supposed to be given meat once every five days. The scribes accompanying the Egyptian army in Syria reported how every effort was made to maintain this standard.[80] In military hospitals, meals were composed of rice, broiled meat, and bread.[81] In the mid-1860s, Egypt's military conquests expanded southward along the east African coast to include Massawa and Harar along the African Horn, the principal area for exports of livestock for slaughter in the Red Sea region.[82]

Cash-cropping and the adoption of long-staple cotton: Starting under Mehmet Ali and increasingly during the cotton boom of the 1860s, Egyptian agriculture shifted from flood irrigation and animal muscle power to coal-fueled perennial irrigation and mechanical power. The move from subsistence agriculture to cash crops, and especially the exponential increase in land allocation for cotton cultivation, stimulated imports. The most significant development lay in imports of food, especially of a higher quality than that produced domestically.[83] The years of the British occupation saw another complementary trend toward cotton and away from subsistence agriculture. While the amount of land devoted to vegetable protein such as beans decreased only slightly between 1879 and 1913 (from 616,317 to 478,187 fedans), the proportion of beans among the country's overall agricultural produce declined dramatically (from 14.1 to 6.2 percent).[84] Considering the population increase, this indicates that the table was cleared and reset for a major shift in the Egyptian diet.

Another facet of animals' transformation from a farmer's companion into a key source of protein was the emergence under British rule of an animal welfare movement. With slaughter performed abroad, at ports, or elsewhere out of sight,[85] legislation and institutions for the prevention of cruelty to animals appeared in Cairo as well as in Istanbul, Jerusalem, and elsewhere in and outside the region.[86] As pets, objects of compassion, and bearers of ethical rights, animals were humanized; or at least their treatment was expected to be humane. This bipolar understanding of the animal—swinging between objectification and anthropomorphism—still characterizes debates about agency, whether of humans, other living actors, or machines.[87]

William Cronon has shown how the development of the US meat packing industry, which many Egyptians followed admiringly in the scientific press, depended on ideas of convertibility, that were, I would add, inherently thermodynamic. Such convertibility presupposed a mechanized body: a hog came to be seen as a way to bring corn to market, and slaughtered animals began being perceived as passing through a "disassembly line." These transformations were enabled by a system of railways and by refrigeration from the 1840s on. The nexus of coal-fired refrigeration and transportation led, in turn, to a significant saving on cargo space, as animals could be transformed into pure "meat" rather than wasting space on transporting skeletons, tails, hooves, heads, and other redundant parts. Moreover, mass transportation and economies of scale allowed clustering together and shipping separately certain animal parts according to particular demand in different markets, further obliterating the unity of the animal into the commoditization of meat.[88] Frozen meat was not simply a euphemism for animal, as beef or poultry are for cow or chicken. It was a commodity form that depended on evaporating and removing animality altogether. The growing coal-fired gap between "animal" and "meat" was arguably an enabler of the transformation that concerns us in this chapter, from engine

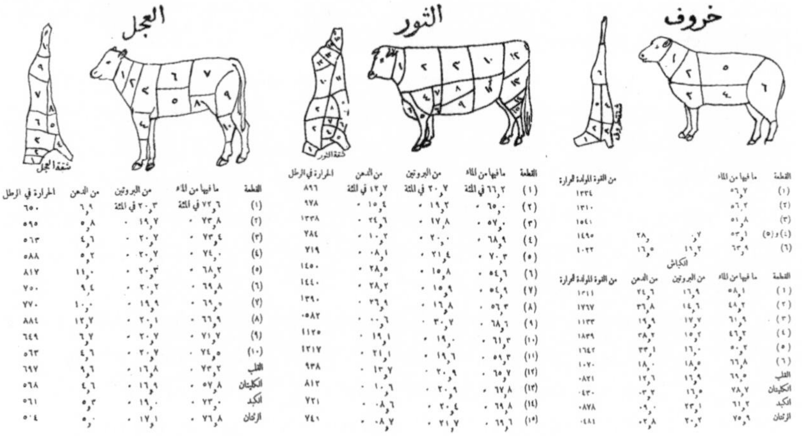

خروف التور العجل

FIGURE 8. "Meat and What's in It Nutritionally," *al-Muqtataf* 47, 1915.

into fuel. This fossil-fueled cryogenic revolution was a global transformation.

Like the press campaign that helped make frozen Australian meat palatable for British consumers, and the way the new language of "nutrition" provided scientific justifications for its consumption across Europe, in Egypt too the daily press and popular scientific journals helped inform urban readers of new ways to handle meat and their own bodies (see figure 8). Just as in other spheres, processes that mainly took the shape of class politics in Europe (where one could draw social fault lines on animals' carcasses and distinguish between the cheaper parts, consumed by the lower classes, and the more expensive ones, consumed by elites)[89] were reflected in the non-West where similar distinctions were articulated as imperial tensions. Thus, the transformation of animals into fuel occurred in tandem with the emergence of a thoroughly new perspective on the human body as a motor governed by the laws of thermodynamics. However, this happened concurrently with fractures in this episteme.

Articles about meat consumption, its nutritional benefits as well as health hazards and discomforts like diarrhea associated with spoiled fresh meat, started appearing in the Cairo-based scientific journal *al-Muqtataf* in the late 1870s and have been commonplace ever since.[90] The topic of frozen meat was addressed regularly from the 1880s, including historical reviews of the transformations in refrigeration technology and expert assurances that the chemical composition of frozen meat was identical to that of fresh.[91] According to one article, by the mid-1910s Egypt's annual meat consumption had reached 750,000 heads of cattle, goats, and sheep, of which 400,000 were imported from abroad. Most of the meat was consumed in Egypt's large cities. The article contrasted local eating habits to those in America, where sixteen times more meat per capita was consumed. In addition, whereas Egyptian meat eaters usually chose the cut based on taste and cooking method, Americans and Europeans valued and paid varying prices for different parts of the animal, based mainly on nutritional value and on water and calorie content. *Al-Muqtataf* sought to promote similar consumer sensibilities in Egypt using texts and images (see, for example, figure 8).[92] The analogies between Egyptian and Western carnivorousness revealed both similarities and differences, and thus the dietary habits of other peoples were at once expressed as external and aspirational, but also as benchmarks for indigeneity and cultural specificity. In other words, they politicized meat-eating in a wide spectrum of ways, explicitly making it into another arena for the colonial encounter.

Evidently, as in Europe and the United States, discussions about meat were entwined with a broader trend toward nutrition. As part of *al-Muqtataf's* agenda to inform readers in the Arab world about contemporary scientific progress and its application in daily life, early articles showed an interest in the nutritional makeup of specific food items ("The Nutritional Values in Eggs," "The Nutritional Values in Fish," "The Nutritional Value of Sudanese Fava Beans").[93] Towards the beginning of the twentieth century, articles adopted a more educational stance, informing readers about subjects such as nutrition for children and nutritional value for money.[94] These articles were part of a burgeoning global discourse in which food was often conceptualized as a function of its caloric value rather than parameters of taste or local availability.[95]

More broadly still, the topic of meat consumption, new indices like the calorie, and discussions about refrigeration were the "cold" version of the emergence in the Middle East of the model of the human engine: if food became fuel, the body that processed it could be seen as a machine. In this, the human body was just one element in a comprehensive system in which all of nature could be mechanized. In the early days of refrigeration, observers were astonished by the capacity of the novel Australian engines to replicate the most extreme of natural conditions and produce "artificial Arctics."[96] Nature was seen as an early model for the frozen meat industry (just as animals were the inspiration for steam engines): the refrigerator was even capable of preserving a giant

Siberian mammoth in ice in perfect condition. However, as Woods shows, by 1882 the familiar story of the mammoth had been inverted such that "nature" in those icy parts of the world was now doing "the work of a refrigerating machine."[97]

Notably, such general trends were uniquely inflected wherever they emerged. Seyma Afacan has shown this to have been the case in Istanbul by attending to the Ottoman articulations of the fusion of the metaphors "man-the-machine" and "man-the-animal" in this period.[98] This was true in Egypt as well. For example, a 1920 article in *al-Muqtataf* recounted the frozen mammoth story as an example of nature's "expert mummification" of the gargantuan beast.[99] Egyptian writers, at that time under the strong spell of Pharaonism, demonstrated that the pull towards cultural and scientific homogenization was colored by local issues. Indeed, another part of the article, calling for the import of frozen meat from the Sudan, demonstrated the prevailing tensions over sovereignty there between Britain and Egypt.

In tandem with meat's differential impact on the individual body, carnivorousness affected the collective body as well. Food in general, and specifically meat consumption, has always presented a bone of contention, offering a familiar way of articulating communal affinity and difference. A place like Aden, almost a tabula rasa which new communities were making their home, witnessed tensions around issues of animal slaughter from its first years as a multiethnic and multispecies coaling station. In 1840, for example, the entire Gujarati merchant community in Aden threatened to leave the port because of the killing of a goat, most likely by Muslims, outside a local sacred shrine.[100] It was around this time that the aforementioned ʿAli Abu Baker and Menahem Moshe competed with one another over meat provision for the British troops. Meat consumption resulted in more than economic competition between rival butchers or between Parsi, Jewish, and Muslim meat merchants. While there is ample evidence of efforts by members of different religious communities in Aden and its vicinity to cater to one another's dietary restrictions,[101] even these efforts reveal how carnivorousness reinforced intercommunal boundaries. As a local proverb indicated, "Nothing differentiates us [Jews and Muslims] other than butchering and marriage."[102]

The tensions flaring in Aden around meat would reverberate in the surrounding region in the following decades. While politics around meat consumption took different forms, ranging from collaboration to confrontation, they were more often than not communal. In the subcontinent, meat consumption consolidated groups at least as early as the Indian Rebellion of 1857, when Hindu and Muslim Sepoys' cooperation was oiled by rumors that British Enfield rifle cartridges were greased with cow and pork fat. In 1858, after the suppression of the uprising, Muslim elites established the Aligarh movement, which promoted Islamic reform across North India. In 1868, one of the movement's luminaries, Sayyid Ahmed Khan, published the book *Risala-i ahkam-i ta'am-i ahl-i kitab* (Ordinances pertaining to the question of eating with the people of the book) in response to questions on the permissibility of sharing food, and especially meat, with non-Muslims. Sayyid Ahmad explicated the reasons for such conviviality, as he had a few years later in his travelogue about a journey to Europe via Aden.

The passage through Aden has been a junction where, like the young Gandhi, other travelers also have had to make ideological decisions about meat. When docking in this port, Islamic reformers like Sayyid Ahmed Khan and later Shibli al-Nu'mani devoted considerable portions of their travelogues to questions of animal slaughter and the ways it worked to facilitate communal solidarity, each coming to opposite conclusions on the matter.[103] Reformers' different approaches to slaughter were also reflections on Muslims' ability to travel together with members of other persuasions, including Hindus (in the 1890s Sayyid Ahmed called on Muslims to avoid cow slaughter to promote intercommunal friendship), and especially Christian Europeans with whom they often shared steamer voyages, meals, maritime anxiety, and free time. Intercommunal solidarities forged around meat were of significant strength and transformative power, thus mirroring meat's more familiar role as the emblematic communal differentiator.

Islamic reformers devoted a great deal of attention to the infrastructures of steam and especially to issues of food and drink in this context. Alongside questions of animal slaughter on board steamers, they were also concerned with water provision on these vessels, both for ritual purification and for drinking. Sayed Ahmed Khan even gave several detailed accounts of how water was desalinated on steamers to reassure his readers that its use was permissible.[104]

Similar dilemmas characterized the halachic debates of the Jews of Aden. Faced with the increased importance of meat and with the uncharacteristic refusal of Yemen's Zaydi Muslims to eat meat slaughtered by Jewish butchers in the port city, the early years of the colony saw a proliferation of printed texts, mass-produced on Calcutta steam presses, that promoted a more lenient halachic interpretation of kosher slaughtering.[105] Aden's place between India and Europe and the importance of steam and meat in this intersection again proved to inform and, in turn, take its cue from various spheres.

Thus, Aden reveals many of the local inflections of the supposedly standardized passage of Australian meat to Europe from the 1880s onwards. Gandhi first experimented with vegetarianism-as-identity-politics when passing through this fueling station. In later years, his thinking and model had developed and was circulated in various languages, including Arabic. Roy Bar Sadeh has shown how *Al-Manar*, the Cairo-based journal and the main platform for "Islamic reformism" which published a series of translations of Gandhi's *Book of Health* between 1926 and 1928, helped proliferate ideas such as the view of one's body and eating habits as sites for political contestation in the Arab world. It also gave Islamic reformers a platform on which to argue and disagree with Gandhi and to modify his program for their own unique contexts.[106] Indeed, as early as 1903, just a few years after the founding of the journal, Muhammad ʿAbduh—Egypt's High Mufti and the main religious authority behind al-*Manar*—published one of his most famous fatwas permitting Indian Muslims to eat meat slaughtered by Christians in the Transvaal. With particular regard to the Hindu emphasis on vegetarianism, Islamic reformers of different stripes sought to forge carnivorous intercommunal bridges, just as Hindus sought to forge vegetarian ones.

Alongside Gandhi's thinking and intellectual production, the model he offered for anti-imperial resistance and especially his captivating ascetic figure circulated more widely still in the Arabic press of the 1920s and 1930s. Egyptian poets celebrated his self-denial ("Resembling the Messengers in his defense / Of truth and in his self-denial / He has taught us about / Truth and patience and commitment"), and when his steamer docked at Port Said in 1931, his presence stirred spontaneous demonstrations.[107] If Egyptians were especially taken with him, this was a much wider trend in the Arab world. A Lebanese cartoon among many even claimed bluntly that "we have all become Gandhi," joking that while the mahatma "had a goat's tit to suckle on, we do not even have a tit" (see figure 9). Gandhi's modular body and diet traveled as text and as image across the Middle East, connecting different ethnic and religious communities.

The 1920s was also the decade when the Khilafet movement, the largest transregional Islamic movement of the day, attempting to sustain the caliphate after the end of the Ottoman Empire, and one led by

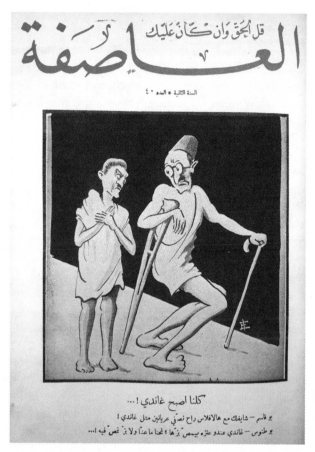

FIGURE 9. "We Have All Become Gandhi," *al-'Asifa*, 20 Iyar, 1933.

a Hindu mahatma, demonstrated that the connections created by British imperialism set unexpected forces and coalitions into action. Faisal Devji directed our attention to surprising affinities between the seemingly extreme opposites of Gandhi's nonviolence and al-Qaida's terrorism (and also humanitarian environmentalist movements).[108] Like the 1930 Salt March, which challenged the British salt monopoly and the tariff structure that encouraged Red Sea (and Aden) salt, this convergence is historically rooted in colonial modes of bodily comportment and resistance. We tend to view the Middle East from the West and from the perspective of oil. Observing it also from the East and through the perspective of coal and meat reveals quite another history.

DIFFERENTIAL CLIMATES

Alongside its role as fuel, meat was thought to heat the body in cold climates. The fact that much of it had to be delivered frozen represented a fire-and-ice combination that was now characteristic of a new global condition. The all-weather refrigeration steamships that brought frozen meat to London via Aden and Port Said benefited from the fact that Australasia's pastoral industry was subjected to the seasonality of the Southern Hemisphere. Ewes giving birth in the Austral spring did so only a few months before London's Christmas market.[109] Frozen meat was not alone in reconfiguring seasonality under coal. The intersection of oasis horticulture and steamship transportation from the 1860s onwards brought Muscat dates to Europe and North America, making Arabian dried fruits a staple Christmas delight.[110] During the same period, citrus agriculture along Palestine's coastal plain expanded enormously with steam pumps that gave access to deeper aquifers and steamers that carried "Jaffa oranges" to European winter markets. This was also true for Gaza barley, which promoted year-round beer drinking.

While the coal-fueled cultivation and transportation of all these foodstuffs contributed significantly to what we now call global warming, their histories also reveal that the process of climatic standardization simultaneously depended on retaining, even enhancing, climatic difference and variation. Clearly, dietary proclivities were key in promoting the material infrastructures and geopolitical arrangements of the age of steam; and this fact can also be demonstrated with coffee, tea, and opium, which shall be our dessert in this chapter. Situating widespread carnivorousness and the developing taste for meat amidst other new dietary habits helps tease out the habit-forming aspects of the coal-fired democratization of meat, and thereby the bodily aspects of adopting fossil fuels.

Coffee first traveled from Ethiopia to Mocha in southern Arabia, reaching Mecca by the sixteenth century. One reason for its proliferation was the fact that it enabled Bedouins to consume brackish water in the desert and to share water with camels and other animals.[111] Until the 1730s, most of the coffee in the world came from Yemen; thereafter it was brought from the French Antilles and then Java. Nevertheless, a century later South Arabia was producing even more coffee and exporting it to new locations like the German and American markets.[112] The takeover by the British of the coffee port of Mocha as their first Red Sea coal depot during the 1820s, and the shift to Aden a decade later, were both informed by the desire to restore the British monopoly over the coffee trade, and simultaneously by the intention to use this trade to justify the projection of naval power in the Red Sea.[113] The passage of the Arabian coffee trade via Egypt was cited in 1837 by engineer Alexander Galloway as one of the incentives for extending the first line of the Egyptian railway.[114] As we will see in chapter 4, about the establishment of the maritime coaling system, occupying new coal depots combined military strategy, economic opportunity, and technical feasibility, among other factors.

A more significant crop that informed the development of global trade and its technological aspects, as well as its coal-fired infrastructures, was tea. In the nineteenth century, this stimulant and the sugar added to the hot beverage made from its leaves became crucial for keeping the wheels of the Industrial Revolution turning by enabling British workers to tirelessly attend to their machines for long hours. By the 1820s, Chinese tea became the chief export item to Europe.[115] More so than coffee, tea transport was informed by seasonality, so in addition to energizing the human body it also galvanized commercial and technological development. Losing its freshness and hence its monetary value by the day, the tea trade and the famous tea races of the 1860s stimulated the shift to steam engines at sea instead of wind-powered clippers.

The British thirst for Chinese tea also animated another process, the opium trade. Opium was forcefully introduced into China in order to balance the trade deficit that tea had created. While the interests of the East India Company lay principally in the trade between India and Europe, it relied on local Indian merchants (including Indian Jews, Armenians, and Parsis) to ship Indian opium to Canton in exchange for Chinese tea brought to India for resale in Europe. The Parsi merchants were especially important: the opium trade transformed the Parsi community into a major trading diaspora. By 1837, eleven Parsi trading companies operated in Canton compared to only nine American and four European ones. Parsi families operated their own fleets, numbering at this time about fifty vessels, which constituted a third of all the ships plying this trade.[116] The China trade transformed Parsis from minor hawkers to transmaritime merchants. The same trading families that started their global careers in Canton would soon dominate the coal depot of Aden, where Mocha coffee replaced tea and coal replaced opium. We have already encountered their prominence there in water desalination and in the ice, salt, and meat trades. From the 1840s, they were also charged with providing liquor to sailors and soldiers in Aden and opium to the town's "natives."[117]

Licenses to sell ganja and opium in Aden were granted in 1847 to a Parsi merchant, with the stipulation that they could only be sold to the town's "natives."[118] Examining opium and meat together reveals that coal dependence formed together with other dependencies through stressing, creating, or strengthening various sets of corporeal, group, climatic, and geopolitical difference. The strictly enforced regulation of drugs in the microcosm of Aden, where opium was sold only to non-Europeans, encapsulated a much broader division of consciousness-altering in which the stimulation of the colder industrial parts of the planet was balanced by the sedation of the hotter agricultural regions.

Unlike tea, a seasonal crop whose freshness and price declined in direct relationship, opium was more lucrative in its processed rather than raw form. Cooking and drying opium made it more potent by reducing impurities and liquids, thus also reducing its volume, transforming it into the ideal high-value low-volume shipment cargo. However, because of the northeast monsoon, only two opium shipments could reach China annually, even on the fastest sail ships from India. A voyage by clipper from Bombay to Canton could be made in forty to fifty days in fine weather, but took twice that time when caught in the monsoon, incurring great risks and insurance costs. These conditions imposed a seasonality on a trade that benefited most from regularity of supply and prevented investors from receiving returns on their capital within the same year. These conditions were important driving forces for the development of steam navigation in the Bay of Bengal, and key reasons that steamer companies like P&O adopted the China opium trade as early as the 1840s. At the beginning of the century, the East India Company held two public sales of opium in Calcutta annually. Shortly after P&O carried its first cargo of opium to China, there was a sale every month.[119]

Opium was so important for steam navigation that even shipping company owners who had moral qualms about the trade soon realized that they could not avoid it.[120] By the 1860s, even before the inauguration of the Suez Canal, Aden became an opium transshipment hub allowing P&O to hold its Bombay-Aden-Suez line and the British India Steam Navigation Company (BI) to establish an Arabian coastal line between the Persian Gulf and Aden, along which its ships carried Persian opium to Aden then continued to China on P&O steamers.[121] BI's men on the ground in Aden (as well as in Jeddah and Hodeida) were the experienced Parsi representatives of Cowasjee Dinshaw & Bros.[122] What came to be called "fossil fuel addiction" after the 1973 oil-price shocks can actually be traced back more than a century earlier; more importantly, it can be identified as part of the literal adoption of various habits and habit-forming substances including tea, coffee, sugar and other stimulants, meat, salt, and opium. This is similar to how Matthew Huber describes the deep inscriptions of oil on the American way of life that deepened oil dependency, and the embodied practices through which subjects come to experience oil in daily life as acts of social reproduction.[123] "Coal dependence" involved the remaking of nature and human nature, but this happened in tandem with the introduction, maintenance, or reinforcement of imperial difference in realms ranging from planetary seasonality to corporeal habitus. That the differential coalonial body emerged as a site of fossil fuel addiction is significant as it allows us to join food and energy politics with connective tissues that are less familiar. The conventional perspectives of "food miles" and "global warming" implicitly presuppose a uniform planetary rise in temperatures. By contrast, the modern fossil-fueled diet involved a long cultivation of different political climates and various physical planetary microclimates. It also enables us to connect these realms in terms of their difference as well as in new ways.

Numerous other examples could be offered to show how nature and human nature were rapidly co-constituting one another; and since both were now being seen in terms of machines, human taste and dietary preference became ways to directly engage the world created by fossil fuels. Yet from the human body, to one's immediate community, to the global position of that community's abode, difference was a regular companion to standardization. As meat and other foodstuffs reveal, temperature variations informed the structuring of the global political economy into a network in which warmer places of agricultural cultivation were grafted onto colder places of industrial processing. If climatic differences were the meat and potatoes of classical racial theories, in the age of steam the all-weather links between different climates offered new thermodynamic grounds for the connection, comparison, and contrast of communities and bodies. The current planetary distribution of weather turbulence and the differential public care afforded to different regions are rooted in this late-nineteenth-century moment.

The abstractions of steam energy obscured the extent to which coal-burning technology was informed by the realities, environments, actors, and apparatuses it extended and eventually decentered. What is at stake here is a better grasp of the potential and conditions for future change. If the machinery of the age of hydrocarbons is informed by local conditions, then change must also be attuned to particular circumstances and to the global canvas on which they unfold. We have seen how coal inserted itself into existing systems and changed them from within. In the process, cooperation was as important as competition. Instead of a replacement of energy sources and regimes, we see a rearrangement of existing powers. Steam, water, and animal power were becoming ever more commensurable with, and dependent on, one another. Modeling itself on rivers, horses, and camels, the steam engine was transformed into a force of nature, and nature, in turn, was reborn as a machine. Such rearrangements followed a pattern of, or at least a tendency towards, intensification. Since any account of energy shifts has an eye on the present and future, and thus to questions of environmental sustainability, emissions, and economy, such intensifications must yet remain part of the picture. Moreover, identifying these sites of intensification and the common language they generate is the first step towards an effective politics that might curtail and resist their thrusts.

A key intensification mistaken for energy transition was that of human labor. Steam engines were assumed to have obviated this form of energy, but in fact they multiplied it while decentering and pushing much of it under the radar, as we will see in the next chapter. For our purposes here, what is noteworthy is the fact that coaling stations were key sites absorbing new manpower (in which we should include the growing "security" sector that accompanied energy), all in need of feeding. The rise of meat consumption in such places is one result of this fact. It is also part of the multispecies dynamic in these settings. Decisions about livestock were part of the calculations of the carrying capacity of a given territory. In Aden, the Somali option was chosen after local water and grass resources were stretched to their limits and deemed uneconomical and wanting in other ways. Large-scale livestock production made little environmental sense per se.[124] But in the context of strategic thinking, proximity to imperial transportation infrastructures, military habits, coal, and salt made reliance on meat widespread and led to the establishment of slaughterhouses inside steamer ports. Coal depots were places where movement and fluidity often made up for local shortages. Water was desalinated with coal brought from England, cattle (meat) were watered and grass-fed in nearby Somalia or Malta, and workers, as we will see in chapter 3, were shipped from other coaling stations. Aden and Port Said were tumors or polyps feeding off the arteries that passed through them.

Wherever meat became ubiquitous in the coaling system, it enflamed communalism and sectarian tension along with (often successful) attempts to curtail them. This created a vicious cycle justifying more troop deployment and, in turn, more meat. Much like coal, which during the nineteenth century animated both energy and empire, meat (and thus violence to animals) was at once a fuel and a cure for human violence. The rise of Gandhian vegetarianism as a form of nonviolence in the human sphere was the mirror image of this process and a reaction to the aforementioned weaponization of meat and militarization of the stomach.

Imperial peripheries were also settings to which European assumptions about the thermodynamic body and nutrition came late. When they did arrive, they operated alongside racial assumptions about nonwhite Arab bodies and non-European modes of togetherness. The Middle East's position as a relay station on the road to India also increased its importance for Indians confronting the imperative to eat meat in Aden, or the encroachment of the Red Sea salt sector. Part of the Indianization of this region thus entailed the inflow of resistance repertoires from the East. In this context, Gandhian and Islamist uses of the body as a site of resistance are revealed as far-reaching in destabilizing broader systems. In various places, the success of introducing new, large-scale infrastructures hinged on seemingly small acts of persuasion or conversion. If remaking nature involved reshaping human nature, human nature is also where these transformations can be resisted or negotiated. In this, meat joins other drinks and food items. As we saw in the previous chapter, the people of Jeddah, Port Sudan, Aden, Kuwait, and Cairo, and their animals, had to develop a taste for condenser water; and drinkers across the British Empire had to take up India pale ale, Chinese tea, and Mocha coffee instead of their old brews. These were individual acts of consent or resistance, but also processes of group formation and cohesion. Also, with regard to water, we saw that consumption habits and social differentiations between "natives" and "foreigners" were mutually constitutive.

Further, we saw how in London, Cairo, and Aden, discursive and material processes were linked to one another like the daily news stamped on a piece of steak. The delays, translation work, and other inflections of the disentanglements of the epistemological and ontological in peripheral settings make these places especially potent sites of subversion. Arabian nomads, Egyptian meat eaters, and members of various religious communities on the move, localized between borders or spread in vast expanses beyond nation-states, often stressed their differences from British steam travelers or American meat eaters.

Industrialization could not have materialized without picking up steam as well as meat in imperial peripheries, or without incorporating and simultaneously marginalizing various human and nonhuman powers. Similarly, resistance to its ongoing legacies has much to gain from attending to intended and arbitrary impediments to the march of steam.

CHAPTER 3

Humans

MANPOWER

The globalization of steam power made coal into "the universal aid."[1] This was facilitated by, and simultaneously promoted, a universalization of new understandings of the working human body and new conceptualizations and practices of manual activity. Among these, a new notion of "labor" was the most prominent, and this chapter seeks to understand how its gradually attained hegemony and form were related to the planetary spread of fossil fuels. Labor emerged from the age of steam like a diamond from a lump of coal—ostensibly symmetrical and untarnished by the messy historical props and pressures that molded it. Examining the process historically, and on the actual scale of this fuel's global reach, reveals its multiple and enduring asymmetries.

Steam engines offered the most powerful demonstrations of thermodynamic laws, testing and taking these principles around the world. This science, which encompassed the conversion of work, heat, and motion, aided the universalization of labor by presenting concrete phenomena as modulations of a single, invisible entity—energy. It allowed for the abstraction of concrete phenomena from specific material, as well as political, economic, and religious affairs. Beyond this deconcretizing, thermodynamic abstraction entailed reconcretizing natural as well as social phenomena, articulating them with a new terminology and metrics that allowed for the development of new operational protocols.

Abstract labor was such a de- and rematerialized category. It drew attention away from certain bodies and physiological conditions, refocusing it on others. While labor was articulated in nonconcrete ways before the nineteenth century, thermodynamics allowed it to be reimagined as energy. Energy, in turn, was increasingly discussed in terms of productive waged labor and seen less and less in terms of heat, motion,[2] or physical toil and pain.

This will be demonstrated in the three sections of this chapter. Each section examines a group of workers involved with steam engines, coal depots, and coal mines, and thus also with one another, even across a vast space: stokers (focusing on working conditions near the sweltering Gulf of Aden), coal heavers (especially in Port Said, one of the world's largest and most expedient depots), and miners (in England, Wales, and Zonguldak on the Ottoman coal coast). Together, these men formed the labor pool of a new, global "energy sector." Though geographically dispersed, the work and struggles of each group impacted the others, something they and others became increasingly aware of, but which historians have yet to fully acknowledge.

As we saw in chapter 1, steam engines and many of the machines they propelled were citational; they did not drive an undisturbed transition between clearly delineated "energy regimes." Rather, in their design and nomenclature they made use of those powers they had unseated from relative prevalence in terms of horsepower, but never actually supplanted them. These powers included waterfalls, animal power, and human muscle. Especially as quasi-human proxies, steam-propelled devices with apt names like the Iron Man (the spinning machine that displaced most workers in the British textile industry) promoted and lent conceptual shape to the shift from bonded to free labor. They also mediated the related shift from work, understood in terms of physical pain, to a quantitative concept that reduced labor to monetized time (hours of waged work). In both processes, the more machines mimicked living things and performed human tasks, the more humans and their labor could be measured against machines.

In particular, the machines that came to replace unruly English textile workers and break their unions were frequently fantasized as various kinds of quasi-humans, particularly slaves. Benjamin Disraeli, who in the 1870s would become one of the greatest promoters of a British imperial coaling system (also within and around Ottoman-dominated regions), captured this zeitgeist early on in the 1844 novel *Coningsby:* "A machine is a slave that neither brings nor bears degradation: it is a being endowed with the greatest degree of energy and acting under the greatest degree of excitement, yet free at the same time from all passion and emotion."[3] *Degradation* was a loaded word during this period. On the one hand, it had a functional referent: fatigue and depletion of energy in the working body. Fatigue emerged as a significant problem in an industrial free-labor regime across western Europe and then North America. The 1840s saw the development of a thermodynamic model of the worker's body, understood as a "human motor," and eventually the appearance of an elaborate "science of labor" meant to curtail fatigue and energy depletion in order to make the interface between man and machine ever smoother.[4] On the other hand, "degradation" referred to disgrace stemming from new moral sensibilities and abolitionist stances that were traveling in both directions between the British metropole and its existing and former colonies. These were sustained by the prospect of the labor-free society that the steam engine helped advance by replacing slaves with "energy slaves."[5] Among other industries, such sensibilities reverberated in the budding "energy sector." If earlier campaigns to abolish the "slavery" of British miners aided the campaign against chattel slavery outside England,[6] from the 1830s and 1840s the degradation of colonial slavery was repeatedly evoked in order to reform working conditions in the metropole, including, again, the mining sector. This was evident, for example, in the findings of an 1842 Parliamentary Commission stating that conditions in English coal mines were "as bad as the African slave trade . . . a wretched and slave-like mode of labour," which led to reforms in the mines.[7]

Thermodynamics and slavery were connected by a third meaning of "degradation." Its racial connotations were used to explain physical and civilizational differences that seemed to contradict scripture-based assumptions of common human origins.[8] Such terminology helped elevate the Victorians over non-Europeans, but it also had disturbing implications. Intersecting with "degeneration"—a polyvalent medical notion anchored in a theory of climates—degradation suggested that enslaved Africans and Asians who produced cotton or sugar in the colonies and the swelling hordes of forced laborers in England apprenticed without pay to process these raw materials in the metropole were unnervingly similar.[9] Such associations tended to float on the hot air of the tropical microclimate in factories and especially in coal pits where temperatures reportedly reached up to 100 degrees Fahrenheit (or almost 40 degrees Celsius), leading the men to work naked and conjuring up scandalous images of topless women evocative of exotic slavery.[10] Labor reforms in mills and mines aimed at distinguishing between metropolitan workers in industrial environments and colonials in "tropical" workplaces.

Whereas a lack of civilization, cleanliness, and suitable clothing symptomatized the degenerating effects of temperature, heat itself changed during the middle of the century from a physical substance to a transfer of "energy." In the process, both degradation and degeneration were becoming synonymous with the move towards entropy—the thermodynamic prospect of disorder according to the Second Law. Together, these aspects of degradation reveal the multiple dimensions of the problem steam engines simultaneously resolved and exacerbated. Colonial slavery and metropolitan labor inspired and negated one another; they shaped conceptions of technology and were shaped by steam power and thermodynamics.

In both settings, bondsmen were now considered more expensive and less productive than free workers who were turned into better laborers by the incentive to work hard and make a living.[11] Free will was a problem; it animated workers' uncontrollable impetus to unionize and claim a share of the employers' profits, and even to break machines. But as the mechanism of discipline-by-incentive developed in the second half of the century, with considerations of "career," it also became a solution.[12] Machines and the factory environment provided a model of "self-motivation," and a steady work tempo that free workers were meant to internalize. Steam engines helped create a situation whereby passage through the factory's gates could be presented as an exercise of choice by "bringing the power to the workers" in crowded urban centers. The right choice was assured by a set of dependencies that pushed men and women to work hard as a proletariat whose only means of livelihood was the factory.

Yet proletarianization, career, and the model of the self-motivating machine had less purchase among the pre- or deindustrialized British colonies and semicolonies. Here worker control was cruder and other mechanisms of fabricating choice were required, as we will see below.[13]

Alongside heat, freedom became another key context where distinctions between metropolitan and colonial deployments of labor took shape. But as in the case of bonded work, divisions were forming within this conceptually unified sphere encompassing cases at home and abroad, and creating an internal hierarchy. The issue of the connection of labor practices and ideologies between the center and peripheries of the British Empire is not new.[14] Yet the study of the universalization of such concepts seldom followed those channels that connected groups of workers in remote corners of the empire—connections which did not necessarily emerge from consciousness, ideology, or solidarity. None-theless, miners' turmoil in Wales or Newcastle sent ripples to Port Said, Aden, and the Ottoman coal coast (and less often, in the opposite direction) in the form of price fluctuations, newspaper and consular reports, and sailors' rumors. These underground, seafaring, and dockside settings contextualize the development of steam and labor power.

The spread and shape of this global network of carbon fibers was informed by geographical and social realities in the British Isles. Water-wheels, which until the 1830s were the driving force for British textile mills, were dispersed across remote locations. This exposed production to the fluctuations of water flow and to demands by workers who could defect from the remote "labor colonies," thus sinking the funds invested in recruiting them from cities, then transporting, housing, and training them. In contrast, steam engines allowed British capitalists to "take the power to the workers" in urban centers, bringing about a transition from hydropower to fossil fuel. Andreas Malm shows that during the early decades of the nineteenth century, steam engines were inferior to water-wheels in almost every way. Their only advantage, and it proved a crucial one, was that they allowed factory owners to better manage their unruly workers.[15]

In the 1840s, steam engines emerged to counter the specter of organized labor in England. Skilled workers who could unionize, make forceful demands, and disrupt production if denied their claims could now be replaced with unskilled, feminized, and juvenile labor in urban centers where competition depressed both wages and militancy. Weavers, spinners, metal workers, cooks, and book printers were all substituted by machines, and new man-, and especially woman- and child power were absorbed into factories with far less agency as machine "hands."

In certain key industries, then, factory labor was increasingly universalized and standardized. But as production became more reliant on steam power, so it depended more on coal. The mining sector, which did not see significant technological innovation until the 1890s, was reliant on the power of human muscle.[16] Like the waterwheels, coalfields were often far from industrial centers and skillful miners were not easy to come by or replace. Therefore, the more coal was needed to fuel the Industrial Revolution in the British Isles, and increasingly across its global reach, the more significant miners became, and the deeper they had to descend underground for high-quality coal.

This increased the risk to which miners were exposed, but it also honed their skills and bargaining power. The energy sector—mining and transportation of the fuel in England and beyond—became the bottleneck where the disruptive powers of workers were concentrated. Steam-enabled textile production became a global industry from the 1840s, relying on raw materials from the colonies that were increasingly irrigated by steam pumps, then harvested and transported by train and steamship; thus the powers of disruption, too, became global. They were now not only in the hands of British miners, but also at the disposal of coal heavers in Port Said and firemen in the Gulf of Aden. Connecting the British minefields to these hubs of the maritime imperial network through which coal entered colonial interiors allows us to examine the tensions and hierarchies in what became a global energy labor market, with arguably the most agentive global workforce.

Disraeli's engine-slave illustrates the point: it was "a supernatural slave," which his conjurer likened to the *jinn* and '*afarit* of Arabic folklore. The novel's protagonist encounters it in Manchester's factories, which in turn are likened to the chambers in Arabian fables. Disraeli would later transform the Conservative Party into "the imperial party" by connecting "the coal question" to "the eastern question" in the 1870s, and thus his literary imaginings during the decade of the birth of thermodynamics are rather telling. The novel also reveals how questions of energy and labor spilled out of the confines of the factory and, indeed, Europe itself, divulging the role of Orientalism in the process.[17] While reflecting a general zeitgeist, Disraeli's Orientalism drew on specific contexts: Eastern and especially Arabic mythology was the most common reference in a "Victorian steam fetish," in which steam engines were likened to slavish yet laborious genii. Disraeli was joined in this trope by Elizabeth Gaskell, Sir Walter Scott, Michael Angelo Garvey, and others in valorizing (or demonizing) the steam engine with references to Arabic lore.[18]

While Arabian genii were stirring Victorian steam, they were under attack in the Orient, where belief in supernatural entities came to be labeled "superstition" and was contrasted with the technological rationality of Europe. Oriental fetish and European "factish"[19] could not coexist in these settings precisely because of their eerie similarity, and the next chapters will explore Islamic theological aspects of this clash. It is noteworthy that this clash, too, was a force field that shaped labor and its abstraction (or in this context, secularization). Bracketing non-European metaphysics and the universalization of "labor" is usually ascribed to the European social sciences. The most familiar critique of this was articulated by Dipesh Chakrabarty.[20] Chakrabarty challenges the ill-fitting superimposition of Marxian universal terminology of "labor" on activities practiced in India involving sufi *ziker* rituals, or the spirit world. Similar claims were made by labor historians of the Middle East.[21] Chakrabarty critically revisits key works by Indian labor historians, including his own.[22] He ascribes such reductions to a universal notion of "power" rooted in European history, yet functioning as a general sociological yardstick against which other historical cases of such reductions are revealed.

However, although the way labor historians define their object of analysis, the Marxist legacies that guide them, and the universalist assumptions of the European social sciences are all part of the problem, at its core the issue pertains more to the imperialism of thermodynamics and the natural sciences than to these more marginal causes, as a brief discussion of Marx's theory of labor reveals. As is well known, Marxian thought is rife with ghostly aspects, romantic and gothic connotations, and theological ploys.[23] Perhaps the most famous is the definition of capital as "dead labour which, vampire-like, lives only by sucking living labour, and lives more, the more labour it sucks." The spectral aspects of labor and vampiric aspects of capital extend also to "commodity fetishism," the desire to live through the dead, and "the necromancy that surrounds the products of labour."[24] The universalization of "labor" might indeed seem to be a European triumph in a clash of fetishisms.

But Marx's views of the underworld and its inhabitants were also grounded in contemporary scientific thought and materialism.[25] At various points in Marx's writing, "labor" is reduced to the expenditure of energy. Human labor becomes exchangeable with and translatable into other motive powers, for "it is purely accidental that the motive power happens to be clothed in the form of human muscles; wind, water, steam could just as well take man's place." In Marx's thought, energy's equalizing powers are seen as "spectral," metaphysical, or "phantom-like."[26]

Philip Mirowsky has argued in the aptly titled *More Heat than Light: Economics as Social Physics, Physics as Nature's Economics* that neoclassical economics emulated physics, and especially thermodynamics' notion of energy, in its theory of value. The argument extends to key aspects of Marxism. While Marx's writings are studded with warnings about the deceptive nature of physical analogies in political economy, in his advocacy of the labor-embodied theory of value Marx falls short of his own standard. Organic analogies and the famous vampire metaphor, Mirowski argues, are central in understanding the Marxian conception of labor. He situates Marx's language of "dead" and "living" labor as a continuation of Leibniz's theory of the dual character of force and a distinction between *vis mortua* ("dead force," existing only as a potential) and *vis viva* ("living force," actualized and in motion) which is at the heart of the modern notion of energy.[27] In this sense, Foucault was right that "Marxism exists in nineteenth-century thought like a fish in water."[28] What made political economy and the European social sciences into "sciences" during this century was their fidelity to the principles of the natural sciences, whose own imperialism too often escapes scrutiny. Breaking away from coal's amber and from nineteenth-century conceptual tools begins with "energy" rather than with Marxism.

Moreover, recovering the lost abundance of real labor or releasing the repressed metaphysics of non-Western activity might not necessarily empower workers. As we shall see in this chapter, historically, the opposite was often the case. During the late nineteenth and early twentieth centuries, transnational coaling and steamer companies—architypes of the modern powers of capitalism—tended to align themselves with antimodern, religious, and patriarchal forces across the empire in order to circumvent state regulation and deny laborer status to colonial workers who, in turn, repeatedly tried to own and inhabit this universal category.

Elsewhere I have shown that the universalization of notions of empty homogeneous time—the unit to which work was reduced in the nineteenth century—cannot be seen as a diffusion from center to periphery, or as a seamless progression through imperial space. Also, the contradictions and counter-tempos that the process involved were not registered in some external sphere. Rather, the inflections and reverberations that abstract homogeneous time generated as it traveled "in translation" to a place like Egypt were constitutive of its universality.[29] The emergence of a thermodynamic theory of labor seems to have followed a similar trajectory. As the European story goes, abstract labor time—hours at work by "a human motor"—replaced a preexisting understanding of labor in terms of pain, compulsion, and natural life rhythms and fluctuations. In early liberal thought, from Locke to Bentham, pain is a key driving force of human behavior. Similarly, in economic thought, the view of work under both mercantilism and classical economics retained an underlying belief that labor was by its very nature a pain.[30] But in the second half of the century, both economic and political liberalism saw bodily pain as increasingly unquantifiable, and therefore a shaky foundation for any egalitarian economics, politics, or physiology. In bonded prison labor, the organization of industrial environments and schedules, and in understandings of the human body, quantifiable, secular, scientific yardsticks were adopted instead, often under the aegis of energetics.[31] How would this story change were we to include the materially integral yet unrecognized laborers of Aden, Port Said, and Zonguldak?

STOKERS

The steam engine was promoted as an artifact that would shift the load from the shoulders of human and animal workers to machines. But it was far from being the first (or last) "labor-saving" technology that in fact increased reliance on human muscle. The canvas of sail ships, which replaced the oarsmen of a previous age, required all hands on deck to climb up and down masts and operate an elaborate system of sails. Much of this labor was unfree. Impressed seamen were second in number only to African slaves, the largest group of forced laborers in the eighteenth century, surpassing any number of oarsmen in history. In the British Empire, press gangs used violence to supply the skilled manpower necessary to maintain British naval supremacy, affecting the lives of hundreds of thousands of British and foreign sailors. Yet towards the end of the Napoleonic Wars, local resistance among sailors, as well as more scholarly critiques that equated impressment with tyranny and enslavement, eroded the legitimacy of this practice. This turn of mind, which also forged attitudes toward other forms of bonded labor like slavery, serfdom, or corvée, accompanied the gradual development of steam navigation. By the 1880s, impressment seemed so archaic that Herman Melville had to remind his readers in *Billy Budd* that "its abrogation would have crippled the indispensable fleet, one wholly under canvas, no steam power, its innumerable sails and thousands of cannon, everything in short, worked by muscle alone."[32]

Nonetheless, muscle power was indispensable. As ships no longer harvested the wind, fuel had to be loaded on board manually, then trimmed (by trimmers) to sizes that matched the ship's stokehole and arranged below deck according to different principles. Coal was then carried by haulers to the boilers and fed to engines by stokers or firemen. And of course, the industrialization of maritime travel meant that steamships, like other coal-burning technologies, relied on the manual labor of miners to extract the fuel from underground.

Most dockyard occupations existed in one form or another in the age of sail, but the category of stokers/firemen was new and significant. One out of every two sailors in the age of steam was employed to shovel coal into the ship's furnaces to maintain the steam pressure.[33] Firing an engine was one of the hardest, dirtiest, most dangerous, undesirable jobs for a sailor, and was usually left to non-Europeans or nonwhites considered racially suited to the furnace-like temperatures of the engine room.

Non-Europeans were deemed fit for this work for other reasons, too. On the early liners, European firemen drawn from the ranks of sailors commanded relatively high wages, and posed a disciplinary challenge in their ability to delay a ship's schedule with strikes or drunkenness.[34] By contrast, Muslim workers did not drink, were deemed unfamiliar with the traditions of protest characteristic of shore laborers, especially the strike, and when other sailors began enjoying new social and legal protections, they were legally more vulnerable to control. Moreover, due to a special legal regime that organized their employment, certain non-Europeans were unable to sign off at a port of their choice and were compelled to return to their Indian Ocean port for rehiring, effectively trapping them in what soon became a low-waged labor pool.

Indian seamen shared white sailors' aversion to stokehole tasks and work below deck. Therefore, from the 1840s, British employers started relying on a new workforce composed of two main groups: Asians and Africans who had not previously served as sailors, and freed slaves whose manumission was part of the introduction of British steamers into the Indian Ocean.[35] The system of coal depots and steamer lines spawned and suckled on a particular geography of emancipation and free labor in the Red and Arabian Seas and Indian Ocean.[36] Royal Navy steamers patrolling this theater created slavery-free corridors, rescuing and sometimes also employing escaping slaves.[37] However, this resulted in raising the price for slaves, and amplifying slave traffic via dhows along nearby routes. Before British regulation, human trafficking in the Arabian Sea was small-scale and part of a heterogeneous circulation of people and goods; but anti-slavery campaigns, ineffective in stemming the existing flow, actually gave rise to a slave trade reminiscent of that in the Atlantic.[38]

Ex-slaves could find legal protection against reenslavement at coal depots, especially in Aden (under formal British control), and heaving and stokehole jobs for wage seekers were offered at coal stores onshore and on board ships. A feedback cycle thus formed during the second half of the century: the more Royal Navy steamers patrolled the Red and Arabian Seas, the more emancipated "coal coolies" gathered in coal depots, pushing down wages and subsidizing the growing commercial steamer traffic. This increased slave prices, enslavement, and hence also escapes and emancipation, pumping new blood into the coolie labor pool. Abolition allowed the burgeoning coal-based infrastructure to tap into local labor reserves in a foreign environment under terms preferable to those at home.[39]

Using emancipation to subsidize a nascent system of steamer lines in inhospitable environments was not a premeditated policy. Rather, it was a result of the introduction of British assumptions about, and experience with, Atlantic slavery into the worlds of Ottoman ports and the Indian Ocean where human trafficking and bondage were anchored in very different social relationships; thus "abolition" and subsequently "free labor" assumed a universal character. In Aden, as in Ottoman port cities in the Mediterranean basin, state-driven "emancipation from above" and dispensation of formal freedom papers by British, Ottoman, or Egyptian authorities disrupted a set of relations anchored in manumission by slave masters, which often retained the freedman's affiliation with the master's household. While individual emancipation continued until the end of the Ottoman period, freedmen who had gained their liberty via this new form of manumission tended to flock to port cities.[40]

Stokers were known to complain of maltreatment and physical hardship wherever steamships took them (Franz Kafka's 1911 novel, *Amerika,* opened with such an account during an Atlantic crossing), especially during the summer monsoon when the Red Sea was one of the most challenging crossings, chiefly for stokers. Despite their ostensibly suitable racial profiles, stokers were the first to suffer from hot weather. For example, in October 1894, as the troopship *Malabar* was leaving Aden to cross the Red Sea during the monsoon, fourteen of its stokers were put on the sick list due to exposure to intense heat.[41] In August 1888, several British steamers reported that near Aden temperatures in the dynamo rooms reached 140 degrees, and 170 degrees (60 and 77 degrees Celsius, respectively) in the hydraulic pumping rooms. Troopship *Tamar* had to ease down, as her stokers could not keep steam; many of them could not even work and were put on the sick list. Two crew members died as a result of heat stroke, and, but for the large stock of ice which had been loaded at Suez, things might have been even worse.[42]

Passage through the Red Sea from Suez to Aden was usually the hottest leg of a maritime trip anywhere in the British Empire. In 1881, sailors were said to have described the steamer's engine rooms as "the nearest to the temperature of a certain mythical place of residence which I need not name."[43] A few years later, another account compared the engine room to Hades, and the stokers who emerged "glistening with perspiration" to cool their black bodies on deck were described as "human salamanders,"[44] the mythical lizard-like creature which lived off the flames of elemental fire. Thus, Adeni and Somalis were considered physiologically best suited to coping with the furnace-like temperatures of the engine room. During the first decade of the twentieth century, between 1,000 and 1,500 Arabs from Aden were hired annually to work as firemen on steamers which docked at this key refueling station.[45] By World War I there were 296,000 men in the British Merchant Fleet; 52,000 of them (16 percent) were identified as *lascars*, of mostly "Indian" origins.[46] The largest groups among them were stokers from Aden and Somalia ("Aden's butcher shop" was gradually also found useful for its human fodder). Yemenis were deemed unfit to work in colder climates, and their activity was limited to between latitudes 60 degrees north and 50 degrees south.[47] The logic of race reinforced that of economics, segregating the below-deck workforces of the Indian and Atlantic Oceans, depressing the wages of those in the former, and assigning stokehole work to Eastern Europeans in the latter.

While the Adeni were considered the stronger firemen, Aden's importance as a resource for stokehole workers can be attributed to the fact that it combined the two most likely channels of vulnerable muscle power. First, it connected the ocean to the famine-struck Yemeni hinterland, and therefore to a pool of casual laborers whose ignorance of maritime work (as well as their repertoires of contention) was a boon to employers. Second, as a British-controlled port, Aden became the largest magnet in the region for escaping and manumitted slaves.[48] The 1869 opening of the Suez Canal brought a sharp increase in both the number of Adeni firemen and of steamer companies using them. In the 1870s, the latter included Dutch, Italian, Austro-Hungarian, and German companies, and the provision of cheap coal labor gradually became a feature of Aden. What the British identified as exploitative hiring and employment practices was intentionally ignored so the station would retain its attractiveness for foreign steamers, avoid delays, and keep its labor costs competitive.

Aden's labor-contracting system relied on brokers (*muqaddam*s) connected to the big steamer companies who recruited workers, housed them, and paid their wages, after deducting a share for themselves.[49] Considered by Europeans as "bribery," an elaborate system involved tribal chiefs, hiring brokers, foremen, ship engineers, and sometimes Aden bureaucrats and policemen, all living off workers who sometimes lost half their salaries in the process.[50] The "*muqaddam* system" demonstrates how transnational coaling companies sought to affiliate themselves with local mediators (and vice versa), and circumvented official and statist forms of regulation.

Beyond stokehole work, cheap and submissive migrant workers from the agricultural and maritime hinterlands were also the main coal heavers in Aden, enabling this major depot to satisfy the demands of the increasing steamer traffic. If, prior to their 1839 takeover of the port, British officials had feared that the dearth of local labor would make Aden inadequate,[51] in subsequent decades the expanding port city attracted farmers, escaped and freed slaves, and the sons of Arab fathers and African bondswomen to the swelling ranks of coal heavers and stokers. These workers were often less demanding than the port's Arab heavers. In the early 1840s, for example, when Arab heavers left their jobs in protest when water supplies gave out, migrant heavers kept going. Men who loaded coal onto steamers found it a short step to working with coal below deck.[52] We will see that this fungibility of coal work between land and sea and between freedom and coercion played a role in eventually bringing the coal economy to an end.

Red Sea crossings and their hardships led to new ways of controlling the workforce and maximizing its potential. In 1891, for example, stokers complained they had lost 50 pounds (22 kilograms) in weight due to hard work between Suez and Australia, that sentries armed with rifles and bayonets prevented them from leaving the stokehold, and that stokers who fainted were revived with water on deck and sent back to work. Until the twentieth century, such complaints—including this one—were usually dismissed.[53]

Forced confinement in the stokehold echoed the general legal and extralegal tactics that kept stokers on board and prevented their "defection." Until after World War II, employers were legally permitted to insert a special clause in "lascar agreements" depriving these sailors of the customary shore leave. Lascars could not be hired for a one-way trip, and their employment had to end in a "British Indian" port, to further limit "defection." These arrangements reflected the construction of the racial geographies discussed in the next chapter. Correspondence between different branches of the British colonial government reveals the constant categorization required to determine, for example, that "Arabs, born at Aden, are natives of India," and so issue the right labor contracts and situate individuals in their allocated places on board, and more broadly within the depots system.[54]

The shipping industry and colonial administration prevented any regulation of lascars' working hours, or limits on the owner's control over their person, including physical violence. Lascars risked being charged with breach of contract (thereby forfeiting any payment and being subjected to criminal charges) for failing to be at their employer's disposal day or night, at sea or on land. As a representative of British shipowners put it in 1903, lascars were "more completely the servants of the shipowner . . . than any other group of men doing similar work."[55] With compensation between one-fifth and one-third of that of white sailors,[56] strict limitations on movement, exposure to physical violence, and the hardships of their tasks, stokers' work bordered on bonded labor. The nominal freedom created by abolition campaigns and the freedom papers and work contracts that formalized it were crucial for steam navigation in the Red and Arabian Seas. But the nature of stoke-hole work made it impossible to adhere consistently even to this mini-mal freedom.

Steamers are seen as all-weather vehicles. Around the Indian Ocean this signified crucial independence from the monsoon conditions that so constrained the itineraries of sail ships. While this was largely true of exposure to the wind, steamer sick lists and stokers' complaints about diminished body mass reveal that before refrigeration and air-condi-tioning, iron steamers were actually more susceptible to intense heat than sail ships, due to their weakest link—the fragile bodies of their stokers laboring under the metal deck (see figure 10).

The data compiled from sick lists and carefully analyzed by Royal Navy medical experts allowed them to calculate, for example, the annual average loss of service days per sailor.[57] Statistical abstractions were coupled with experimental observations. For instance, the *London Daily News* reported such findings in 1909, in an article titled "Heat in Stokeholds: Interesting Experiments in the Red Sea." The piece was based on a report by staff surgeon Oswald Rees, who had subjected himself to two hours of work in the stokehold during his ship's Red Sea passage and lost 2.5 pounds of body weight. Integrating these findings with observations of other workers and the racial physiology of his day, he stated that "the small animal loses heat much more rapidly than a large one—not only does a mouse not sweat, but its vital functions must be carried out at rapid rate to keep up its body temperature. The inter-est in this lies in the fact that a small, wiry stoker will lose heat more rapidly than the fat one. . . . On the other hand, the fat man will lose by heat [sic] by water evaporation at a much greater rate than the thin man." In terms of skin color, Rees found that "the Negro not only has an advantage over the white man in the sun, but also in stokehold, for the lesser diathermancy of his dark skin is an advantage in sunlight, and in the stokehold it radiates heat better than a white one."[58]

The language of heat loss seems unseasonably inappropriate when used in this sweltering environment. Yet it anchors the experiments in the notion of the calorie, which was gaining traction during the second half of the nineteenth century. Introduced in the 1820s by Nicholas Clément in lectures on heat engines, a calorie was defined by physicists as the amount of heat required to raise a gram of water by one degree Celsius. In nutritional science and physiology, a calorie was seen as the amount of food which, when combusting in the body, provided a specific amount of heat as designated by the physicists. In the Red Sea, too, the human body was seen as an engine, which may have contributed to the disregard for the pain workers felt and reported. Yet in this environment, the universal thermodynamic body was also affected by another heat-based theory—racial physiology—also evident in Rees's language. From the late eighteenth century, such physiology was based on climate theory, which helped replace polygenic assumptions about different human origins with monogenic claims about common roots. Evident differences in physique and civilization could be attributed to a degenerating exposure to sun and heat.

As Anson Rabinbach has shown, during the second half of the nineteenth century European physiology and labor science, among the fields most inspired by thermodynamics, regarded the working body as a particular kind of machine. If this century inherited a moralistic conception of work as activity measured in terms of "industry" and "idleness," it was replaced from the 1840s onwards by the metaphor of the "human motor." (Moralistic attitudes to non-European and especially Oriental races embodying the vice of indolence were more persistent.)[59] Thus, workers' fatigue could be curtailed and eventually even eradicated by proper diet, scheduling, and ergonometric arrangements. Yet the beginning of the twentieth century was characterized by the deterioration of this optimism as new recognition of the inevitable limits of the "human motor" was manifest in different ways and spheres: again according to the observations of Rees, one such method was a fine-tuning of racial and other physiological typologies to better match stokers and their tasks.

The racial inflections of labor physiology helped externalize human limit and distress—features made legible only on Rees's own white body. While complaints about the severe heat in English factories and coal mines echoed repeatedly as early as the 1830s, and although the detrimental effects of such heat on (white) workers' bodies were famil-

FIGURE 10. Lascars at the stokehole. Photo by Reinhold Thiele/Getty Images.

iar,[60] nonwhite bodies were subjected to much higher temperatures before their limits were recognized. Thus, even after abolition decommodified human bodies, racial considerations offered another way to engage their materiality through size, skin color, heat resistance, body mass, etc. Via the mediation of a white body, the insights of labor science and physiology gained entry to the engine room, albeit in an anachronistic fashion. If nonwhites complained and suffered pain as universal bodies, losing weight and collapsing from the heat, their entry into the sphere of universal labor science was nonetheless delayed and truncated. Their pain, compounded by the mechanisms that veiled it, was part of what made steamers move.

COAL HEAVERS

Loading coal onto a steamer was a job that could be performed by a ship's crew. But coaling was despised by sailors, especially in hot climates, where it was considered dangerous. We have already seen that the boundaries between coaling onshore and below deck were not clearly drawn. Commercial companies, the British admiralty, and colonial authorities considered natives of tropical and arid climates racially more suited than British sailors to work as heavers in these settings.[61] Again, such racial logic made economic sense, keeping the wages of non-European heavers well below those paid to European laborers.[62]

The costs of coaling were usually factored into the price per ton of coal at any given fueling station (often including calculations of the cost of human life, according to which, commonly, one man died for every one hundred tons of coal loaded).[63] Shipowners and captains thus had to include in their itinerary decisions the proximity of fueling stations, the types of coal available at a certain harbor, the regularity of its availability, the distance of the coal pile from the ship, the mode of its storage, and the manner and rapidity of loading. If speed of motion between points was a decisive factor in both commercial and military steam navigation, the potency of the coal and the time it took to load it, steam, and labor could be considered as complementary factors. The evaluations of local heavers' efficiency appeared frequently in ships' logs and travel accounts, and in handbooks like *Coaling, Docking, and Repairing Facilities of the Ports of the World* (printed in various editions between 1888 and 1909).[64]

In this genre, Port Said was regularly commended as the "acme of coaling ports, as coal can be brought on board here much faster than at any port in the world."[65] In another ship's log, such rapidity was attributed to a "strength that is not human" of "sturdy niggers . . . of small stature, and often mere boys" carrying the coal baskets on their heads, and to the verbal and physical abuse by their overseers (see figure 11). As with the Adeni stokers and heavers, British sailors and officers found the ability of heavers in Port Said to work in such an inhospitable climate especially impressive.[66]

From the 1830s onward, the coaling process at each of the key coal stores was expedited with great efficiency. During the 1830 virgin voyage of the *Hugh Lindsay*, it took four days to load one hundred tons of coal in Suez; in 1834 the same quantity took two-and-a-half days to load.[67] By the turn of the century, on average the task took one hour.[68] Such acceleration required a concentration of heavers at ports and new schemes for controlling and coordinating them, greatly facilitated by the fact that depots were hubs of information and connectivity. Telegraphs and later telephones enabled regulation of the flow of workers on the macro- and micromanagement levels, respectively, for example, by facilitating the conscription of seasonal workers and telephonically informing sleeping heavers of a ship's arrival.[69] Indeed, coaling also took place at night, when heavers used braziers to light their way.[70]

While connectivity provides one explanation for coaling's accelera-tion, a more significant reason was the steady increase in the number of local heavers. In Port Said, heavers were organized into a new guild in 1870, when the state employed a sheikh, elected by the majority of guild members, to allocate labor, and officials started taking stock of this workforce.[71] In 1872 they numbered 501; in early 1875, 600, and later in the same year, 815. In 1878 there were 1,104 coal heavers in the port city. The guild's numbers kept swelling rapidly, and in 1894, the number rose to 3,500.[72] As British official Thomas Russell put it, "Port Said has the reputation of being the fastest coaling station in the world, with its thousands of Sa'idi labourers swarming like ants up the ship's gangways and tipping the coal into the bunkers."[73]

Harbors were contact zones between land and water and had to cope with the unpredictability resulting from storms, shifting winds, naviga-tion variations, and other factors that complicated the synchronization of ship traffic and human labor. Oversupply of workers was therefore one of the most typical features of dock work in harbors from London to Bombay. Employers encouraged it in order to meet the irregular peaks in demand, without incurring the expense of a permanent work-force. Abundant, cheap labor disincentivized stevedoring and coaling companies and dockyard employers from investing in new technologies, and the labor-intensive nature of dock work persisted until the con-tainer revolution of the 1960s.[74] Dependent on the arrival of ships, workers experienced bouts of labor followed by periods of waiting.[75]

Perhaps due to their connections with the Egyptian nationalist move-ment, the transnational and transmaritime aspects of coal heaving and of the heavers' labor militancy have been largely overlooked.[76] These aspects, especially the fact that heavers could disrupt transnational trade, communication, and production, came to the fore around the 1882 British occupation of Egypt, when these characteristics of heaver militancy trumped the more localized, intraguild, provincial, and statist politics that had characterized it during the previous decade.

Unrest in the Egyptian coal-heaving sector characterized this occupa-tion almost from its inception. In 1871, workers accused their elected guild leaders of assigning nonmembers to fuel ships while members were left idle.[77] Although the guild was only a year old and decreed from above, its members were quick to adopt it as a framework for disputation, and hastened to take advantage of the logics of officialdom that sanctioned it; this included submitting a formal written petition ('ardhhal) to the provincial authorities.[78] In 1873, they approached the Port Said harbormaster with complaints against the guild shaykh that included issues rooted in pietistic epistemologies and religious moral economies, such as the shaykh's engagement in usurious money lending. But most of their gripes—about miscalculations of the coal quantities they loaded, or arbitrary deduction of wages—were aligned with the rationale of state officialdom. Heavers found accommodation to the new protocols of labor management quite fruitful: while the judicial committee addressing the matter eventually decided to keep the shaykh in his job, he had to commit to better regulating the guild inspectors and address the complaints of its rank and file. Moreover, guild members were given special receipt books to document the sums paid them.[79]

Although such state interventions could better regulate the relationship of workers and middlemen, they could hardly transform the basic conditions generating unrest. A bloated, underpaid workforce was essential for expedient and inexpensive coaling. The transnational coaling companies, with their inherent tendency towards large-scale comparison, measured the wages of Port Said heavers against those in Malta, Marseilles, and London, ignoring the higher (and rising) cost of living in Port Said itself and the labor unrest in these other ports.[80]

A strike in April 1882—part of a wave of nationalist-worker-military unrest known as the 'Urabi Revolt, which prompted the British occupation of Egypt in July—became a source of concern among shipping companies worried about the interruption of passage via the canal, or fearful that the strike would inflame interimperial French-British tensions; shipping companies therefore pressed for its quick resolution.[81] One proposition was to use foreign strikebreakers: in April 1882, British companies tried to convince the Egyptian authorities to bring Maltese or Armenian replacement workers, but were refused.[82] The strike did not completely demobilize steamer transportation, but it forced ships to use their own crews to coal, creating long queues at the harbor. This was enough to force coaling companies to agree to a raise in wages.[83]

If in April international dimensions worked to the advantage of the strikers, the tide turned after July.[84] As many RN steamers rushed to join the Egyptian campaign via Aden at the height of the summer, when lots of heavers left to tend to their fields during the planting season, the remaining workers took advantage of the labor scarcity and the dire need for warships to start one of the most intense rounds of strikes and unrest in the port's history.[85] At the same time, as the military campaign was rolling across Egypt and triggering a stream of refugees leaving the bombarded city of Alexandria, many evacuees boarded ships departing for Malta. At that point, Maltese coal heavers announced a strike, forcing P&O steamer crews to coal their own ships and triggering the idea of importing Chinese laborers to replace the strikers.[86]

Worker unrest and its suppression were part of a self-reproducing translocal cycle: even the availability of Chinese labor resulted from the steam-fueled opium wars that weakened the Chinese state and caused the Taiping Rebellion from which a stream of refugees could be "freely engaged" on the cheap by mine and plantation owners, railway and steamer companies worldwide.[87] In Yemen, China, Egypt, and Malta, the British policy of opening up new markets and routes led to weakening state protection and deregulation of local labor markets. These markets were now integrated into a transmaritime labor pool from which steamer companies could quickly mobilize Chinese workers to crush a strike in Malta, or Maltese or Adeni coal heavers to break a strike in Port Said, as we shall see more than once below. In October 1882, now with official colonial backing, coaling companies used such tactics, locking out all local heavers from Port Said's bunkers and forcing them to accept the old piece rate.[88]

The transnational aspects of collective action in Port Said characterized subsequent strikes there too. A coal heavers' strike in April 1894 attracted to Port Said a British cruiser, under orders to land its men in the event of riots.[89] This scenario was repeated in June, at which point both British and French cruisers took positions at Port Said, each with orders to dispatch a landing party if the other side were to do so. As in 1871 and 1882, the surplus of workers ("about twelve hundred more men that can possibly be employed") was seen as one cause of the strike, and British commentators suggested resolving the crisis by removing these workers to other hubs of the coaling system.[90] This was the other side of expanding systemic control: the ability not only to borrow, but also to offload superfluous workers.

In both strikes, Maher Pasha, the town's governor, was accused of "ill-advised interference with the chief industry of Port Said, which has hitherto been successfully regulated by the co-operation of European firms and the Arab Sheiks."[91] The "interference" included passing a set of regulations allowing him to appoint and dismiss the coal heavers' guild shaykhs. The coaling companies rejected these regulations, thereby triggering the strike. In the aftermath, the regulations were revoked and the limited autonomy of Egyptian officials further curtailed.[92] This fusion between the local and transnational, and the successful circum-vention of the national sphere, characterized other ports under British control. In semicolonial Port Said, the antipolitical aspects of the proc-ess were thrown into sharp relief against the backdrop of a previous decade of similar intervention by a relatively independent Egyptian state, due to attempts by the governor and the language of the workers. Another factor made this all the clearer: the 1894 strikes were part of a larger wave of labor unrest in the Canal Zone during that year, includ-ing a prolonged strike by Port Said dredgers. These workers, many of them Greek, managed to get the Greek Consul to negotiate with the steamer companies on their behalf.[93] Managing to transform the strike from "a riot" into "politics" entailed getting a nation-state involved.[94]

In the 1890s, coal heavers again resorted to the language of Egyptian independent state-formation, imbued with an unmistakably transna-tional tone, including the comparative yardstick of "other kingdoms."[95] They directed their complaints against employers, and especially against the guild shaykhs, as they had done throughout the 1870s. John Chal-craft has documented petitions to this effect from this and the following two decades, again revealing that a supposedly premodern vocabulary of grievance was interlaced with appeals to government regulation, correct enumeration and bookkeeping, and the injustices of monopoly, bribery, and, notably, slavery. (In 1896, they complained to Lord Cromer, Egypt's de facto ruler, that contractors "buy and sell us like slaves.")[96]

Such petitions, directed at Egyptian and British officials who dis-missed them, suggest that heavers were deploying a universalist discourse of labor, demanding order, standardization, and regulation. Employers and officials, for their part, fought and delayed this process by aligning themselves with the parochial forces of shaykhs, "bribery," and deregu-lation. The petitions help historians reverse the received wisdom about how universalism worked: local actors adopted a language that tran-scended their particularities and attempted to speak to power in its own dialect (sometimes actually writing in broken English). But the system's stability was predicated on keeping these universalist demands at bay.

The Port Said heavers' guild already had electoral protocols and grass-roots forms of participation in 1870. As historians have shown, along with other forms of electoral politics—of village shaykhs in the country-side, or guild heads and deputies in towns and ports (all involving ballot-counting by the police and widespread animated participation)—this guild democracy formed the electoral basis for the Egyptian Consultative Assembly of Representatives, connecting local and national politics.

FIGURE 11. Coal heaving in Port Said, 1911.

But the Assembly was abolished less than a year after the British occupation, and a new electoral law replaced direct elections with truncated indirect ones. Elections for guild heads were completely abolished soon thereafter.[97] From this point onwards, Port Said joined other locations where local struggles could be global without engaging on the national level.

After the 1882 occupation, and without the protection of an independent state with international clout, the repeated strikes in the coal heaving sector did very little for the Egyptian workers. But this is not to say they had no historical impact. A major spoiler of heavers' labor action was the fact that steamers that reached a striking dockyard could be coaled by their own crews. During the last decade of the nineteenth century and the first decade of the twentieth, this happened increasingly in Port Said, Malta, Gibraltar, and elsewhere, eventually changing the basic conditions of maritime labor for British sailors. The *Impeccable* reported in its log for a 1901–4 commission that "one of the advantages of laying at Malta—native labour is employed in coaling ship," a rarity or "luxury," as the logs of other steamers put it.[98]

Strikes and the resulting intermittent availability of local heavers in many coaling stations were linked to a growing use of colliers for coaling steamers under their own labor, rather than by jetty or open lighters in ports (see figure 12). The constant increase in the size of ships and engines, and their capacity for coal, which more than tripled during the three decades after 1880, also promoted the shift towards colliers and towards tapping the free labor force on board.[99] When coaling could no longer be delegated, complaints, injuries, and tensions arose among

British steamer crews and the yearning for alternatives—"oh for oil fuel!"—became more audible.[100]

Indeed, by the end of the first decade of the twentieth century, there were ports where even simple sailors could compare the rapidity and especially the ease of fueling a steamer with oil as opposed to coal. In Suez in 1908, for example, a ship could be fueled with oil pumped from the quay at a rate of one hundred tons per hour.[101] This, on average, was also the rate of manual coaling during this decade (in this particular station on the quay or by lighters). The higher potency of oil somewhat compensated for its higher price per ton, $13.75, compared to the Cardiff coal available at Suez for $9.50 per ton. Under the right conditions, the human element could become an important variable, perhaps a decisive factor, in considering oil as an alternative to coal.

As a problem of white British sailors (one suggested that on "nearly every ship some poor soul lost his life" heaving coal),[102] coaling was finally politicized through the "proper" and familiar statist channels. Here the right kind of body interfaced with the right kind of body politic. The issue even found its way to the British Parliament, as the subject of a Parliamentary Paper addressing serious coaling accidents during the years 1910, 1911, and 1912 and as a set of questions posed to the First Lord of the Admiralty in 1909 and 1913. When the newly appointed Winston Churchill responded to the latter, he stated that "the risks attached to coaling are great, but in many cases accidents occur through want of care at a critical moment, and this, no action on the part of the Admiralty can prevent."[103] But in January 1913, Churchill already knew what every sailor intoxicated by the smell of oil fuel knew: that there were ways to limit the dangers of coaling, which indeed he opted for later that year.

As with stokers, the human body—both of white sailors and of unruly striking dark-skinned ones, whose suffering was invisible but which nonetheless helped expose physical vulnerabilities—again emerged as a bottleneck, this time in the coal-heaving process. Wallerstein and Balibar have claimed that racial hierarchy and the ethnicization of the global workforce justified paying lower wages to a segment for work that was sometimes harder than that done by white Europeans.[104] Rather than precluding the possibility of a universal sphere, such differential practices actually enabled it. As far as the transnational coal system was concerned, these practices allowed for reducing labor costs, often below subsistence levels, in a manner akin to the ostensible ignorance of the physical limits of stokers. As an owner of a coaling com-

FIGURE 12. Coaling from a collier, circa 1910. Canadian War Museum. George
Metcalf Archival Collection, CWM 20030174–011.

pany in Port Said put it, "it is hardly necessary to point out to you that
there is a vast difference between the requirements of an Arab compared
with those of an English labourer. Thanks to a genial Climate and a low
order of civilization, the wants of an Arab are comparatively almost
Nil."[105] But the arrangement was unstable and eventually had to be
replaced. Arab and other labor militancy played an indirect role in promoting this transition.

MINERS

Similar to other features of the coal system, workers' strikes had interconnected and often synergic effects. It was not uncommon for strikes
to erupt concurrently in faraway parts of the system. Coal miners in
Wales launched a strike in April 1882, on the heels of the strike by Port
Said's heavers.[106] Welsh miners (whose steam coal was the global benchmark for "best quality" and was sold in Port Said) were constantly told
that this coal constituted "foreign trade," connecting labor wages in the
coalfield to overseas fluctuations of supply and demand. This was
indeed the case. Coalfields were more tightly integrated into the world

market than the British economy as a whole. Nearly one-half of all coal mined in South Wales was exported by the turn of the century, at a time when little less than a quarter of the output of the UK economy was comprised of exports.[107] The sudden 5 percent reduction in wages that prompted the work stoppage that April might have been connected to the situation in Egypt.[108]

More often, however, labor action in the far larger British and other European mining sectors reverberated in the Ottoman Empire. These disturbances were registered as price and wage fluctuations, as depleting reserves in depots, and as congestion of ships in harbor. Ottoman ambassadors in Europe frequently sent reports about strikes such as those in Ruhr and Westphalia, which in early 1889 caused price increases in the Empire and which were reported in the Arabic and Turkish press and in the Hebrew press in Palestine.[109] These strikes inspired a major dockworkers' and coal heavers' strike in London later that year, similarly echoing from Aden to Australia (we saw in chapter 2 how Gandhi supported it). This was also the case in 1893–94, when strikes erupted simultaneously in the coal mining sector of Colorado, Illinois, Ohio, Pennsylvania, and West Virginia in the United States and in the Tankersley colliery in England. The strikes erupted in tandem with the aforementioned strikes by Port Said coal heavers, which were closely related to the unrest among the Suez Canal workers which, in turn, radiated "east of the Canal" to provoke unrest in Aden and Perim in the Red Sea region.[110]

Mine and ship owners as well as striking workers of different occupations were well aware of connections between coal mining and coaling overseas. During a Welsh miners' strike in the summer of 1893, for example, interested parties closely followed the quick depletion of Welsh coal in Port Said, as well as attempts to restock this and other eastern coaling stations with French and German steam coal (from the mines which went on strike in 1889).[111] Speakers at a miners' assembly convened after the strike had ended analyzed the depletion of Welsh coal in Gibraltar and its effects on their strike's success.[112] The Ottoman government, which also followed these interconnected strikes closely in the press and through informants at the coaling stations, was similarly well aware of their effect in its territories.[113]

The effects of intersecting waves of labor unrest differed between the exporting British Isles and the importing Ottoman Empire, and were characterized by several imbalances. Ottoman mines provided only about 20 percent of the empire's total coal needs; the rest was imported, mainly from the British Isles.[114] If fluctuating coal prices, which mining companies tried to offset by reducing workers' wages, were found to be the key factor in miner militancy in England and Wales,[115] it was even more the case in the Ottoman Empire, where coal was more expensive and subject to more extreme price fluctuation.[116] Beyond the increase in price as one moved away from the source and incurred transportation costs, other factors informed coal prices in the Ottoman Empire.[117] Significantly lower coal tariffs for foreigners within the empire made price differences a bone of contention and frustration between locals and foreigners.[118]

The great British coal strikes in the years leading up to the 1913 decision to adopt oil (and the peak in British coal during that year) coincided with a wave of labor unrest in the Ottoman Empire, which saw the fiercest struggles in the production and transportation of coal. Global historians who noticed the concurrence of labor struggles have associated them with cycles in the world economy, and have tended to ascribe their diverging aspirations, tactics, and outcomes to historical contingency rather than connectivity per se.[119] What follows is an alternative account, which unpacks the implications of connectivity.

Whereas Aden was torn from Ottoman Yemen and officially made part of British India, and Egypt was an occupied Ottoman province informally annexed to the British Empire, the Ottoman coal coast can be seen as another point on the same semicolonial continuum, under full Ottoman political control but subjected to intense foreign economic intervention. Transnational mining and coaling companies that affiliated themselves with local middlemen in the two former locations applied pressure on the Ottoman state, and on the sultan in person, to mediate in mining-sector labor disputes. In the latter half of the 1910s, settlement meetings organized by local government officials were likened to the sultan's gift-giving ceremonies, where large crowds expressed their loyalty to the ruler.[120]

To understand the effects of this strategy vis-à-vis transformations in the British energy sector, consider the Ottoman career of a major Welsh mining company. After a British House of Lords verdict in the case of *South Wales Miners' Federation vs. Glamorgan Coal Co.* severely restricted union officials' ability to call a strike and workers' willingness to embark on one, the British Parliament passed the 1906 Trade Disputes Act, a key piece of legislation immunizing participants in industrial disputes from tort actions for damages and injunctions. Meanwhile, outside the United Kingdom, transnational coal companies could still get away with using practices that were quickly falling out of favor at home. Glamorgan Coal Company (GCC) was one such firm which had been

FIGURE 13. A military camp set up outside the Glamorgan colliery after the Tonypandy riots. Getty Images.

involved in Ottoman coal mining at least since the 1890s, after the Ottomans opened up the sector to foreign companies in 1891.[121] In the following decade, GCC also involved itself in steamer coaling in the region. When the right to unionize was officially recognized in the Ottoman Empire in early 1908, this was soon reversed with the adoption of the Provisional Strike Act on Associations in October.[122] The act was adopted due to intense pressure from foreign capital (in the mining sector, the proportion of foreign capital was exceptionally high, climbing from 60 percent in 1907 to 72 percent in 1909). A new 1909 Strike Act restricted the establishment of associations and limited the workers' weapon of strike by establishing obligatory compromise periods.[123] In 1909, therefore, GCC could use its time-tested union-busting strategies in the Ottoman Empire, bringing in blackleg coal heavers as replacements with official sanction.[124] As in previous heavers' strikes, such as in Port Said in 1882, 1893, and 1894, coal companies could resort to heavers from neighboring stores (in those cases from Malta and Aden).[125] Official Ottoman intermediaries sometimes proposed resorting to blackleg heavers for ending labor disputes, offering official reassurance that the security of companies using such strikebreakers would be guaranteed.[126]

Back home, things were quite different. In 1910, GCC and another Welsh coal company invited the police to break a miners' strike in Tonypandy, South Wales, paying the policemen's expenses and accommodation (see figure 13).[127] The invitation triggered a chain of events that involved the army and tarnished the reputation of the young interior minister Winston Churchill, seen as the official responsible for sending soldiers against strikers.

Whereas the reversal of the House of Lords decision in the House of Commons was a triumph for participatory representative politics, those at the Ottoman end of these global disruptions did not enjoy the same benefit. During the second half of 1908, 111 strikes erupted in the Ottoman Empire's coal sector and other industries—about twice the number of strikes recorded throughout the empire during the previous half-century.[128] The transnational character of these strikes was clear to all involved. The governor of Ereğli, for example, attributed strikes in his region to "foreign coal workers who were aware of the labor movements in European countries," as did the Ereğli Company later that year.[129] The success of labor militancy in Europe and its impact in the Ottoman Empire were the main reason for passing the restrictive strike law and curtailing the spread of universal labor. The Ottoman minister of commerce and public works, Ali Bey, pointed out during the parliamentary debate of the law that the category "laborer" should not be given legal status in the Empire.[130]

What came to be called the Young Turk Revolution reinstated the Ottoman Parliament and constitution that had been suspended by the sultan in 1878. However, the Young Turks deflected popular unrest away from representative parliamentary politics and toward an understanding of government as controlled by an enlightened elite (more a House of Lords than a House of Commons). Manifestations of labor politics were often similarly rearticulated by nationalist middle classes as anticolonial or anti-imperial in ways that absorbed their energies and deflected their agendas toward national and hence bourgeois goals.[131] In August 1909, an addition to the strike law was passed that prevented workers in essential public services from striking without official sanction. The coal sector was included in this definition around 1910.[132]

While strikes were forcefully suppressed everywhere in this period, it was more easily done on Europe's periphery. The securitization of coaling meant that the British army and navy were increasingly used to break labor unrest across the empire: at Port Said in 1882 and 1893, Gibraltar in 1898, 1890, and 1892, and in India in 1908; and if we include the navy crews used as coal heavers-cum-strikebreakers—as workers in the various coaling stations saw it—the number of incidents rises considerably. Fierce police and paramilitary action and the arrest of striking miners and heavers were also common in the Ottoman Empire.[133] This was the case in 1906, 1908, and 1913 with the striking Zonguldak miners, and in 1911 with the striking coal heavers of Istanbul.[134] These strikes coincided with miners' strikes across western Europe, particularly in England and Wales. But, as we have seen, when Churchill merely garrisoned troops in preparation to break the Tonypandy strike of 1910, the myth of sending the army against miners nearly cost him his political career. In England, by the 1850s, the use of troops to break labor militancy, so common in the eighteenth and early nineteenth centuries, was a thing of the past.[135] The "war capitalism" of the colonies was successfully kept at bay from the "industrial capitalism" back home.[136]

For Churchill, this was a constitutive experience, one enhanced by the work of Herbert Stanley Jevons (nephew to W. S. Jevons), who argued in 1909 that the only limit to the coal regime was organized labor and strikes, which in the first decade of the twentieth century continuously disrupted the stability of the global coal system.[137] Theory and experience pushed Churchill, now First Lord of the Admiralty, to convert the RN to oil.[138] Breaking, and breaking away from the unruly miners and heavers of the coal system, Churchill advocated their replacement with foreign oil, pipes, and fewer men more susceptible to pressure.[139]

The coal strikes in England and Wales were a direct result of what Marx called "proletarianization": miners were farmers transformed into full-time workers, completely dependent on their wages and on the mine, lacking any other skills or means of subsistence, and subsequently fighting as a unified class for better pay, shorter working hours, and reasonable employment conditions. Their new power and limitations were a result of this process. By contrast, from the activation of the Ereğli mines in the late 1840s to the strikes in the years around 1910, Zonguldak miners retained their connections to their villages and to agricultural modes of subsistence, making them less dependent on wages (which at any rate were constantly "in arrears").[140] In this they were not very different from other Ottoman workers who took up coal-heaving when the agricultural season was over. Like heavers in Aden and Port Said, they too had used the language of the day to protect their rights or protest ill treatment, repeatedly complaining that they were treated "like slaves in colonial countries."[141]

Scholars from the subaltern studies school have argued that European-style proletarianization cannot be seen as one of the globalizing features of capitalism.[142] Surprisingly, however, the 1910 strikes in England and Wales resulted in a push toward proletarianization in Ereğli, just as decision-makers in England were devising new ways to weaken the rebellious working classes at home, in part by diversifying imperial fuels with oil. The fate of "fully" proletarianized Ereğli miners was thus completely tied to coal just as it lost its political potency and could no longer serve as a driving force toward participatory politics. Examining the timing and circumstances of miner proletarianization from a systemic perspective helps wind up the account of the universalization of free labor that began with abolition. Proletarianization combines the political and economic meanings of liberty in "free labor." As Marx put it, it creates "free labourers, in the double sense that neither they themselves form part and parcel of the means of production, as in the case of slaves, bondsmen, &c., nor do the means of production belong to them, as in the case of peasant-proprietors; they are, therefore, free from, unencumbered by, any means of production of their own."[143]

The wave of Welsh and English strikes of 1910–12 reverberated violently in the Ottoman coal sector.[144] "Because of the strikes of the European coal workers," states an alarmed 1910 report to the Ottoman Ministry of Interior, "all merchant ships in the Black Sea and even in the Mediterranean need to purchase coal at Zonguldak port, and today 14–15 ships wait there for coal, including three big military transportation steamships, which would take 12,000 tons of coal for the Ottoman Navy." The report, wired during the high agriculture season, complained that miners had left for their villages just at the hour of urgent need. "To ensure the production of necessary coal for the navy and in general," it added, "it is necessary to guarantee a sufficient number of workers, who themselves cannot recognize the real interests [of the empire] and are currently engaged in useless work in their villages."[145] The situation in the global coal sector sensitized all parties involved to the circumstances of production in the coalfield. As in 1908 and earlier, the possibility of systematic transnational influence was again on the table. Even when investigations proved that strikes were not precipitated by "foreign agitators," subtler transregional mechanisms were at work.[146]

The path to proletarianization in the Ereğli coalfield passed between the competing forces of agricultural work and military conscription. Workers often preferred to maintain the connection to their villages and lands, but around 1910 farmers and miners in the Zonguldak region were forcibly conscripted into the Ottoman army. However, the strikes in the British coal sector boosted the demand for local coal, which allowed the Ottoman navy and the Ereğli Coal Company to negotiate an exemption for mine workers. Taking advantage of this new agreement, the company and the governors of Zonguldak and Bartın began encouraging locals to adopt mine work as a way to escape military service.[147] Finally, in 1911, increased demand for Zonguldak coal and the creation of a proletarianized workforce resulted in the coalfield's shift to foreign management.[148] While strikes continued in the Ottoman coal sector in subsequent years, after World War I workers found it increasingly difficult to force the hand of a corporate employer backed by both the regional authorities and the central state.[149]

If the previous two sections demonstrated how the bodies of stokers and heavers were gradually integrated, albeit in a mediated and belated fashion, into a labor physiology via sick lists and coaling accident reports, the same was true for proletarianized miners in Ereğli. Analyzing mine accident reports between 1870 and the 1910s, Donald Quataert found that the "post-1912 era" reports were scientific, personalized, and formulaic.[150] And Nurşen Gürboğa showed that, after the war, the accident rate increased significantly, along with quantitative and scientific risk-management schemes that eroded existing forms of communal aid and solidarity.[151]

How do we put together the three cases presented above? We may begin by revisiting a platitude about steamers: that they were relatively labor-free. Stating what now must be obvious, steamships relied on ample human muscle. But if they depended on manpower, they also relied on its invisibility. We have seen three variations on this theme: stokers represented the point where the individual human body posed a limit to the functionality of the engine. Simultaneously, their case reveals how human limits became significant only when demonstrated by a white body. The invisibility and pain of nonwhites helped steamers move. The same applied to coal heavers, but as a human collective. The joint action of heavers helped demonstrate the physical limits of British sailors engaged in coaling at striking foreign harbors and helped politicize those limits. The third case, of British miners connected to various Ottoman coal workers, demonstrates the same point on the larger scale of relationships between multiple collectives and nation-states. Only British workers reaped the benefits of interconnected global convolutions, and again their suffering gave them access to the political, while other suffering bodies were ignored. Thermodynamic notions of labor had a global impact in an uneven manner.

Tumultuous workforces made the operation of the coal system far from optimal in the eyes of policy-makers who eventually opted for oil. The importance of bodily harm and material breakdown will be probed in chapter 5, where I examine how accidents acted as forces for technological change via protocols of insurance. At this point, we might regard the strike (like nineteenth-century capitalists) as belonging to this category of disruption, pushing for techno-fixes to human-element collapse. Strikes recalled other kinds of malfunction where broken bodies were tantamount to machine failure. In this sense, the writings of Jevons Jr. demonstrate the cross-fertilization between the natural and social sciences as much as those of his uncle.

The lesson is not only that Middle Eastern workers were key for the operation of the global coal system, industrialization and trade—that the Industrial Revolution happened on a global scale—but also that their presence was written under erasure. The analytical task was not simply to make something visible, but to flesh out the importance and mechanisms of its invisibility, which might account for the enduring received wisdom about steam engines as labor-saving devices. Alongside Britain's "ghost acreage" or "phantom carrying capacity"[152]—the vast spaces outside the British Isles from which British industrialization could draw raw materials (thus freeing up space at home for food production, for example)—there seem to have been foreign ghostly bodies whose transparent life-blood nourished global capitalism. In part, the success of the more familiar specters haunting Europe depended on the invisibility of these bodies.

What helped to highlight these factors in the above discussions was thinking about workers in England, Wales, Zonguldak, Port Said, and Aden as sharing a labor pool interconnected by the same resource and material infrastructure, subjected to similar natural rhythms, and exposed to related fluctuations in price. This perspective showed how earlier proletarianization in certain parts of this system, where more of the process's benefits could be gained, came at the expense of other parts. Instead of a diffusion model whereby European history travels to the colonies, in our examples European history actually precluded such diffusion.

Following thermodynamics' spread and blind spots, moving from the individual to the collective, from the body to the body politic, can contribute to political theory, a path opened by the notion of "carbon democracy." Timothy Mitchell's framework explains broad political participation as tied to a limited historical window of opportunity connected to a particular energy regime. By using the unprecedented vulnerability of an industrial system dependent on coal, British miners forced elites to grant them access to the political arena and the right to vote. The shift to oil came in order to close this window.

Regarding political participation as a question of global labor power under the sign of energetics complicates this argument. First, democracy cannot be attributed only to one fuel, coal, that allowed it and another, oil, that limited it. What enabled political participation in Britain simultaneously prevented it in the Ottoman Empire already under coal. Inclusion under the umbrella of labor science and physiology, trade unionism, protection from the more brutal kinds of state violence, and eventually the right to vote were developments gained in Britain on the backs of colonial workers, often by transforming them into "workers."

While these workers were prevented from recruiting their weakened states to their cause, the opposite was the case in the metropole. Let us conjure up Disraeli's *Coningsby* again to illustrate how. The novel is set at the time of the 1832 Reform Act extending the franchise in Great Britain to major urban centers, and for the first time connecting industrialization and parliamentary politics. Published in 1844, *Coningsby* steps back in time to situate a beginning for Victorian political identity founded on what would later be called "Tory democracy." Novelist Disraeli, hailed as the architect of this vision, laid its legal foundations in another reform act in 1867 as prime minister. This Second Reform Act extended the vote to "labourers"—men defined as virtuous workers in contrast to the multitude of the unskilled poor—further tying labor to parliamentary politics. In 1884 and 1885, the franchise was extended again in the Third Reform Act and the Redistribution Act to boroughs with large miner concentrations, constituting coal miners' major entry into electoral politics.[153] It was during the latter year that Lord Randolph Churchill defined the process as "Tory democracy." This was the legacy and framework with which his son Winston contended.

The gradual extension of the franchise was always a belated and reluctant response to effective public pressures, often directed toward other goals, such as labor rights. It reflected, and simultaneously shaped, the growing powers of workers and the consolidation of ties between them. But different industrial sectors were not necessarily aligned as comparable components in the homogeneous space of a nation-state. We have seen how both the energy and textile sectors were truly global, in ways that connected them to people and places overseas more directly than to geographically more proximate settings. Around the turn of the twentieth century, the employment of two hundred thousand English miners depended on foreign coal exports.[154] Workers did not naturally consider themselves as members of the demos, the citizenry, or the proletariat of a nation-state; these categories emerged in tandem with making and breaking potential alliances.

Democratic politics was such an instrument for making and breaking coal-itions. British miners' most significant successes were in forcing the state down the mine shaft, and in increasing state regulation and supervision of their bodies and worktime. They forced the state to intervene between them and employers in managing the safety of their persons both inside and outside of the mine in terms of nutrition, subsistence, insurance, schedule, and wages.[155] They fought for the franchise not for its own sake, but in order to secure these "labor" rights. In this light, the franchise can be seen as a mechanism for being governed in a new, thermodynamic way. If steam engines were union-busting devices in the mills, they energized specific groups of miners, allowing them to beat employers at their own game. At the same time, they were absorbed into this statist game. Understood as thermodynamic, carbon democracy appears to be tailored to fit one kind of body politic with one kind of body.

CHAPTER 4

Environment

ARTIFICIAL ARCHIPELAGO

Between the 1820s and the 1890s, a system of coaling depots metasta-
sized between the Mediterranean and the Indian Ocean, mainly along
what would be called "the Middle East" a decade after that system had
stabilized. Even if some of the coal stored in these depots soon found its
way into the interior, in the eyes of their British architects and operators,
the main importance of these depots remained as maritime coaling stations. In contrast to industrial western Europe, most of the coal exported
from Britain to the Ottoman Empire (constituting 80 percent of the total
coal burned there) was used for transportation, mainly by international
steamships.[1] This was generally the case with coal leaving the British
Isles: "The great bulk of our export," wrote Herbert Stanley Jevons
about coal in 1909, "is for the use of steamships."[2] Britain's global coal
trade extended Britain's reach and mobility overseas. Whereas we associate the adoption of fossil fuels mainly with industrialization, I will show
that it was the thrusts of transportation that made a major contribution
to the global coal economy. This chapter focuses on the extension of
British military and commercial power through interconnected maritime
footholds and dryland enclaves, and the construction of the "artificial
archipelago" that enabled it—a chain of depots on natural islands and in
littoral zones on the mainland, the consolidation of which involved massive shifts of land and water to simultaneously connect and insulate these

stations from their surroundings. The extension of this archipelago was two-directional and multifaceted. The powers that promoted its construction came from both east and west, from London and the East India Company capitals of Bombay and Calcutta, aided by local forces and dynamics at various midpoints along this corridor. So, too, with the coal stored in these depots: initially, much of it came from England, but whenever possible, local sources were tapped. The maritime uses of coal soon stimulated mining in India, Japan, South Africa, and China. Ottoman coal mining was an early instance in this dynamic, which in some respects is another case of extending a British "extractivist" mindset, both overseas and underground.

Coal never moved alone. Its progress along this corridor was intertwined with the circulation of other materials, species, and political institutions simultaneously Europeanizing, Arabizing, and Indianizing this network. So, too, with the connective tissues that joined coaling stations to one another: carbon fibers were reinforced by other connectors, sometimes organic or else the stuff of age-old Islamic family ties. Understanding the emergence of this corridor helps identify key structures of the global entrenchment of the fossil fuels economy. Although this process obviously had many other tentacles, following the one that connected the main global coal exporter to its key colony and crown jewel through a region that a century later would become a key global oil producer is a reasonable point of departure.

What follows examines different elements of the spread visualized in figure 14, and some of its consequences. The first section stresses the importance of movement for the creation of this stationary network. Unlike sail ships, which could stop at any harbor and stay long periods at sea, steamships regularly refueled at designated hubs, on average once a week, which made them semiterrestrial technologies. The name given to these hubs in British parlance, "coaling stations," is reminiscent of the railroad and, just like that overland infrastructure, fossil-fueled ships depended on groundwork and disrupted existing understandings of land and sea.

The second and third sections attend to the importance of engine and ship design and then weight in informing naval movement and the terraforming that resulted from the different challenges of taking earth matter—coal and other pieces of England—abroad. The earthiness of the depot system was shaped by coal's weight and materiality, and resulting water ballasting schemes, as well as by processes of land reclamation, dredging, and mineral (especially salt) extraction. Connections

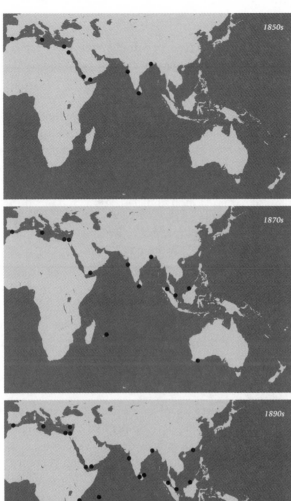

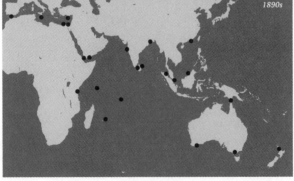

FIGURE 14. Three snapshots of the British coaling system: 1830–1850, 1850–1870, and 1870–1890. From left to right: 1850s: Gibraltar, Malta, Alexandria, Suez, Mocha, Aden, Bombay, Galle, Calcutta. 1870s: Gibraltar, Malta, Alexandria, Port Said, Aden, Mauritius, Bombay, Galle, Calcutta, Tebing Tinggi, Singapore, Labuan, Albany. 1890s: Gibraltar, Malta, Alexandria, Famagusta, Port Said, Zanzibar, Perim, Aden, Seychelles, Mauritius, Diego Garcia, Bombay, Colombo, Trincomalee, Calcutta, Tebing Tinggi, Singapore, Labuan, Hong Kong, Albany, Thursday Island, Melbourne, Auckland.

along this archipelago involved signaling systems like telegraph cables and lighthouses whose deployment was politically and legally fraught.

The sections below, which explore how the archipelago was populated, reveal the artificiality of the new system as transient. Large-scale, man-made transformations resulted in unintended processes that transformed the newly formed material, social, and legal environments into a second nature along similar lines as meat and opium, described in chapter 2. The steam-boosted mixtures of matter and species created new "natural" environments on land and sea. Artificiality thus stimulated processes of naturalization, indigenization, and racialization where the previously scarcely populated terra nova of the depot system sprouted natives and local traditions. It animated the spread of extraterritorial legalities as well. As also revealed below, this naturalized environment was characterized by specifically human features, such as anxiety about shipwreck, that fueled a constitutive "energy insecurity." The last section connects the maritime corridor of coaling stations and the global spread of this anxiety under conditions of a global maritime arms race to the activation of coal mining in the Ottoman Empire and elsewhere.

TRAILBLAZERS

To underscore the importance of movement, take stock of the forces and processes that shaped the development of the coaling system, and begin fleshing out key factors in how depots were chosen, abandoned, or reconditioned, I advance this and the following chapters' argument by following several steamers, examining how a maritime corridor congealed around them. We begin with the *Hugh Lindsay*, constructed to assess the practicability of Red Sea steam navigation, combining a body built in Bombay and an engine sent from England, thus also revealing the bidirectional thrusts, from India and England, which made up the depots system. The ship left Bombay in March 1830, so heavy and low in the water that it was nicknamed "the water lily" by skeptical observers. It carried coal piled on deck, in the saloon, and even in the cabins to ensure sufficient supply. After eleven days, and hourly entries in the ship's log describing the wind, waves, and each of the engines, boilers, and side wheels, it arrived in Aden with only six hours' supply remaining. The need for coal was, of course, anticipated, and a collier loaded with Welsh coal, accompanied by a sailing brig, the *Thetis*, was sent around the Cape to the Arabian coast. But the collier was wrecked on a reef, a fate the *Thetis* narrowly avoided, on the north Arabian coast.[3]

Given these outcomes, the assessment was mixed: it showed that the distance between Bombay and Suez could be crossed with a steamer, marking Aden as the optimal point for a depot as well as one of the only suitable harbors. However, this was only barely feasible: when it left Bombay heavy with extra coal, the *Hugh Lindsay*'s draft was depressed from eleven-and-a-half to almost fourteen feet, which severely strained the paddles and slowed her engines to eleven-and-a-half revolutions per minute, or five knots an hour. Before reaching Aden, when it was much lighter, the engines accelerated to nineteen revolutions or eight knots.[4]

Furthermore, in attempting the voyage's second leg and sailing directly from Aden to Suez with all the coal in Aden on board, the *Hugh Lindsay* was still lacking supply because of the accident, and in order to reach his next station the captain had to improvise by wetting and reusing the coal ash.[5] As we will see, such technical lessons would be key in the development of a new generation of compound engines that would reuse the same steam two and later even three times.

This voyage allows us to catalogue some of the main issues pertaining to the formation of the depot system: one was weight, the coal being too heavy for effective engine function at the start and too light and risking capsizing at the end; another was shipwreck, which endangered life and property as well as the regular provision of fuel. These two elements intersected in the reef-studded Red Sea with its dangerous proximity to rocks and diminished maneuverability exactly when it was most needed. Compound engines, hydrography, lighthouses, and other signaling schemes were all responses to such challenges.

The ship's weight also informed the *Hugh Lindsay*'s search for an alternative to Mocha, the first coaling port. In 1820, EIC agents had been granted land and exemption from anchorage fees at Mocha, which proved significant for traders and sailing ships stocking this store with coal or moving it from there to Aden, and later for the steamers that used this coal.[6] Yet evidence collected by James Wilson, who captained the *Hugh Lindsay*, indicated that "no vessel drawing more than 10.5 feet of water can go into the inner anchorage at Mocha and the outer anchorage is exposed to so great a swell . . . that the taking [of] coals on board is a matter of great difficulty and often impossibility."[7] Harbor depth, its relation to ship draft and coaling protocol and to possibilities for steam dredging and land reclamation proved vital during the following decades.

Alongside these technological and environmental aspects of the passage, human factors emerged as well. First, with so much coal on board, there was little room for anything else; thus communication, mail, and later passengers were key in financing the development of early steam navigation. In 1832, the voyage had stimulated debate on the punctuality of Arab and Turkish coal suppliers, and whether a private coal contractor would be better than an appointed governor.[8] Along its course, the *Hugh Lindsay* changed the social landscape. For example, to expedite coaling its captain replaced the Ottoman official Hussain Agha with ʿAlim Yusuf, an Armenian merchant, as the EIC representative and coal supplier in Jeddah, on the assumption that a private individual would perform better than the tardy official.[9] The hybridity resulting from such privatization and reliance on non-Muslims is also revealed in the fact that while it was named after a British naval captain and chairman of the EIC, the *Hugh Lindsay* was built by the Wadia family of Parsi shipbuilders in Bombay. While the British engaged with various groups in and around Aden, including Jews, Greeks, Armenians, and various local and foreign Muslims, Parsis—who had accompanied the British troops almost from their arrival—held a special place in the establishment of the coaling station. Seen as combining familiarity with Islam and an entrepreneurial disposition geared towards hard work and a commercial spirit, anchored in a Protestant-like religious ethos,[10] Parsis were key enablers of Aden's development materially and socially. Following Aden's Parsi community also explicates transformations of local and Indian Muslims in and around the depots. The *Hugh Lindsay* thus charts the key topics addressed in the following sections: technology, weight, shipwreck, new legal and social ecologies, and the surprising origins of reform movements in the Islamic world.

DRIVING FORCES, LITERALLY

Coal depots are usually seen as energy stores, and energy certainly played a significant role in shaping this system. Under sail, the location of Red Sea ports was informed by the region's wind regimes and the fact that due to the monsoon it was much easier to sail south and out of the Red Sea than northward into it.[11] Steam navigation made this principle partly irrelevant—"partly" because existing powers like winds or animal caravans still played a significant role in transporting coal to depots and provisioning them with food. Significantly, though, steam integrated what, under wind, had been a divided space into a single Red Sea, which, in turn, was divided again according to principles such as the distance between coaling stations. However, as the first part of this book repeatedly demonstrated, highlighting energy regimes and transitions often obfuscates more than it reveals. What this perspective erases is the fact that the marine steam engine, the ships it propelled, and the environments they traversed were all in the making during the nineteenth century. Engine, ship design, and environmental exoskeleton were interconnected; the following explains their gradual co-constitution.

As exemplified by the ingenuity of running the *Hugh Lindsay*'s engine with the same coal twice, and the problems with its side paddles, one way to read the maps in fig. 14 is through marine engine development and the shift from side paddles to rear propellers. Early steamers with heavy single combustion engines required vast amounts of coal to turn their side wheels and advance slowly, resulting in frequent stops at the small coal dumps proliferating along the way. (Figure 14 shows the main stations, deployed at equally spaced intervals. Near them, smaller depots were created and then abandoned—e.g., on the Arabian coast of the Red Sea in places like Mukalla, Perim, al-Hudaydah, Kamaran, Jeddah, and Yanbuʿ—to allow steamers under British, Egyptian, Ottoman, and other flags to fuel, and also to mitigate buoyancy problems resulting from burning coal.) By the late 1830s, engine technology had evolved sufficiently so that a ship fueling in Suez could complete its journey to Bombay, stopping only once to fuel in Aden; at this point, this harbor was occupied by Britain. As the first map illustrates for 1830–50, this early period is defined by equidistant coal depots along a single line stretching from Gibraltar via Aden and Ceylon to the Indian ports of Bombay and Calcutta, the two capitals of the EIC.

The second period in question (1850–70) saw technological innovation in both ship design and engine efficiency (mainly screw propulsion and iron and later steel shipbuilding). In the 1860s, the development of compound steam engines that used the same steam twice (expanding steam via multiple cylinders of progressively lower pressure) proved more energy-efficient and allowed engines to sustain higher pressures. The 1869 opening of the Suez Canal drove the infrastructural explosion evident in the third period in figure 14. Technologically, steamers could now carry cargo as well as mail and people. By the 1890s, the triple expansion engine, which worked at two hundred pounds per square inch (psi), had come into use, taking engine coal-consumption efficiency to its pinnacle during that era. By now the depots system had stabilized and the last of the major coaling stations was established during this decade.

Besides shortening the route to the east, the Suez Canal evened out and reduced the distance between coaling stations. This allowed a European steamer en route to India to take on coal at Gibraltar, Malta, Port

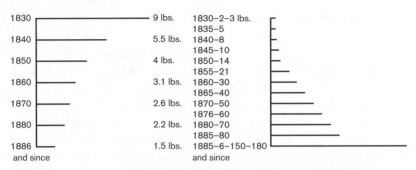

FIGURE 15. Decrease in expenditure of coal per indicated horsepower per hour based on good average practice" (left); diagram showing increase in steam pressures (right).

Said, and Aden, reducing the size of its coal holds and increasing its cargo capacity.[12] Ship design and technical features were also informed by other characteristics of the east-west voyage. For example, since its inauguration, the canal's narrow waterway promoted screw-propelled steamers and demoted side-wheelers, which hit the banks with their propellers, and sail ships, which depended on costly towing services for the entire stretch between Suez and Port Said. The canal itself kept changing: during its first decades it was continuously widened and deepened, a process which both suited and informed the changing size of the steamers that passed between its banks.[13] Ships' shape and geography recreated one another.

Other features of the canal also changed the shape of ships; for example, there were significant price differences for coal east and west of the canal: during 1911 and 1912, the price per ton of Welsh coal nearly doubled when passing between Port Said and Suez. As long as the canal's depth forced heavy ships, when fully loaded, to replenish their supplies east of the canal, light steamers had the advantage in terms of money and coaling time.[14] As the canal was gradually deepened, it propelled an increase in the size of steamships, just as the ships' size encouraged its deepening.

Changes in the Suez Canal and the growing number of depots and maritime traffic also highlight the fact that the growing efficiency of the maritime steam engine during the nineteenth century (evident in figure 15) did not reduce coal-burning. Quite the contrary. The amount of coal stored in the major depots increased exponentially during the second half of the century (see figure 16), feeding the growing steamer traffic and increasingly sustaining other uses inland. Here was another instance in

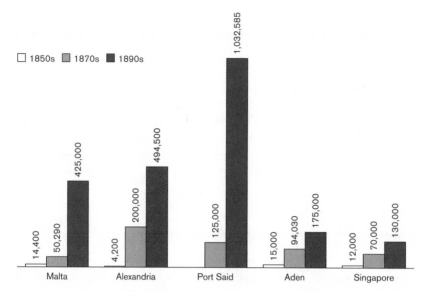

FIGURE 16. Coal depots, average yearly coal volume in metric tons. Based on multiple sources.

which energy saving and efficiency meant more rather than less circulation, expansion, combustion and emission. The following sections retrace this intensification in terms of ship and harbor size.[15]

WEIGHTY MATTERS

In 1850, the world's merchant fleet had about 9 million tons of carrying capacity. By 1910, it had 34.5 million.[16] We tend to think about such leaps in terms of circulation. Rapidity, turnaround, and regularity were no doubt important elements in this story, and had much to do with fossils being used as fuel or energy sources, and coal's combustibility. Yet exclusive focus on combustion obscures the crucial importance of weight itself, which is revealed by the fact that it was the guiding principle for almost anything at sea or in harbor. It determined how maritime insurance premiums were set, the permissible number of passengers on board steamers,[17] and the calculation of most ship and harbor fees, from pilotage to lighthouse charges.[18] Weight affected labor arrangements in and around a ship,[19] the ship's shape, the size and shape of ports and waterways steamers passed through, and the shape and composition of the depots system itself. What follows unpacks several key aspects of coal's weight—beginning with makeweight.

On the open sea, a ship must contend with lateral forces of wind and waves; ballast regulates buoyancy and affects a vessel's stability and maneuverability. While this has always been obvious to mariners, historians have usually been blind to the prominence of weight, stressing other aspects of freight such as monetary value or levels of processing. Traditionally more interested in dynamics on land than maritime connections and water's pushback, they largely overlooked the importance of weight and its role in enabling European industrialization in the now-conventional story of the Industrial Revolution, tailored to fit a world-systems model in which the colonies were underdeveloped agricultural peripheries that provided raw materials to the mechanized metropole.

The world-systems perspective tends to stress the difference between agriculture and industry. The industrialization of England made it increasingly dependent on imported raw materials such as Black Sea grain from southern Russia and the Ukraine, or Egyptian cotton: enter cereals and fabrics, exit finished goods and textiles. These neat divisions are disrupted, however, by coal—a key raw material shipped from England. While coal may appear as an exception to the rule, it was in fact the exception that *enabled* the rule. Most imports, particularly those from the Black Sea and Mediterranean regions, were low cost and bulky, and most exports were far less bulky but expensive (such as drugs or dyes, which were highly valuable but small, or heavy rather than bulky items like expensive iron products such as anchors, chains, and later, machinery, much of which was coal burning); therefore, ships would have to leave the British Isles half empty or in ballast.[20] Coal export solved this problem, at least for ships traveling to the Black Sea and eastern Mediterranean.[21]

The difference between raw and processed items was manifest largely through bulk (processing meant reducing bulkiness and weight), and coal equalized this differential by enabling a significant return trade. In a British trading system focused on carrying homebound cargos, coal was a readily available item that at the very least paid a shipowner's expenses and could increasingly find customers around the world. As the nineteenth-century economist William Stanley Jevons put it, "coal is to us that one great raw material which balances the whole mass of the other raw materials we import."[22] By the 1860s, British coal was consumed in almost every sizable port in the world. Beyond the excellent quality of Newcastle and Welsh steam coals, a more important factor, according to Jevons, was the fact that "coal is carried as ballast or makeweight and is subject to the low rates of back carriage."[23] In 1903, Welsh industrialist David Alfred Thomas estimated:

> More than four-fifths of the weight of our exports consists of coal; without it the great bulk of the shipping bringing corn, cotton, wood, wool, sugar, &c, to our shores would be compelled perforce to clear without cargo, and in ballast . . . Indeed, it is hardly conceivable that our foreign trade could have reached its present dimensions had it not been for the outward freight provided by coal.[24]

Coal's weight informed the shape and, even more so, the depth and underpinning of the environment it helped traverse on sea and land, as the development of water ballasting demonstrates. Many dissimilarities between sail ships and steamers resulted from their disparate sources of energy: harvesting the wind forced sailing ships to zigzag or tack, whereas steamers sailed straight. Another often-ignored, key difference concerned vertical rather than horizontal movement: steamers were ships on which the draft—the vertical distance between the waterline and hull bottom—changed during a voyage. In the first years of the 1910s, just before the RN converted from coal to oil, a cargo steamer of average tonnage rose about 1.3 inches per day owing to coal consumption.[25] Several decades before, fluctuation in the water had promoted the development of water-ballast based on pumps connected to the ship's engine that linked the burning of coal to the inflow of sea water, compensating for the steamer's rise owing to its depleted coal. Other reasons also applied: although wind was free and coal cost money, a sail ship's reliance on earth ballast balanced things out somewhat. Loading ballast (like shingles or gravel) required resources to purchase the makeweight and pay the workers; it was also time-consuming and required cargo space. Moreover, off-loading earth ballast clogged river mouths, and when this was banned, ballast was dumped in mounds on river banks, taking up fertile agricultural land. By contrast, liquid ballast and water-dumping was seen as not having any financial, labor, or environmental costs at all.[26]

From 1844, the steam engines in new screw-propelled ships pumped seawater into a compartment directly beneath them to compensate for weight loss during the voyage. Linked to the abandonment of paddle wheels and the adoption of screw propulsion, new iron colliers designed specifically for British coastal coal runs were the first to adopt water ballast; oceangoing steamers followed suit in the 1870s.[27] From the early twentieth century, it was clear that water made coal redundant by replacing it as ballast. In 1921, the journal *Coal Age* stated, "it should be understood from the outset that coal exports can never again mean to any country quite as much as they meant to England around the beginning of the 20th century. At that time there hardly was a portion of the globe where British coal could not be found. But since then important changes have occurred," referring primarily to double-bottomed ships that used water ballast and made coal redundant for this purpose.[28]

Concurrent with the increase in ship tonnage was the need for port expansion and improvement on a grand scale as larger, heavier, and more powerful steamers wielded an ever-greater strain. Coaling facilities had to be accommodated in deep, wide basins of still water, where there was adequate depth alongside the quays. Even the best natural harbors of the wind era were soon totally inadequate.[29] In 1910, an engineer described this as "a race between engineers . . . On the one hand, we have the ship designers turning out larger and larger vessels; on the other is the harbor engineer, striving vainly to provide a sufficient depth of water in which to float these large steamships."[30] If the *Hugh Lindsay*'s voyage was a search for a deep, natural harbor, by the 1850s steamers had become so large and heavy that only man-made ports could accommodate them. Several coal depots in figure 14, including Gibraltar, Port Said, Aden, Bombay, Singapore, and Hong Kong, were either created or significantly transformed by land reclamation and dredging. These two processes of earth-removal were often closely connected, and involved similar dynamics to ballasting. It was coal-fired steam dredgers that allowed British engineers to reshape the sea. Meanwhile, copious quantities of industrial by-product—gravel and shingle—facilitated dredging and land reclamation. Increased water depth and breadth in canals and ports correlated with expansion and construction of infrastructure, including wharves, jetties, depots, warehouses, offices, etc.

Both processes were used in Port Said and Aden, and much of Port Said's acreage was gained by land reclamation. After the 1865 erection of the city's western wharf, and the resulting sand deposit, more land was added to the city area under Suez Canal Company jurisdiction. An average yearly increase of 50 meters to the municipal sand beach was measured between this year and the late 1880s, when the company formally claimed the additional land and started leasing it to non-Egyptians. Thus, beyond the special legal status and tax exemption of non-Egyptian subjects in the canal zone, the area was part of a special real estate arrangement that excluded it from normal Egyptian property regulations. The town's municipal border kept moving, and the canal's architects could compellingly argue that they had produced the territory to which they now laid claim.[31] In Aden, by the 1880s, the town's natural harbor—previously considered one of the finest in the world—had gradually become unsuitable for bunkering. The merchant community of Aden led by Parsi traders[32] had communicated the importance of dredging the harbor to the India Office and had been submitting concrete proposals for this since 1859, but it took three decades to carry out.[33]

The entanglement of land reclamation, ballast, and coaling, as well as the importance of Parsi mediators, was also evident in the emerging salt-production sector at depots like Port Said and Aden. In these and other Red Sea stations, salt was extracted both by quarrying and by solar brine evaporation (for example by "digging holes . . . into which water at high tide filters through the soil and the salt is thrown down in crystals as the water evaporates.")[34] By transporting Red Sea salt, colliers from both Europe and India could avoid returning or continuing empty. The ballasting function of coaling-station salt lowered its price, making it competitive with salt produced in India.[35] During the 1880s and 1890s, several coal companies diversified into salt production in various Red Sea coaling stations including Aden and its vicinity, with such calculations in mind.[36]

SHINING AND INVISIBLE CONNECTORS

Exchanging coal, water ballast, and salt allowed steamers to regulate their cargo and buoyancy as they traversed an increasingly integrated network of depots. These stations were also connected through other means, often also informed by weight. For example, the tonnage of a ship determined the "light fees" it was charged in ports of call. Heavier ships which sailed closer to reefs and the seabed benefited from signaling and illumination systems proportional to their weight: thus, more weight translated into higher light fees and more revenue for expanding the lighthouse system, as figure 17 reveals.

The all-season, 24/7 movement of the steamship was not intrinsic to steam engines; it required a supporting framework. Lighthouses were important components of this framework, providing light at night and demonstrating yet again how earth and water were redefined and renegotiated during the age of steam. These towering structures at liminal points where water met land provided architectural challenges that stimulated the development of civil engineering and the use of new building techniques and materials. Hydraulic lime, for example, a mortar that set under water, eventually led to the development of Portland cement and

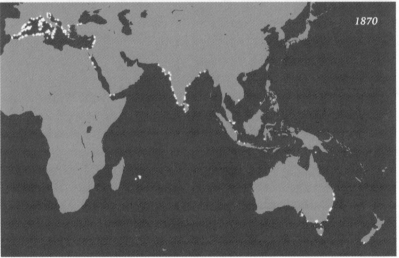

FIGURES 17A (this page) AND B (next page). Mediterranean and Indian Ocean lighthouses in 1850, 1870, 1890, and 1914; based on multiple sources.

to the use of concrete in modern construction.[37] The pioneering Port Said lighthouse, completed a few days before the festive inauguration of the Suez Canal, was one of the first major structures built with reinforced concrete by the inventor of this building system, François Coignet.[38]

When promoting the establishment of a lighthouse at Cape Guardafui, near the mouth of the Gulf of Aden, the British Board of Trade listed the

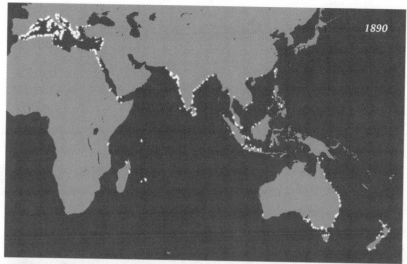

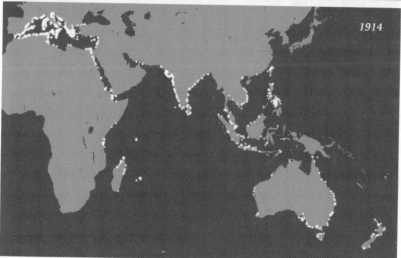

FIGURE 17B.

nearby shipwrecks between 1872 and 1885. All sixteen British steamers, mostly of one thousand tons or more, were tabulated with their cargo value and type (much of which was coal), and losses in property and life.[39] Ship size and external environment were again co-constitutive: increased tonnage financed lighthouse construction in the Red Sea, which could now be traversed in all seasons and all hours of the day and night.

As these maps suggest, weight and light fees incentivized the clustering and spread of lighthouses near coal stores and well beyond them. If this helps fathom the volume of maritime traffic, then the maps reveal the great intensification of maritime circulation and the importance of direct maritime movement for commerce. This was also evident in the illumination of the Red Sea rather than the Persian Gulf, and in a comparison of the 1850s—when the Egyptian overland rail connection was already operating—to the 1870s. Lighthouses came with new trends of point-by-point navigation, allowing steamships to fulfill their potential to sail straight. The bird's-eye view of the maps reveals the global and transmaritime nature of coal-fired interconnectivity. Finally, the scale of the maps makes separate light sources appear to blend with one another. This was an actual synergy, but it was also the result of quite different, not to say competing, approaches to illumination, safety, and responsibility of the various political entities involved.

Like other features of the coalonial infrastructure, lighthouse construction and operation in the Red Sea and Mediterranean were rife with interimperial tension, especially at the Ottoman-British-French-Egyptian nexus. A lighthouse was usually a fortified structure, often with a garrison (or armed steamship anchored nearby) tasked with defending it from pirates or violent locals. In these waters, and around the British Isles, coastal communities often lived off shipwrecked vessels, which they regarded as a gift from the sea or God. Captains were regularly warned to "be on your guard against false lights, which are often shown by them to allure vessels."[40] They were also tasked with preventing defection by lighthouse keepers, usually ex-mariners who found manning a lighthouse lonely and tedious.[41] This made lighthouses emblematic military strongholds whose deployment in foreign territory constituted an imperial foothold. During the closing decades of the nineteenth century, the British were backing Egyptian schemes to construct such posts in the Red Sea, whereas the French insisted that a French concessionaire working for the Ottoman government across the empire continue its work in this territory too. The sides made little effort to hide the fact that the semimilitary capture of this key maritime route was at stake.[42]

In 1836, the British government abolished private ownership of lighthouses in the British Isles and entrusted the upkeep of such facilities to Trinity House, a corporation already responsible for ballast regulation, thereby further bolstering the link between weight and light. Rather than justifying the shift in terms of safety at sea, however, reformers used economic rationales and presented the public interest in terms of business. During the following decades, Britain extended this logic to the Red Sea (which, as we will see in the next chapter, became a sphere of quantifiable "risk management"). With the most ships at sea at any given time, the British had the greatest interest in lighthouse proliferation. However, due to free-trade ideology and French involvement in Ottoman lighthouses, Britain usually opted for minimalism instead, whereas the Ottomans pushed to expand and enhance the system's capacity.[43]

By contrast, in the Ottoman Empire, the provision of light at sea and the attendant acknowledged safety considerations were not configured as an abstract public good, translatable to weight and financed by the volume of trade. Rather, light provision was seen as the basic, personal obligation of a just ruler (whether with regard to sufficient quantities of candles at the market, or cultural representations of the radiance of the sultan's power itself).[44] While foreign pressure was not uncommon, requests for lighthouse construction often came from coastal communities of traders and fishermen.[45] Plots for such buildings were given for free, and local authorities were expected to render any assistance necessary to the lighthouse engineers.[46] Especially after the inauguration of the Suez Canal and the related activation of maritime Hajj routes at the expense of those over land, providing light to pilgrims was increasingly important.[47] In the early 1860s, the Porte franchised this task to a French firm, Collas et Michel, founded by two French navy officers. The company built almost a hundred lighthouses in the first twenty years, where only a few had existed before, and continued through the empire's demise and beyond.[48]

Lighthouses clearly illuminated questions of sovereignty and interimperial tension. Moreover, they yoked such struggles to the logic of cartography. In both the Red Sea and the Mediterranean, the Ottomans, British, and other powers became aware of the need to deploy lighthouses as part of charting and mapping, and of the need for better maps as part of selecting sites for lighthouses.[49] Lighthouses came to the Red Sea with new trends in cartography and point-to-point navigation, befitting vessels that could sail in straight lines rather than tack.

Contrasting the British economic rationale of "what the trade would be willing to finance" with Ottoman attitudes toward lighthouses demonstrates that as a light source, coal shone in different ways for different beholders. Similarly, during the second half of the nineteenth century, lighthouses were introduced into various local mythologies around the Gulf of Aden. Mahlal Ha-ʿAdani (whom we encountered in chapter 2), related how Aden's Jews' narrative of the British occupation connected this event to the Sabbath candles and to lighthouses:

> The Jews of Aden . . . lit candles every Friday in glass lanterns hung with a rope to the ceiling. As the wind would blow and rock the suspended lantern, the candle would flicker continuously through the window. These flickering candles, like a lighthouse in miniature, attracted the attention of the Captain of a British ship that happened to pass by Aden on a Friday night . . . When [the captain] returned from India, he brought hundreds of soldiers, who arrived at the spot from which they saw the flickering lights, boarded small boats, approached the shore of Aden, and planted the British tricolor flag there.[50]

On the coast of neighboring Somalia, lighthouses were said to host jinn who would sometimes interfere with Islamic devotion.[51] The exhilaration of Indian Muslim travelers passing through these waters at seeing Arabia for the first time blended with the sight of the Aden lighthouse.[52] It is no wonder that the most important periodical of transnational Islamic reformism was called *al-Manar*, the Lighthouse. The name was taken from a hadith according to which "[Prophet Muhammad] said that there were for Islam landmarks and a lighthouse, like a lighthouse of the path." This beacon was said to connect east and west, linking Arabs, Indians, Turks, and Persians.[53] A few years after *al-Manar* was first published, the foundational guidelines for a utopian pan-Islamic society were similarly described as "rays of light shining from between the articles and lines, rising above the lighthouses, illuminating the dark sides and blinding the eagles."[54]

Sayyid Fadl bin 'Alawi, a Hadrami whose career we will examine more carefully later, used similar tropes when comparing the Prophet Muhammad to a "ship of rescue," a person "successful for the brightness of his light." These metaphors drew on older analogies between prophetic power and light, but also on the *manari* symbolism and, in turn, on actual coastal infrastructures which Fadl knew well. Indeed, the book where they appeared, *Idah al-Asrar al-'Alawiyya wa Manahij al-Sada al-'Alawiyya*, was published in 1898–1899, just as *al-Manar* was first printed, by the pan-Islamic *al-mu'ayyad* press, which also published the writings of Muhammad 'Abduh and other *manarists*.[55] Such inflections of fossil-fueled light and legitimacy helped forge new group identities and modes of belonging; and, as the maps in figure 17 demonstrate, they also combined to create a synergy that brought integrated passageways into being.

Together with these coastal connections, the depots network was also connected invisibly underwater (see figure 18). Early schemes for "steam communication" in the Indian Ocean were designed primarily to expedite and regularize the transfer of letters rather than goods.[56] Before submarine telegraph cables connected the subcontinent to Europe in the 1860s and 1870s via locations—Aden, Suez, and Malta—that were also key coal depots (unsurprisingly, as their generators required an energy source), steamer lines were the only means of guaranteeing regular communication between the metropole and its key colony. This changed with submarine telegraphy, an innovation made possible by the steady, linear motion of steamships and their ability to lay cables evenly. As the steamer was seeking out new functions, the large cable-laying steamships (large enough to carry long cables that could connect continents) led to ever larger vessels.

As well as their implications for ships, telegraph cables also created new realities on land. In British dossiers concerning sovereignty in the Red Sea, telegraph cables were filed together with lighthouses, both "opening the question of sovereignty over the coast and islands in that sea."[57] Especially from the late 1850s, when telegraphic connectivity between Aden and India was negotiated, British diplomatic activity, including the request of several Ottoman sultanic *fermans,* sought to ensure the safety of the cables at landing points on shore. But such assurances, without which the technical work could not start, came at the political price of potentially acknowledging Ottoman rights in territories near Aden. The more generous the wording of the *ferman,* extending licence "as far east as the straits of Bab-el-Mandeb," the bigger the problem.[58] On the other hand, British steamers surveying potential telegraph routes in the region could be summoned to intervene militarily in commercial or other disputes.[59] The British thus weaponized telegraphy, not only in terms of information transmission, but also in justifying the presence of gunboats in foreign waters.

Operating, negotiating, purchasing, or capturing territories for the telegraph, lighthouses, and various other requisite constituents of coaling often involved non-European actors. Obviously, the telegraphers had to be polyglots and often hailed from diverse backgrounds. So, too, with the more mundane aspects of operating these infrastructures; for example, the financial memoranda of the Eastern Telegraph Company reveal the dependence of this service on Welsh coal as a main source of fuel for electricity, on desalinated water provided by the Parsi company of Cowasjee Dinshaw & Bros., and on different steamer companies for regular travel between the offices at Aden, Suez, Zanzibar, Perim, Bombay, Malta, Mozambique, etc.[60] Even if the telegraph connected Britain and its key Indian colony, and Welsh coal generated the signal, it still required

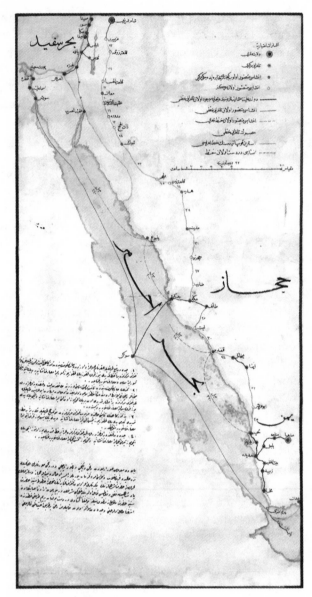

FIGURE 18. Land and submarine telegraph lines. HRT, MAP
O476, BOA.

amplification from midway stations. Egyptian users supported the enterprise with their fees, the Ottoman sultan supported it with his blessing and *fermans*, and Parsis provided potable water. Like the lighthouse system, telegraphy's entanglements created social imbroglios.

ENERGY AND INSECURITY

"When in this country," wrote Daniel Defoe about his visit to Newcastle during the 1720s, "we see the prodigious heaps, I might say mountains, of coals, which are dug at every pit, and how many of those pits there are, we are filled with equal wonder to consider where the people should live that can consume them."[61] In Defoe's day and for about a century to follow, most of these people lived in London and the south of England, and they largely used coal for domestic heating. At costs ranging from one-fifth to one-half the price of wood, coal was essential for survival during the English winter, supporting a population growth that enabled the Industrial Revolution. Its distinct smell and murky cinders were among the most evident signs of British domesticity.[62]

However, in the nineteenth century, coal's heat-generating function was fundamentally reconceived. Coal was increasingly used for locomotion, gradually becoming "the means of transporting itself . . . coal carries coal, by rail and by boat."[63] Even when leaving the British Isles, coal carried with it a piece of home. By the time Defoe's more famous book, *Robinson Crusoe*, was translated into Arabic (as early as 1835, spurring other Arabic and then Turkish translations), the Ottoman Empire also saw the arrival of Newcastle coal, a confluence demonstrating Ralph Waldo Emerson's understanding of coal export's civilizing mission ("every basket is power and civilization").[64] *Crusoe's* phenomenal success has been attributed to the way it captured the mixture of British imperial aggression and anxiety about the risks and ordeals of leaving the British Isles.[65] It was part of a long tradition, beginning perhaps with the *Odyssey*, that dealt with Europeans shipwrecked on foreign shores.[66] Such literature expressed and transformed these fears into a particular kind of expansionism that also applied to steam power: *Crusoe* tells of a shipwrecked British gentleman who transformed a desert island into a piece of England, aided by Friday, a native collaborator. The famous Mercator projection maps in which imperial holdings were depicted in the same red or pink color as the British Isles were similar anxiolytics; another way to carry "home" to the empire. Coal's progress in the nineteenth century via reddish dots on non-British territories (often aided by non-European Fridays) followed a similar arrangement, establishing a system of depots and locals through which British ships could sail safely and feel at home. I claimed that before "energy," fueling was often an excuse for a land grab. However, it would be hasty to reduce this motivation to a sinister desire for domination and ignore the constitutive angst that animated or justified it. As we have seen in chapter 1, fear and weakness often propelled infrastructure expansion, and this was also the case here: prior to any land grab, "energy" and "energy security," there was a constitutive maritime insecurity.

Many coaling stations, from the very establishment of the network, showed how the need to manage widespread instances and fears of shipwreck blended with captivity and schemes of territorial expansion. The first British coal depot in Ottoman lands was constructed in Mocha, the coffee port that Mehmet Ali, Egypt's powerful governor, had secured for the Ottoman sultan. A case of manhandling the local agent of the East India Company in 1817 was used to justify a British bombardment from the sea in 1820, after which the company was granted land and exemption from anchorage fees. In the following decade, when Mehmet Ali volunteered to create a similar coal depot for the British in Aden, London flatly rejected his proposal. In 1839, capturing the port by force and establishing a fueling station was an explicit way of curtailing Egypt's own encroachments in the Yemen. The general aspiration to block Egyptian expansion was combined with a particular pretext for the British attack on Aden: the shipwreck and plunder of the *Daria Dawlat,* and grave ill-treatment of its passengers.[67]

These imperial dynamics as well as interimperial anxiety animated British coaling expansion. In the mid-1850s, the British captured Perim, a small volcanic island dividing the Bab el-Mandab strait into two channels, to establish a coaling station and lighthouse under the noses of the French who were thought to be planning a similar takeover. They eventually adopted a strategy of co-optation, allowing the French and other European powers to maintain coal bunkers in territories under British control "instead of giving to the government of France a plausible pretext for acquiring possession of an independent position."[68] Protection of British personnel, vessels, and commercial and imperial interests was key in the capture of all these locations. Throughout the coaling system, intra-European rivalries were being extended eastwards.

Like the European powers and the Egyptians, Istanbul also used coaling to justify expansion. Aden's takeover indicated to the Ottomans that the establishment of foreign coal depots might result in nonthermodynamic projections of power, and they attempted to answer in kind. For example, in 1879, through their Basra governor, they gathered information about local shaykhs, tribal loyalties, and religious pedigrees in the Persian Gulf to found a coal depot commanded by a charismatic "coal officer" who could establish rapport with local tribes at Bahrain, and then secure the port as Ottoman territory.[69] Such maneuvers fueled action by the British who thwarted this plan by taking over the port in 1889 and, from 1890 onwards, imposed British extraterritorial jurisdiction over British Indian subjects and foreign merchants there.[70]

The imposition of British extraterritoriality and its promotion by British Indians against Ottoman designs (which also relied on local non-Ottoman actors) in the Persian Gulf and the Red Sea (and in between) is a nod to the need to gain traction and safety. It is no coincidence that key eruptions of violence that led to the establishment of Red Sea coal depots during the 1820s, 1830s, and 1850s involved Indian merchants, often Muslim, ships under the British flag coming and going to India, and EIC coastal factories, or other extraterritorial sites like consulate buildings with legacies in the Indian subcontinent. Seeking safety, the EIC was often drawn into military conflict by its local collaborators. Such was the case when local Yemeni forces attacked the EIC factory in Mocha in 1817, after a dispute about loading a Muslim-captained India-bound ship, "under the British flag, which had for ages afforded protection and commanded respect for the British character."[71] Similarly in 1837, the EIC ships' inability to offer timely assistance to survivors of the *Daria Dawlat*—and attendant fears that this might shake the confidence of South Asian merchants and pilgrims and result in Indian traders adopting other arrangements for their security—led to the occupation of Aden. EIC officials believed that due to the "insults . . . offered to the British flag," Aden should be seized for use as a coaling station to ensure security in the region.[72] According to Scott Reese, Aden's small, diverse group of elites wasted little time in accommodating themselves to the new political and economic realities. Sayyid Zayn b. ʿAlawi, the effective leader of Aden's Muslim community, for instance, declared his family's loyalty and desire for protection under the British flag. British Aden emerged as the preferred port of Indian merchant houses, which also sought to minimize the hazards of shipwreck that were compounded by often whimsical Ottoman customs policies. With British control secured, Muslims from across the western Indian Ocean put down roots and began to coalesce into a community in Aden.[73]

So, too, in 1858 in Jeddah, Hadrami merchants were singled out by the British as the main perpetrators of the violence that spread from the port to European consulate buildings. A key context for the events and their aftermath was the commercial rivalry between members of this trading diaspora and Europeans and their protégés. Sayyid Fadl b. ʿAlawi was blamed for invoking anti-British sentiment and for provoking incitement between the Hijaz and India.[74] The catalyst was an argument about whether an Indian ship should raise an Ottoman or a British flag. Military enforcement of British extraterritorial sovereignty triggered a violent popular response and, in retaliation, a bombardment by a British steamer meant as a spectacle of British power. Several similar instances will be discussed below.

Another culprit was the Ottoman Egyptian ʿAbdallah Agha, a tax collector who was also involved in the monopoly of the salt trade—salt being a reexport from nearby Suakin and an important ballast on the voyage from Jeddah to India. A further context for the violence was his constant rivalry with European consulates (established in the context of multiethnic trade in the Indian Ocean) who tried to shield their protégés from the unfair levies he was said to extract.[75] Clearly, the need to pacify and secure an arena where natural hazards were compounded by multiple trading traditions and interests prompted British territorial capture and aggression.

From the 1860s, steam navigation along our corridor blended security with another major source of British anxiety—sanitation—a connection captured by the widespread fear of the "twin infections" of cholera and Islamic radicalism. The "twin infections" combined the traumas of the 1857 Indian Rebellion—seen by the British as a jihad resulting in "fugitive Mullahs" spreading to Arabia, Istanbul, Egypt, and elsewhere—and the 1865 global cholera epidemic that resulted from the mass congregation at Mecca. In both cases, Muslims' reputed fanatic disregard for their own bodies had grave implications for the safety of the bodies of Britons. British observer W. W. Hunter put it succinctly in 1872 when defining hajjies as a "squalid pilgrim army." Though they might "care little for life or death . . . such carelessness imperils lives far more valuable than their own."[76]

Steamers shortened the vector of contagion by allowing vibrio cholera to reach Arabia from India, where cholera was endemic, before it killed its hosts, thus facilitating its spread around the world. The British refused to explicitly acknowledge that steam navigation might be a contributing factor, but neither could they completely ignore its role in spreading disease. The blurring of infection and security hazard pointed the way to an awkward solution: monitoring the steam hajj through the shadow of

Islam. In order to regulate flows of potentially contagious bodies along the hajj routes, a network of inspectors was extended from British India into the Arabian Peninsula. British extraterritoriality, which is the subject of the following section, was not driven by a malicious desire to capture Islam's holy cities, as many Ottomans, Arabs, and Indian Muslims believed, but mainly by fear that resulted from the coal-fired circulation and a reluctance to depend on Ottoman quarantine to curtail it.[77] Instead, the British used "Mohammedanism" for entry and action in the Hijaz, Muhammad's birthplace and an age-old Islamic space of refuge and limited temporal sovereignty. In the process, they helped shape Islam as a natural religion befitting the task of regulating the body.

EVERY MAN IS AN ISLAND: HUMAN RESOURCES AND EXTRATERRITORIAL ECOLOGIES

The language used in the British press to frame the Jeddah violence is revealing: "the Mussulmans," it was claimed, were pushed to violence "following the instinct of their lower race and lower creed."[78] The British refusal to address maritime safety, as revealed above in the discourse about lighthouses, reared its head in the context of human conflict, too, revealing the centrality of "race" and the notions of "native," "nature," and similar environmental determinisms when coming to grips with man-made environments. These blended the general racial thinking of the period with more specific contexts, to which we now turn.

As the above incidents also reveal, much of my "artificial archipelago" was rooted in the institution of the EIC factory. This was not a place of production but a permanent fortified warehouse and ship-maintenance facility. EIC factories were established in clusters of fishing villages in Madras and Calcutta, and on the islands near Bombay, seamlessly integrating with the coaling archipelago of the nineteenth century. They were created by special dispensation of land by local rulers and became "white islands" in "brown towns" (which might have prompted literary scholars to see them in the context of Defoe's desert island).[79] Factories were sites of extraterritoriality, where the EIC administered its own civil and criminal law under the British flag, attracting foreigner and Indian profit and asylum seekers. Among these, Parsis were a common fixture; indeed, in the first factory, founded in Surat in the early seventeenth century, members of the Wadia family (who built the *Hugh Lindsay* two centuries later) worked as shipwrights and chandlers. Parsis were as much a part of the scene as Friday on Crusoe's island.

Coal depots and the port cities that grew around them constituted a legal archipelago where regimes of exception and extraterritoriality flourished. During the late nineteenth century, these made-in-India extraterritorial schemes blended with the capitulations, which initially served Western Christians in Ottoman lands, and now started serving non-Christians who carried British interests to the heartlands of Islam via the physical and legal enclaves of steamer ports. These were sites of intersecting, partial sovereignty related to the more durable features of the system: the deployment of depots, lighthouses, and telegraphs, as well as the fact that land was created ex nihilo and newcomers flocked with their different bacteria, documents, and claims, creating multiple and unending disputes and crises.[80] The Arabian coast of the Red Sea and its hinterland, especially after the vacuum left by the Egyptian withdrawal in the 1840s, was such a space, an Anglo-Ottoman borderland crucial to the flow of the steam hajj and thus for British safety. Such social, legal, microbiological, and political realities were (literally) organically connected to the depots system.

A good example of how nature became a category covering both the artificial and human components of the architecture of the depots network followed within a few years of Aden's occupation, when the British authorities started experimenting with the colony's capacity to grow fresh vegetables.[81] They were responding to an unhappy predicament: while the European troops dispatched to South Arabia were in relatively good health, "the native troops," that is, South Asian sepoys, were not doing well at all. Alongside the regular ailments familiar in India (fever, dysentery, and liver complaints), they were affected by the Arabian climate and resistant to conventional treatments that had been effective in India. In response, the British first resorted to "only sending mussulmans and low caste men on the Aden detachments, men who would not object generally to eat meat, and to the use of stimulants." But it was later assumed that the cause of their ailment was proximity to the salty, damp earth, and that these conditions similarly affected people of all religions and castes; thus, they opted to improve living conditions and supply better nutrition.[82]

Many of the improvements came from another non-European group: Parsis were the first to grow cabbages, turnips, and radishes in Aden; having tasted them, British commanders authorized an experimental vegetable garden to complement the sepoys' diet.[83] In the construction of the *Hugh Lindsay* and in previous chapters, we witnessed their central role in meat, salt, and water provision, as well as various drugs.

Parsis were one of several non-Christian diasporas populating the coaling system. What made them so resourceful?

The answer lies in the Parsis' historical relationship with the EIC, and with another group, namely the Muslims. As Parsi lore would have it, between the eighth and tenth centuries, after the Islamic conquest of the Sassanid Empire, a group of Zoroastrians refusing to accept Islam fled Persia and settled in Sind and Gujarat. This legacy and Parsis' estranged intimacy with Islam were well-known among members of this community and in EIC circles, and might partly explain the company's tendency to rely on Parsis in mediating its relationship with the subcontinent's Muslim rulers.[84]

From the early 1850s, Parsis organized in religious reform associations to defend Zoroastrianism against Muslim and Christian challenges, transforming it into a scripture-based faith adhering to European models of religion.[85] They also created a flourishing educational system and entered en masse into the legal profession, where their multilingualism and diverse cultural know-how made them sought-after mediators across the subcontinent and on both shores of the Red Sea by the end of the century. In Aden, for example, Parsis were the first to occupy the position of court registrar from its creation in 1873 until the next century. In the early twentieth century, Parsi registrars became the fiercest opponents of Aden's qadis and "traditional" Islamic authorities. Whereas the latter regarded their legal role mainly in terms of arbitration, the former advocated "Anglo-Mohammedan" law—a British attempt to rationalize and codify Islamic jurisprudence as a system of "natural law" that gave precedence to textual rules over social contexts.[86]

Using Parsis to promote an understanding of Islam anchored in the fixity of nature recalls another legacy of their association with the EIC. In the early-modern period, Parsis mainly acted as chandlers to European companies, supplying wood for ships, foodstuffs, and alcohol.[87] The origins of formal British-Parsi relations are rooted in the space of the factory and in the native agency system of the EIC, in which brokers were needed to connect European merchants to local markets and supplies. As early as the turn of the eighteenth century, the chief broker for the company's Surat factory was the prominent Parsi merchant Rustum Manock. He was responsible for promoting EIC affairs with the Mughal court, which in turn recognized him as the company's ambassador. Gradually emerging as a comprador elite and relocating their center of activity to Bombay, progenies of Rustum Manock assumed the newly established role of political agent after the chief broker position was abolished.[88] Utilizing these resources, Parsi brokers started trading along the west Indian coast, and in Persian and Arabian ports.[89] When the EIC sought to extend its Indian reach to Aden, they naturally relied on the Parsi material and legal technologies developed in its factories. Whereas locals in foreign environments were usually reduced to malevolent forces of nature like the "twin infections," Parsi technical, logistical, and legal understanding was seen as the cure.

Parsi improvement schemes greatly impressed those Islamic movements seeking models for communal reform. Sayyid Ahmed Khan, for example, could not contain his admiration for the Parsi community, which he encountered in Bombay, and was completely taken by the Parsis' economic success, education system, and language skills, as well as their dress, hygiene, and more.[90] Through his writings and those of Christian Arabs like the Syrian Christian traveler Athnasius Ughnatiyus Nuri, the Parsi model was exported to the Arab world. Nuri, who visited India during 1899–1900, regarded Parsis as "the most advanced people," praising their education, their charity, their high positions in the British government, and their European dress.[91] In 1912, Muhammad Rashid Rida, pan-Islamist founder of *al-Manar*, visited India and expressed similar admiration. In his view, Parsis were the most developed community in India due to British cultivation resulting from the latter's reliance on them.[92]

Rida offers an incisive assessment of the genealogy of Islamic educational reform in India vis-à-vis Parsi and other similarly "advanced" communities. His acerbic account of colonial divide-and-rule strategies states that the British only allowed Sayyid Ahmed Khan to found the Aligarh College after realizing that education reforms among India's pagans had reached a threatening level that justified co-opting the subcontinent's Muslims as potential allies in case others adopted an anti-British stance.[93] A mutual feedback cycle seems to have emerged in which Islamic reformers drew inspiration from the Parsis for combatting Muslim ʿulama; this, in turn, provided the legal ammunition in places like Sayyid Ahmed's Anglo Oriental College for Parsis to apply Anglo-Mohammedan law against similar Muslim ʿulama in places like Aden.

Parsis and sepoys were not the only Indians from the subcontinent frequenting the Red Sea coal depots. Muslims of both Arab and non-Arab descent represented another wave of incomers, and this was often a counterpoint to the Parsi advance, and one which was mutually reinforcing. Parsi legal knowledge was often required to deal with such itinerants who spread along the commercial, family, and religious networks that preceded the age of steam.

Aden was "Indianized" under steam; this was preceded by the migration of Arabs to India which, in the nineteenth century, the British were trying to curtail. "Stoppage of Arabs along the Sea Coast into the Interior of India" was the frequent title on official documents about these attempts, and this was high on the British agenda in the Indian subcontinent, so they routinely apprehended and deported Arabs seeking work.[94] From the late eighteenth century, places like Hyderabad had an important Arab—and especially southern Yemeni—presence. The British viewed these "Indian Arabs" as the "most truculent and dangerous specimens of humanity" and as a "whole race" that should be deported.[95] They were often seen as "fanatics," understood in psycho-physical terms, which, for example, reduced their tolerance to cholera. Typically for the times, Winston Churchill said: "What the horn is to the rhinoceros, what the sting is to the wasp, the Mohammedan faith is to the Arabs of the Soudan—a faculty of offence or defense."[96] Indeed, before the separation between the "natural" and the "social" and the rise of "faith"-oriented ideas of religiosity, religion was seen as part of the former and was, of course, predicated on how Christians saw natural religion, as revealed by the category "Mohammedan" used for Muslims in the colonial censuses and literature.

Yet even as fanatics, Indian Arabs had their uses for the British Empire in interactions with other Mohammedan subjects. We get a glimpse of such uses in Rida's Indian travelogue. After describing Sayyid Ahmed Khan's Aligarh University, he commented on several other new centers of Islamic learning, including the well-funded British-licensed *al-Madrasah al-ʿArabiyyah* (the Arabic/Arab School) in Hyderabad. This school was an example of what Rida saw as a broad British campaign to promote the study of Arabic among India's Muslims. A likely reason for this, according to Rida, was the British government's desire to capture Arab territories in the Persian Gulf, Iraq, and Yemen, for which it required Arabic-speaking Indian Muslim agents. The way to combat foreign encroachment, he concludes, was for the Ottoman government to strengthen its local administration in the Arabian Peninsula and rely on local tribes to block British forays.[97]

Rida was quite perceptive. By the end of the nineteenth century, European extraterritoriality was expanding with trade and transportation. It now extended to non-Europeans, non-Christians, and even to Muslims, mostly those from India, and to geographies closed to Christians, such as key parts of the Arabian Peninsula. What allowed these Indian agents of British imperialism to move into Arabia and enter the holy cities of Islam was their "Mohammedan" identity. This model was exported to the colonies from Christian Europe by colonial officials and scholars, and was the basis on which religion involved a Jesus-like figure and communal relations where this figure mediates and connects the bodies of believers. Even as Muslims repeatedly stressed that their Prophet was a mortal and that their adoration of him was as the perfect believer rather than the object of their belief, what permitted their movement and commensurability in a coaling system was the Christianized model of their religion. Itinerant Mohammedan Indians served as European consuls and vice-consuls, hygiene inspectors, quarantine supervisors, and commercial representatives across the Arabian Peninsula. Without them, it is hard to see how the system could have operated. Moreover, their travels meant they were treated as objects of protection and/or suspicion, and thus the reach of British surveillance, aid, and jurisdiction was extended to new environments. Their portable British subjecthood was combined with their religiosity.

Entangling the worlds of the Ottoman Empire and the EIC, extraterritoriality in the Red Sea and the Hijaz blended the *jus sanguinis* of the capitulations—"blood rights" originally extended to people, based on religion or ethnicity and granting freedom of trade and travel—and the *jus soli* of the factory—rights extended to space or soil. As the commander of the British steamer bombarding Jeddah in 1858 put it, "[the consul] called my attention to several Treaties existing with Turkey, giving to the Consular authorities full jurisdiction over all British subjects in the Ottoman ports 'in the same and as ample manner as if Her Majesty had acquired such power or jurisdiction by the cession or conquest of territory.'"[98] This mélange befitted a coaling system in foreign waters predicated on carrying pieces of home, as well as a notion of religion in which blood is a conveyer belt of soil and territory. Legal historian Lauren Benton has shown that Europeans imagined imperial space as networks of corridors and enclaves. She argued that the pathways that became conduits for law and even corridors of jurisdiction blended with the actual geography of rivers, coasts, islands, and oceans. Benton's early-modern riverine legal geography is extended here artificially by the coaling system and its human mobile parts.[99]

Indian Arabs traveling from the subcontinent to Arabia and other Ottoman territories were by no means easily manageable imperial underlings or torpid predictable components of a British-controlled system. Rather, they and the older trade, family, and religious networks that connected Arabia and India before the arrival of steamers were the indispensable yet intolerable connectors that allowed the system to operate, simultaneously creating a rich repertoire of frictions and tensions. One such itinerant, Abdullah Arab Moonuffer, was a merchant who originally lived in Bombay; he became a Turkish subject in the 1880s and at the decade's end resided in Medina and visited Istanbul. Moonuffer created quite a stir when, using the influence of another Indian Arab to approach the sultan, he proposed that if he received the sultan's sanction he could use contacts in India to convince hajjies from the subcontinent to opt for Ottoman steamers only and boycott the European vessels.[100] His Istanbul contact and deal broker was Sayyid Fadl b. 'Alawi. British intelligence reports feared Fadl was trying to counter British extraterritoriality with steamer lines between Jeddah, Bombay, and Calcutta that the Ottoman sultan would endorse and brand.[101] Fadl was indeed no novice in carbon politics: he had previously been involved in a dispute about establishing a coal depot at Makullah, and in several other confrontations between the British, Ottomans, and locals of the Arabian Peninsula involving shipping and coal infrastructures.[102] In 1879, before his own plans to return to Arabia were aborted, Fadl put together a strategy for Ottoman involvement in South Arabia that combined his lineage to the Prophet Muhammad, Sufi blessing, and the aid of a mining engineer.[103]

Fadl also offered valuable legal expertise. For example, he wrote a memorandum on European impingement into Arabia in the coaling station of Aden, the lighthouse of Bab al-Mandab, the coal depot of Bahrain, and several other sites along the coaling network. Fadl detailed sources and legal evidence for Ottoman sovereignty in these settings, and followed the paper trail through the archives of Hudaida and the repositories of Mecca and Cairo that housed accounts of negotiations, deeds of sale, and bills of rights indicating, for example, that Arab shaykhs who sold territories to foreigners did so without authority.[104] Such maneuvers elucidate not only why the Ottomans relied on such mediators, but also why the British turned to Parsis as legal counterpoints to people like Fadl (especially as counter-experts in Islamic law and history) and as sources of local knowledge about tribes and loyalties around the peninsula.

Repeatedly pushing the Porte to adapt to the age of steam (he was also one of the first to propose the idea of a Hijaz railway), Fadl's actions stirred considerable British concern.[105] He was probably the most visible among a group of Hadrami Indian Arabs familiar with the coal depots around Aden who advised the Ottomans on such issues.[106] The Porte, for its part, integrated these actors and their agendas into

Ottoman politics. Sultan Abdulhamit gave Sayyid Fadl the title of pasha and appointed him as his advisor on Indian affairs (a mandate that covered South Arabia, indicating that the Ottomans also partly subscribed to the Indianization of this territory).

Equally interesting is how such involvement affected Islam. When transcultural brokers like Fadl wrote treatises in support of the caliphate (the Ottoman attempt to project religious authority extraterritorially), such as the 1881 *Tanbih al-'ukalah* (A word to the wise),[107] they biologized Islam by lending Arab genealogical support (and that of Sayyid, through his lineage to the Prophet Muhammad) to a Turkish project, thus helping extend Islamic law and legitimacy beyond the territories where it could be politically enforced via their own extraterritorial networks and bloodlines.

We see how these issues appeared to the Ottomans from Ahmed Midhat Efendi, an Ottoman author of the *tanzimat* and Hamidian eras, who situated British attempts to extend capitulations to the Hijaz in their biopolitical context. Such endeavors, he diagnosed, amounted "not to autonomy (*otonomi* or *muhtariyet*) but to anatomy (*anatomi*)."[108] An anatomy-based ethnic politics was threatening to trump religion, language, and other Ottoman modes of belonging, togetherness, and hierarchy. Ahmed Midhat's diagnosis suggested that the extension of capitulations to Muslims was not so much a problem of the demise of religion, but of its racialization. As Cemil Aydin has argued, "the idea of the Muslim world is inseparable from the claim that Muslims constitute a race"—a race formed with the aid of imperial transportation and communication technologies and around notions of social Darwinism during the last decades of the nineteenth century.[109] Imperial biopolitics, Ahmed Midhat discovered, was extending an anatomical logic towards Islam.

Muslim itinerants between colonial India and Ottoman lands were key promoters of an "infrastructural turn" in Ottoman political discourse, whereby political legitimacy in the inter- and intraimperial domains was measured according to the ruler's ability to meet the challenges of technological modernity. Large infrastructures like railways, telegraphs, and water desalination works were seen as vehicles of an Ottoman civilizing mission in the Hijaz, and this mission also extended to the individual body. As Michael Christopher Low has shown, the Ottomans applied European "environmental orientalism" in this territory, connecting the defective Arabian environment to the defects they sought to correct in its inhabitants.[110] Transcultural mediators like our itinerant Indian Arabs were one possible conduit for this mentality. For example, during the hajj season of 1891, Sayyid Fadl wrote an alarmist memorandum to the Porte about foreigners sending Muslim consuls to Jeddah for the purpose of residing in Mecca. Beyond the obvious point that such meddling might create problems for the Porte, Fadl came up with another, more sophisticated angle: "it is important to confine these consuls and their subjects to Jeddah and prevent them from moving to Mecca because of the risk that travelers from among the brutish Bedouins (al-'urban al-mutawahishin) might hear that Christians are ruling Mecca and Medina, as they do not distinguish a Muslim consul from an infidel." [111] As a real go-between, Fadl was simultaneously deprecating the region's locals like a cultivated imperial official and alluding to their naïve yet potentially insightful point (as himself a part-local). And he was harping on a matter that was a sore issue for about a decade by this point, eventually resulting in Ottoman refusal to extend the capitulations to the Hijaz.[112] The chilling effect this had on non-Ottoman Muslims seeking to travel or purchase property in Arabia created a paradox: it was a British power which promoted transnational Islamic solidarity and the Ottoman caliph who limited it.

Whereas colonial officials were diagnosing the fanaticism of hajis or the "twin infections" of pan-Islam, Fadl and other "pan-Islamists" were adopting the medicalized language of religion, usually in inverted form. Arabian, Indian, and North African pan-Islamists of Sufi and Salafi stripes have all subscribed to such language, although usually in its positive articulations. In his 1899 book *Kitab Idah al-asrar al-'ulwiya wa-minhag as-sada al-'alawiya,* Fadl explained how members of his Sufi denomination approached their disciples:

> The position of shaykhs is like the science of medicine. One who is proficient in it looks at each sick person in terms of the treatments and medicines appropriate to [that person], and the medicine may differ according to the illness. Thus, if the shaykh were proficient like this then he is truly the master [al-murabbi: pedagogue, literally, care provider].[113]

A few years later, when weaving a fictional account of a Salafi pan-Islamic summit in Mecca, Syrian pan-Islamist 'Abd al-Rahman al-Kawakibi explained that the reason for convening reformers from various locales was to heal the sick body of the Islamic community.[114] Maghribi pilgrims on board steamers bound for Arabia sang similar tunes: "My sicknesses no doctor in the world can heal them . . . I reach Taha [the prophet] and see him . . . O, Prophet, heal my body and its members."[115] Pathological or remedial, religion was now located in (and as) the body of believers.

Figurations of body, community, and politics traversed India and the Ottoman world along the artificial archipelago that connected both via legal islands of extraterritoriality and other technologies. Parsis were integral to such processes precisely because their communal success was found in the major hubs visited by Islamic reformers, located as they were in almost every key life-support junction along the way. Muslim travelers and thinkers were specifically attuned to this model as it was both historically and contemporaneously geared to target Islamic institutions on the subcontinent and its Arabian extensions. Moreover, the Parsi presence in EIC sites extending toward the Red Sea, and in the water-desalination, salt, opium and meat sectors, etc., originated in their auxiliary roles as shipwrights and chandlers in the seventeenth- and eighteenth-century factories. They were part of the infrastructure, which is why an infrastructural analysis is so appropriate to explicating their roles.

Factoring human actors into large-scale infrastructures also applies in the Ottoman "infrastructural turn." The hydrocarbon pathways that came to be seen as the key yardsticks for Ottoman effective and legitimate sovereignty in the Anglo-Ottoman frontiers of the Red Sea and the Hijaz also acted as conduits for British extraterritoriality couched in religious terms. Whereas the Ottomans relied more and more on machines that burned British coal (and were often made in England), Britain relied on the "Mohammedanism" of its Indian subjects to extend influence into the Hijaz and on traditions and legal technologies often made in the Ottoman Empire. The following chapters explore the synergy of Islamic and steam powers. This was not a process of secularization,[116] but one in which the technological sublime of these networks interfaced with, changed, and amplified existing forms of sublime and ethics.

That British India was an Islamic empire or, put more accurately, a Mohammedan empire could be seen also in the importance it ascribed to enabling the hajj—a role not very different from that of another non-Arab empire which gained legitimacy from its role as Servitor of the Two Holy Places—the Ottoman Empire. As we have seen, "Islam," or even just Indian Mohammedanism, was a multifaceted thing. To transform it into a generator of inroads to Arabia required relying on the involvement not only of Indian Muslims but also on Parsis and even on Arab migrants to India, with their language skills, legal knowledge, and family and commercial networks. Both Parsis and their Muslim opponents/admirers who manned these legal and material enclaves helped imbue this infrastructure with religious meanings, and also to naturalize it and make it and themselves into parts of the landscape.

PATH DEPENDENCE: TRANSPORTATION AND THE ACTIVATION OF GLOBAL COAL MINING

Highlighting industrialization in the British Isles, the "energy transitions" framework challenged in the first part of this book is by and large Eurocentric and fairly static. It tends to ignore the activation of coal mining outside of Europe and the importance of maritime transportation in this process. Previous chapters demonstrated that while the British Isles were a unique setting for the activation of an industrial revolution due to coal's proximity to water there, British sail ships and then steamboats connected places like India and China to the imperial waterways in which coal could flow in and gradually out. Colonial coal mining was the "gift of the Raj,"[117] and so were the "energy security" and "energy independence" mentalities. By the early twentieth century, India was proudly self-sufficient in terms of coal production.[118] China had the same aim from the 1890s.[119] But if British steam-fueled marine imperialism was a key part of the story of coal mining outside of Europe, the various infrastructural connections between the British Isles and places like China and India, and especially in-between entities like the Ottoman Empire and its Red Sea extensions, must be part of the picture.

Previous chapters discussed how coal and opium together created a fossil fuel addiction. Although commercial coal mining in India started in Bengal about a decade before the First Opium War, the Parsi-led shipment of this drug to China significantly stimulated it. Dwarkanath Tagore (poet Rabindranath Tagore's grandfather) inaugurated Indian coal mining via the triangle of opium, coal, and mail. He pioneered a mail and opium shipping line between Calcutta and Suez via Aden after a thrilling maritime crossing that rivals that of the *Hugh Lindsay* a decade earlier. His ship, the *Waterwitch,* was also built by Parsi shipbuilders. It raced from Calcutta in 1839 fully laden with coal and letters.[120] The *Hugh Lindsay* charted a course through the formation of the coaling system; now, if we follow other vessels, mainly to the Mediterranean but eventually also back to the Indian Ocean, we see the activation of coal mining and the development of dependence—as well as attempts to secure "energy independence"—in the various peripheries of the British Empire.

Questions of energy (in)security like those discussed above were not alien to the Ottomans, who first encountered steam power facing the barrel of a gun, as we shall see below. (As in India, coal itself had been familiar since the late eighteenth century, and failed attempts to exploit it partly accounted for the empire's military reforms.)[121] This happened during the same decades that the Ottomans ceded control in the Red Sea ports of Mocha, Aden, and Jeddah to Britain, which was understood in Istanbul as part of a carefully designed British conspiracy and a cautionary tale that promoted Ottoman energy independence.[122] It was also framed as such by local consultants like Sayyid Fadl, and the implications—primarily the urgent need for what we would today call energy independence and energy security—were clear. The same lessons were taught in the Mediterranean. In this sense the British Empire was infecting rival empires with its angst-ridden approach to coal.

The first steamer ever built and used for naval combat was designed to battle an Ottoman fleet in the Mediterranean under Greek flags during the Greek War of Independence. The ship (along with its coal) was brought from England, but in 1826 it was a novelty even there. When a conservative British admiralty finally introduced fighting steamers to the Royal Navy during the 1840s, a decade and a half had passed since British engineers and soldiers of fortune, as well as colonizing corporations such as the EIC, had begun experimenting with this technology away from the metropole. The construction of this pioneering ship, the *Karteria*, and the squadron of steamers meant to join it reveals some of the conditions for scientific progress in this interimperial environment. The Greeks entrusted the project to the engineer Alexander Galloway, whom E. P. Thompson described as one of the first to deserve the moniker "engineer."[123] Other actors in Europe's periphery were even quicker to identify this pioneer, demonstrating how well they had England's technical landscape mapped: the first student mission from the Islamic world knew how to find him in 1815.[124] Egypt's Mehmet Ali, who at that time controlled Mocha and Jeddah and had designs to expand towards Aden, was also familiar with Galloway and employed his son as chief engineer.

Mehmet Ali's own son, Ibrahim Pasha, commanded one of the fleets fighting the Greek secessionists for the Ottoman sultan. With the Galloways providing engineering services to both sides of the conflict, Galloway senior deliberately delayed completion of the ships for as long as possible. The *Karteria*'s eventual completion and deployment in the Mediterranean was a major reason for the collapse of the Greek-Ottoman ceasefire and for the Battle of Navarino in 1827. With the ability to turn about using pedals, it could fire accurately and rapidly from all sides. The ship's furnaces, its maneuverability, and its stability also allowed for the use of heated shells from the sea. Until then, it had been coastal batteries that usually fired hot shots at wooden ships, rather than the other way around. During her first year at sea, the *Karteria* fired eighteen thousand hot shots, mainly at onshore fortifications.[125] If the *Karteria* fired the opening shot in overturning received definitions of land and sea in the Mediterranean, we have already seen that this reversal was a feature of steamers elsewhere. In 1828, the Ottoman navy purchased its first steamer, the British-designed *Swift*. One of its first tasks was to tow the crippled Ottoman ships that had participated in the Battle of Navarino back to harbor.[126]

These early experiences with foreign and domestic naval steam power were the main context for the emergence of the mining, import, and proliferation of coal in the Ottoman Empire. The first and most significant coalfield in the empire was in Zonguldak, on the shores of the Black Sea. As Donald Quataert has suggested, the story that it was discovered in 1829 by a local returning to Zonguldak from military service in the Istanbul naval yards is probably apocryphal.[127] However, the spuriousness of this myth throws into relief the belligerent maritime context in which the Ottoman state began to show interest in fossil fuels.[128]

Subsequent phases of the introduction of steamers into the Ottoman navy also coincided with the development of coal mining. In the late 1830s, the first steamers were built in the Ottoman shipyards.[129] In 1840, one of these ships explored another coalfield near the Black Sea port of Pendaraclia ("Bandar of Eregli"); it even served as a floating laboratory to gauge the coal's quality, using some of the mineral specimens it had collected to generate steam.[130] Seven years later, the Ottoman fleet formally converted to steam when a new steamer squadron was built, prompting the initiation of active mining in Zonguldak in 1848. In 1865, the mines were officially transferred to the authority of the Ministry of Seafaring.

The *Karteria* and the *Swift* also marked a significant development on the British side. Since the 1730s, patent and tax regimes and border controls had restricted the proliferation of English steam engines and technical drawings. In the 1830s, feeling that its steam engine industry had matured, Britain started changing its policy and sold the technology together with its energy source.[131] Our two steamers, for example, arrived with English captains and engineers who helped to train local officers. Promotion of the British energy sector through proliferating coal-burning technologies and personnel went hand in hand with the reform of tax structures and the abolition of fees on coal export. In 1831, Britain halved its coal export duty, yielding a 109% increase in tonnage exported abroad. A further increase of 320% followed the cancelation of duties a decade later.[132]

Newcomers found it difficult to compete with the mature British industry. While the Ottomans were attempting to master the skills of steamship building, evidenced by the experiments mentioned above, it soon became obvious that a new Ottoman steamer would be obsolete by the time of its launching. This realization led to a policy of favoring acquisitions over production, and as early as the mid-1840s, the Ottoman fleet was regularly buying its steamers, mainly from England.[133] Concurrently, attempts were also being made to achieve "energy independence" through local coal mining.[134]

The Ottoman central government and provincial rulers understood that reliance on foreign minerals and expertise would be costly in various ways. An 1854 report submitted to the sultan stated that "the annual coal production of the Ereğli mines is estimated at about 400,000 kantars [22,580 tons]. Since the annual need of the imperial docks, the armory, the mint, the gunpowder factory, the fez factory, other factories, and steamships totals 1,886,000 kantars, except the mentioned 400,000 kantars, 1,486,000 kantars of coal should be purchased from England."[135] Thus, soon after the activation of Ottoman coal mining, and together with persistent attempts at improving it, the central government and several provincial governors began exploring and developing imperial coal sources that would promote energy independence. This was part of a wider overhaul of energy resource management which included the introduction of scientific forestry in 1840.[136]

As in other spheres—especially the art of war on land and at sea, but also agriculture, education, law, and medicine (all by-products of military reform)—the sultan took his cue from Mehmet Ali, who pioneered geological surveys and the search for coal, albeit with limited results.[137] Subsequent Egyptian khedives also sought opportunities for coal production, even beyond the territories under their direct control.[138] During Mehmet Ali's Syrian campaign, the Egyptian army searched for coal in Mount Lebanon.[139] Mehmet Ali's heirs, 'Abbas and Isma'il, also dispatched geological expeditions and discovered coal in Upper Egypt, but no mines were subsequently developed. In the 1860s, Ottoman coal explorations were conducted on the Island of Semandirek, in the villages of Marmara and Güzelşehir, around Draç in the Tiran Mountains, in Resmo in Crete, and elsewhere in the empire. In official correspondence, the benefits of internal coal production were contrasted with the disadvantages of import. Consider an imperial decree addressed to the governorships of Saida, Salonica, Yanya, and İşkodra, which also captures Egypt's catalyzing role:

> Since there is no coal source around Egypt, the coal for daily use, trains, and factories is purchased at great cost from Europe. The Governor of Egypt therefore proposed that if coalmines are discovered in other parts of the empire, officials will be sent from Egypt, and if the coal proves valuable, it will be produced under their administration. If significant coalmines are discovered in the sea or underground, [Egypt's] Pasha requests local help for his officials. We request to be notified regarding coal discoveries in Çayağzı near Salonica; Draç on the coast of İşkodra; Avlonya in Yanya; on the Aegean Islands; Gelibolu; Crete; and Tekfurdağı.[140]

But scarcity limited the possibilities of empire-wide cooperation, and instead only aroused tension. Cairo's growing political and economic independence from Istanbul over the nineteenth century was manifest in its decreasing dependence on Anatolian wood and, given the dearth of alternatives, its increasing reliance on British coal. In 1831, Egypt's governor undertook a military campaign to wrest control of Syria from the Ottoman government. He was driven largely by the desire to acquire wood for shipbuilding and industrial consumption in the mint, foundries, and armory and to fortify his hold in Egypt.[141] The adventure ended in 1840–41 with the British (steamer-aided) bombardment of Acre. The last scramble of the age of wood was settled by a decisive demonstration of steamer power.[142] During the rest of the century, locally mined nonfossil energy sources and coal supplied only a small portion of the growing demand for energy, making British coal imports more, rather than less, indispensable. The Ottomans responded by allocating their limited local resources to specific functions, prioritizing, for example, military energy independence by supplying a strategically selected network of coal depots with Ottoman coal.[143]

Besides the bombardment of Acre, another spectacular example of steam power was offered in 1840 by the EIC iron steam- and sail-powered ship *Nemesis,* which was the first steamer to sail around the Cape of Good Hope and played a central role in the First Opium War by crushing enemy ships and coastal defenses. The *Nemesis*'s success came as a surprise even to her operators, who saw in the conflict "an extremely favorable opportunity for testing the advantages or otherwise of iron steam-vessels."[144] While Britain ruled the waves long before the advent of steam, its advantage on the high seas was limited by the supremacy of small enemy ships in their native shallow waters. The *Nemesis* helped reverse this limitation: its hybrid construction allowed it to cross oceans under sail, while its shallow draft and steam engine allowed it to maneuver along rivers in order to engage vessels and onshore targets. Like the *Karteria's* ability to fire hot shots—a capacity which until then only characterized onshore forts—the *Nemesis's* construction enabled the conventional division of military distinctions between land and sea to be reversed. This hybridity was informed not only by the precedent of the *Karteria* and the secession of Greece, but also by the conquest of Aden. In 1839, the year of the *Nemesis's* construction, an EIC fleet of sail ships sent to capture this port was severely becalmed and almost missed the battle. The incident was used to promote the introduction of auxiliary steam engines in sail ships, and the builders of the *Nemesis* were among the first to take this into account when they constructed the ship.[145]

Kenneth Pomeranz, historian of China, has stressed the importance of British coal, and coal's proximity to waterways, to the emergence of a "great divergence" between England and China circa 1800. As Pomeranz also acknowledges, England and China were not independent units that could simply be compared to one another, pointing out that narrower sites and corridors make for better comparisons. These two empires were connected via various intermediaries, such as the Ottoman Empire and Parsi shipbuilders and merchants frequenting the ports of Canton and Aden, through which the coal economy could be extended eastward while other influences traveled west.

The *Nemesis's* performance stimulated steam power and a habit for coal by providing a sample of another substance that would create a new dependence. As in the Ottoman case, British military and commercial intervention eventually triggered the development of coal mining in China from the 1880s.[146] In sum, while the different geographies of coalfields were no doubt an important factor in understanding the timing and shape of industrialization in the English, Ottoman, Indian, and Chinese cases, when viewed comparatively, another connective step is required. Such a key bridgehead of the coal economy eastward was the corridor of the Ottoman Mediterranean and Red Sea.

Taking account of the *Karteria*, the *Swift*, and the *Nemesis* together with the *Hugh Lindsay*, the *Waterwitch* and numerous other vessels sailing between London, Bombay, and Canton, we see a complex picture of the origins, conduct, and outcomes of British steam-based naval power. This multifaceted legacy is also a keystone of the current global addiction to fossil fuels, which should be seen not just as a race to industrialize, but also as related to keeping and expanding key maritime corridors.

Nor was this a neat story of technological diffusion from the metropole. Britain's gradual shift to combat steamers from the 1840s was animated by experiments conducted in such frontiers by private and semiprivate actors, who were not necessarily British. This maritime dynamism involved changes in steamer technology and in both the mobile and fixed components of the system, which included a great deal of terraforming and eventually also global coal mining, thus leading to subterraforming, too. The depots network represented in figure 14 was a result of this hybrid dynamics. Superimposing the various maps considered above—including British and Ottoman cartographies and my own renditions of depots, lighthouses, and telegraph cables, as well as geological and mineralogical maps I will discuss in the final chapter— one on top of the other, and adding the ships that traversed this space to the mix, reveals coal depots as synergetic hubs where existing forces were intensified, as we have already seen with water, animal, and manual power. Beyond thermodynamics, coal's weight and anxiolytic properties also shaped this built, social, and legal landscape. These artificial conduits of materials, as well as legal and even religious traditions, were soon naturalized. Recounting such entanglements in all their dizzying complexity and insisting on their multidirectionality and on the agency and complicity of local actors obviates our adoption of simple narratives that allocate full responsibility solely to industrializing Britain, capitalism, or imperialism. Instead, we are compelled to acknowledge that the complexities of imperialism are integral to the global fossil economy.

CHAPTER 5

Risk

RIZQ MANAGEMENT

We have seen that rather than breaking neatly with its predecessors, coal often extended and intensified existing driving forces. The development of steam engines and steamers and the march of the global hydrocarbon economy were animated also by existing forces that were less terrestrial than bunkers and depots and less palpable than water or muscle. If previous chapters exposed the intensification of prevailing concrete powers, steam also required spiritual environments in order to expand. This chapter will show that as with the more tangible forces we have seen intensify by their contact with coal, so was the case with the mutually animating metaphysics of risk management and *rizq* seeking.

Over the past decade and a half, historians have been debunking the "diffusionist" understanding of technological and scientific development as a process that begins at the metropole and only later extends to the empire's periphery.[1] The previous chapter challenged such an approach to the steamship; the current chapter arises out of this revision and shows how this vessel was developed close to the British Isles and in the Atlantic world, but also in the Indian Ocean, and how it was shaped by the social and even spiritual land- and seascapes of this region. However, the point is not to allocate credit, or blame. Examining technological modernization in foreign waters and lands, and focusing on the pilgrimage to Mecca as a driving force for technopolitical innovation, I seek to identify dormant ethical potentials embedded in seemingly familiar fuels and machines and in the amoral epistemologies that propelled these vessels and their energy sources.

Even before they materialized, the prospects of quicker and cheaper passage to Arabia's holy cities—which, during most of the nineteenth century, was as yet a hypothetical sea change—were advertised by steamer companies' agents in many a mosque and saint's tomb. Such "technological sublime,"[2] bolstered by the religious imperative to undertake pilgrimage if possible, created a hype that far outweighed any existing infrastructural capacity to accommodate it. Early steamers, pushed by their limited cargo space to specialize in passenger transport, were attracted to the hajj sector and underwent significant modifications to cater to it in the process. The steam hajj was a self-fulfilling prophecy.

The development of the steamer was informed by the particularities of this sector. These included the climatic and geographic conditions prevailing in the Indian Ocean and Red Sea (e.g., monsoon, heat, abundance of reefs, among others), and the religious proclivities of passengers. A key characteristic was the massive overloading of ships, regularly compared to the Middle Passage[3] and indeed precedented by the traffic in human slave cargo and that of immigrants to the New World—both made possible by the non-European ethnicities and/or lower class status of those transported. One important aspect of what follows is the exploration of such continuities, and the connection between slaves and then indentured laborers to "energy slaves," as steam engines were increasingly called, and to the Muslim pilgrims on board those vessels which they propelled.

The advent of the steamship is often understood through a liberal perspective according to which these were labor-saving technologies where mechanical slaves replaced human ones and where existing modes of transportation anchored in the slave economy were replaced by the abolitionist thrusts of steam navigation. Looking at indenture, pilgrimage, and other kinds of bondage that were not exactly slavery but which nonetheless shared several of its characteristics, this chapter continues examining modes of handling and (re)commodifying human bodies, already evident in the emphasis on the diathermic benefits of dark skin for engine-room work, or in the way abolition licensed a British campaign against dhows—the steamer's main competing means of transport in the Red Sea—feeding freed slaves to vessels' stokeholds and to coastal coal sheds as stokers and heavers. Human trafficking, too, was shaped by steam travel and changed it in its turn. Again, retracing this mutual constitution is significant, not only in order to cast light on the tormented and tangled histories of progress, but also because such histories deposited in the engines and fuels of this passage ethical potentials that enable us to approach these technologies and their progeny with recycled sensibilities.

Overloaded pilgrim steamers created new problems such as breeding grounds for infections and chaos on board, as well as new solutions. The main engine for teasing out this socio-technical process of trial and error in what follows and examining the connection between piety and profit is the famous 1880 maritime accident, "the Titanic of the nineteenth century,"[4] immortalized in 1899[5] in *Lord Jim* by Joseph Conrad (who was no stranger to maritime disaster himself).[6] The accident happened on board the SS *Jeddah*, a heavily insured steamer, owned by the Hadrami al-Saqqaf family, carrying a human cargo of 778 men, 147 women, and 67 children, all Muslims bound for the holy cities in Arabia, in early August. The ship left Singapore but never reached its destination. I will recount its adventures gradually as this chapter unfolds, begging the reader's pardon for the resulting suspense. As will become clear, I resort to this narrative strategy as suspense itself is another of the chapter's themes and handling it in a certain way one of its arguments.

Accidents like the *Jeddah*'s prompted various risk-management mechanisms designed to reduce the chances of such adversities recurring. After explicating key structures that informed the steam hajj, what follows examines some of the technical, territorial, and especially financial preventative arrangements to which the *Jeddah* affair and similar accidents gave rise. The argument is concentric, moving from inside the ship to the physical environment it traversed, to the still broader economic and regulatory environments that propelled this movement. Indeed, new preventative mechanisms in turn stimulated risk-taking and hence further disaster in a vicious cycle. The word *risk* comes from Latin and its derivatives: *riesgo* (Spanish), *risco* (Italian), *risque* (French), "that which cuts," hence "cliff" or "reef" and consequently "risk to cargo on the high seas."[7] The Latin word comes from a Greek navigation term *rhizikon, rhiza*, which means "root, stone, cut of the firm land." The term *rhizikon* transferred to the Arabic-speaking world via the Mediterranean as *rizq*, or "divine reward" or "daily sustenance." This chapter examines risk's role in navigating the challenges of approaching reefs in the Red Sea while transporting *rizq*-seeking Muslims to the coast of Arabia, probing the synergy of fossil-fueled financial capitalism and Islamic piety.

THE STEAM HAJJ OR THE HAJJ STEAM?

Control of the holy cities and organization of the pilgrimage routes have been important markers of political power in Islamic polities. The Ottomans, who gained such powers in the sixteenth century, required overland pilgrims to congregate and depart from either Cairo or Damascus. Under Ottoman rule these caravans transported between ten and twenty thousand pilgrims annually.[8] As in pre-Ottoman times, the hajj maintained and even boosted its significance as a commercial opportunity, impacting the economies of Cairo and Damascus, the route to Arabia, the Hijaz, the empire as a whole, and the pilgrims themselves.

Before the advent of steam, Indian Ocean and Red Sea wind patterns introduced a split in the Islamic global community, dividing it between the "*Ummah* above-" and "below the wind."[9] Pilgrims from South and Southeast Asia usually traveled to Mecca by sea, via the port of Jeddah. Estimates indicate that seafaring pilgrims numbered around fifteen thousand annually between 1600 and 1800.[10] Over the course of the nineteenth century, changes occurred in the routes leading to Mecca, in the means of transport plying them, and in the number and (ethnic and class) profiles of the pilgrims using them. Steam travel brought the territories "below the wind" closer to Arabia, rendering the divide obsolete and making the passage cheaper, and thus significantly changing the annual congregation in Mecca.

Most overland hajj traffic shifted to sea routes, as steamers made the trip quicker and cheaper. The South Asia hajj, too, was changed, with direct passage replacing layovers in the Indian Ocean's many port cities which were also part of the hajj economy. Steam severed the hajj from existing commercial activities that informed regional economies, and from the possibility of combining religious duty with earning a livelihood for many a pilgrim.

The shift in routes also concentrated the hajj in the hands of shipping companies, giving rise to "the hajj trade." Whereas the modern hajj severed existing economic ties and disrupted the economies of entire cities and empires, it also created new commercial opportunities and connections. Expanded participation in pilgrimage invigorated steam travel in various ways. Elsewhere, I described how participation in pilgrimages within Egypt, which were connected to the larger annual hajj, offered an indispensable source of revenue for the Egyptian railways.[11] The same was true for other railways, such as the Hijaz and Trans-Siberian railways.[12]

The expansion of the hajj also supported a second- and third-hand market for older steamships (leaking tubs that were the hallmark of the hajj trade), and thus new ships could be improved or replaced more rapidly. Offering mediation services, financing and group organization, networks of Hadramis connected plantations and debt collectors in Southeast Asia to camel rental agents and pilgrim guides in Arabia.[13] Beyond informing tourism and other business, this also gave rise to a network of quarantine stations near coaling depots, and the development of new medical and statistical protocols as routes changed and pilgrim numbers increased. Many of the infrastructures examined in previous chapters, from water desalination to lighthouses, were similarly stimulated by the hajj. The question "what did the hajj look like in the age of steam?" should clearly come after the more basic question, "what did steam navigation look like in the time of the hajj?"

Yet such a separation is artificial. The steam hajj was both fueling and fueled by the solicitation efforts of the companies' touts and the new financial arrangements (often in the form of debt-bondage) such as those offered by the al-Saqqafs to the passengers of the *Jeddah*. Celebratory voices promoting the ease and affordability of the steam hajj were heard across South and Southeast Asia by pilgrim brokers who now traveled annually between Arabia and their home countries in search of clients. Touts did not shy away from intimidation and misinformation. And they were provided with information by pilgrim guides in the Hijaz. This profession, passed down from generation to generation, concentrated vast genealogical knowledge about who has or has not performed the hajj.[14] These new kinds of transmaritime leverage and pressure were combined with a more epistemic duress: the hajj is a *faridah,* a mandatory duty commanded by God, to be performed at least once in the lifetime of every adult Muslim who is physically and financially able to do so. Both the physical and financial conditions changed significantly as steam propelled increasingly more Muslims to visit Mecca.

To a certain extent, "as the European steamship companies that dominated global maritime traffic took an interest in the hajj traffic, the Muslim pilgrimage became incorporated into this new industrial infrastructure of human transport."[15] Western businessmen have sought to apply to the hajj "the same logic and mechanisms that they developed in their transportation of other populations, or even freight, around the world."[16] The *Jeddah* affair itself was the *cause célèbre* that precipitated several attempts at reforming and streamlining the hajj in a businesslike fashion. In 1885, after this global scandal, which was attributed to the poorly regulated nature of the pilgrimage, the Indian viceroy extended an offer to the Thomas Cook Company to monopolize railway and steamer transportation of pilgrims from India to Arabia, which it did until 1893.[17] The legacy of tourism which came to characterize the pilgrimage to Mecca still animates it to this day.[18] However, such accounts of the "industrialization of the hajj" are incomplete without the mirror image of "the Islamization of industrialization," and they ignore shrewd non-Western actors, such as the al-Saqqafs and their Singapore Steam Company, which owned the *Jeddah*. If the logistics and agencies of tourism were penetrating the hajj, the reverse was also the case, though obviously not in equal measure. Anecdotally, beginning in the 1880s, we first encounter the notion of tourism-as-pilgrimage and must-see tourist sites designated in guides as "Meccas."[19]

The business practices mentioned above depended on careful scheduling and therefore also on the punctuality and speed of all-weather steam vessels. Yet the maneuverability and swiftness of steamers were not simply built into these vessels; rather, they required technical modifications and external support systems such as coal stores, lighthouses, canals, rear propellers, and many other components, which were all in the making during the second half of the nineteenth century. Steamers became all-weather vessels by developing new internal, social, and external environments. This development in steamer technology was surely a global story, much of which unfolded in the Atlantic. Yet the Indian Ocean and Red Sea, with their climatic and geographic particularities that combined high temperatures, shifting winds, strong currents, and multiple reefs, were seen as a testing ground for the practicality of steam navigation from the 1820s, and as key laboratories for its development during the rest of the century.[20]

Thus, rather than assuming a priori that "steamships had two advantages over sailing ships in the Red Sea: they could ignore the prevailing wind patterns and because of their easier manœuvrability they could come closer to reefs,"[21] it is important to inquire how they came to develop these advantages. As an explanatory mechanism, I propose below what may be termed *accidentalism*. Accidents, as I will show, occur as a result of contingency and also the schemes devised to curtail, direct, and use it. In the Red Sea, it was the combination of the fierce monsoon winds and the proximity of reefs that put the *Jeddah* in harm's way, prompting us to rephrase the above-quoted statement as a question: how did confronting, and in many cases succumbing to, accidents involving such dangers produce steamers capable of successfully meeting these challenges?

Alongside climate and geography, the Indian Ocean's human environment, and especially the hajj with its irregular itineraries and schedules, and with the unique spiritual and medical conditions thought to characterize Muslims en route to Mecca, was similarly an important laboratory where experiments in religion, transport, and quarantine, among others, were regularly conducted.[22] Early steamers had little cargo space, leading to their specialization in mail, high-value and often perishable freight, and passengers. The latter could be carried on deck and would load and offload themselves.[23] Historians of transatlantic migration and transportation still debate the relative importance of steamship companies' recruitment agents, risk management calculations, and of travel conditions and fares in determining the fate of this sector.[24] They are unanimous, however, about the crucial importance of the third-class or steerage compartments and of immigrants' agendas as the bedrock for the development of the industry and the liners themselves.[25] This was even more the case with pilgrims in the Indian Ocean. Notwithstanding the modernizing and secular fancies of their middle-class operators (Europeans and non-Europeans alike), as on the Egyptian railway, it was the third-class steamer passengers who kept the business afloat.[26]

What we already know about the *Jeddah,* even before it ran into trouble, allows us to explain some of these issues. The ship's itinerary, course, ownership, and the number of people on board indicate how the hajj and the age of steam were mutually constitutive. The pilgrimage season of AD 1880 (corresponding to AH 1297) began in early November, and hajjis started to advance towards Arabia a few months ahead of that time. The late-July departure date of the *Jeddah* was meant to bring the pilgrims to their destination just in time for the beginning of the month of Ramadan. Indeed, as Southeast Asian hajj travelogues from the period indicate, the month of fast in the land of *haramayn* was considered a much longer and more intensive experience than the few days of pilgrimage. Many Muslims who could afford it tried to combine these two highlights of the Islamic calendar.[27]

The *Jeddah*'s itinerary is an indication of how the hijri calendar informed the timing, trajectories, and different kinds of patronage of steam navigation, demonstrating also some of the related political, cultural, and technical protocols of steamer traffic (as well as those of other coal-burning vehicles).[28] Our vessel was one of 132 steamers that sailed to the port of Jeddah under the British flag in 1880, a fact that informed the insurance premiums of 328 steamships in total that sailed there that year.[29] As the main function of this port city was to serve as a feeder harbor to the holy cities, we may safely assume that most of these ships, and especially those like the *Jeddah* which braved the summer monsoon, were set in motion by the hajj or related Islamic trajectories. (As the sea gate to the Hijaz, Jeddah was also an important trade entrepôt, and much of that traffic was also hajj-related and connected to the need to feed the large itinerant population passing through the port.)[30] The westward passage was met in Jeddah by pilgrims coming via the Suez Canal. The year 1880 saw 35,262 pilgrims crossing the canal in this direction, 14,372 more than in 1879; and, as the British consul providing these figures indicated, the shift of the hajj traffic to the canal represented the abandonment of the traditional overland pilgrimage routes.[31] Eighteen eighty was the first year in which the number of sea arrivals surpassed the number of land arrivals to the hajj. The leap was from 46,000 maritime pilgrims in the year 1876 (and 154,000 land pilgrims) to 60,000 in 1880 (and only 31,500 arriving by land). During the half-century or so we call "the age of steam," the total number of pilgrims counted at the congregation at Mount Arafat near Mecca rose from an annual average of 90,000 during the 1850s to 190,000 during the 1900s.[32] These figures suggest that both in terms of total numbers of hajjies and in relative terms of sea versus land pilgrimage, steamships were quite significant agents of change. This also gives us indication about the other side of the coin—the impact of the hajj on steam navigation in these waters. The remarkable general volume of maritime hajj traffic is revealed when compared to the global steamer volume. Estimated figures for the year 1870 suggest that about four thousand steamships sailed the globe—a number which doubled by 1886.[33] The steam hajj represented about 5 percent of this total global traffic.

Yet the large and growing number of steamers traveling the hajj sea routes fell short of an even steeper increase in demand. Up until 1800, about 15,000 South Asian pilgrims came annually to Mecca; by the 1880s this number had quadrupled.[34] The number of passengers on board the *Jeddah* is thus symptomatic of an infrastructure propelling, and then trying to keep up, with a sea change. Here, too, we see that the demand to undertake the pilgrimage outgrew the capacity to satisfy it, providing a powerful engine for the growth of the sector as a whole. The Court of Inquiry held at Aden after the ship's accident remarked that "nearly 1,000 souls on board a vessel of the tonnage of the 'Jeddah' was a greater number than should be allowed by any regulation, especially for a long sea voyage, as taken by the 'Jeddah,' and at a season when bad weather might naturally have been expected."[35] Overcrowding—

determined in an 1870 act as the ratio of a vessel's tonnage to the permissible number of passengers—had been a constant problem afflicting ships on the Singapore-Jeddah line throughout the preceding decade.[36]

Several factors shed light on the dynamics of overcrowding, including the temporality of the hajj trade and its interface with other global flows; periodic contagion scares (usually regarding cholera) that created a priority situation whereby the city of Jeddah had to be evacuated as quickly as possible (justifying overburdening steamers on the return);[37] the hype around steam travel and advertisements by agents as well as returning hajjies; the charity that Muslim-owned steamer companies, Islamic rulers, and especially the Ottoman Sultan extended to poor pilgrims; and new debt-oriented financing options.

Even when combined with the month of fast, the pilgrimage to Mecca was not a year-round affair. Steamer companies that depended on the hajj trade engaged as many ships as possible to "make a killing" during the season, and reallocated them thereafter. The more passengers they could compress into a small fleet, the less waste they would incur. Congestion also characterized different ports of departure, all the more so at Jeddah, leading to increased accidents and collisions during the hajj.[38]

Representatives of steamer companies, who could now travel back and forth between Arabia and South and Southeast Asia annually, actively recruited potential pilgrims, and were paid a commission per head.[39] Along with the promise of a quick and easy passage, they offered various loans, attracting hajjies with "neither possessions nor wealthy family members, who nevertheless undertook the voyage, knowing how easy it is to obtain the necessary credit in exchange for their commitment to labour." Agents often created a false picture of the hajj, its hardships and financial costs, and then they provided high-interest loans for those pilgrims who underestimated the costs of the trip.[40]

In order to compete with moneylenders, steamer companies such as al-Saqqaf's lent money to "pilgrims who spent their money too quickly or did not bring enough, and who had difficulties getting credit for the funds to cover the return voyage." The debtors were jointly liable for a common debt and committed themselves to repay the full amount or earn the missing sum by working as farmers in the creditor's lands in Cocob, Singapore.[41] According to some estimations, as many as one-third of Javanese pilgrims financed their hajj by entering these contracts with the al-Saqqaf. Such indenture schemes were modeled on the Chinese credit-ticket system, making these pilgrims into so-called *hajji-coolies*, repaying their trip by working between a few months and two-to-three years in Singapore's plantations. Also, the British protective legislation that was meant to protect pilgrims from exploitation was predicated on Dutch legislation targeting Chinese coolies.[42]

Given these new ways of financing pilgrimage and a new economic structure in which pilgrims had become the commodity, the specter of the slave trade was regularly evoked in European analyses of the hajj trade, especially with regard to "Hadjis who, at the end of the Hadj, are forced to 'sell themselves as slaves,'" like the passengers of the *Jeddah*, to whom these analyses referred directly. An 1881 high-profile court case against the al-Saqqafs found that they were keeping pilgrims "in conditions of slavery" before and after the hajj through debt-bondage.[43] Indeed, several Hadrami families, including the al-Saqqafs, were involved in the lucrative Red Sea slave trade into the late 1870s, and even more vigorously in previous decades.[44] The family diversified from the dhow-propelled slave trade into the steamer-aided pilgrim trade at the beginning of the 1870s, at once extending time-tested protocols of overloading that characterized dhows and developing new schemes of bondage and new interfaces between global flows of migrants, workers, slaves, and pilgrims. In this sense, too, coal-fired pilgrimage was more of a continuation than a rupture with slave trafficking.

In addition to various slave-like circulations, the hajj was also an opportunity to ship actual slaves to the slave markets of Jeddah and Mecca with minimal intervention by the British, who were hesitant to interfere in religious affairs. Conversely, the hajj was traditionally an opportunity to manumit slaves and benefit from the enhanced *barakah* (blessing) bestowed at this auspicious time.[45] Steam navigation was integrated into these economies of trafficking and manumission, introducing into the already fraught scene aspects of technological sublime and terror, and the various financial mechanisms of insurance, monetary liability, and indenture. The hajj created a spiritual obligation to God that blended with monetary debts to shipowners that were repaid in Singapore's plantations, for example. The categories of "slave," "indentured laborer," "pilgrim," "passenger," and even "cargo" and "animal/ cattle" were bleeding into one another.[46] Such a confluence rested also on the convergence between "slave" and "Muslim" (derived from the Arabic word for submission)[47] which was not unique to the world of the Indian Ocean or to the age of steam. Locating true freedom and pure piety in complete submission to God has been a longstanding Sufi tradition in which "absolute freedom would lead to absolute disaster."[48] Such ideas had been circulating on ships as far as the black Atlantic.[49]

To contextualize the *Jeddah* affair, it is noteworthy that the 1870s saw the emergence of regular steamer lines to and from Singapore, and the shift to them of the lion's share of migrant traffic. A significant portion of this traffic was composed of "the coolie trade," which had become a mainstay of the local economy. In 1880, 78,196 migrant workers came to Singapore.[50] Indentured labor in Singapore obeyed the temporality of the monsoon and of local agriculture: ships arrived with workers between December and March, during the northeast monsoon, sailing to China during the southwest monsoon between June and September.[51] Al-Sagoff (i.e., al-Saqqaf) was one of several non-European steamship companies specializing in the hajj from the early 1870s, and infuriating the colonial authorities in the same way that British officials had scorned other non-European competing steamer companies like the Bombay Persia Steam Navigation Co., "a gang of Arabs whose chief source of profit is the carrying of excess pilgrims."[52] Had the *Jeddah*'s voyage gone according to plan, the ship would have been able to return from Arabia at the end of the hajj in December, bringing back hajjies to pay off their debts in Singapore's plantations.

BY ACCIDENT

Accidents were driving forces of steam power through almost every constitutive moment of this technology. As we saw in chapter 1, the steam engine was conceived as a solution to the accident of mine flooding. Chapter 3 situated this development vis-à-vis the failures and stoppages—both intended and not—of human elements. Chapter 4 showed how in the Indian Ocean and Red Sea regions, from the 1830 trailblazing voyage of the *Hugh Lindsay* through the occupation of Aden in 1839, the steam engine's career was characterized by shipwreck. In the second part of the century, steamers sought to capitalize on their advantages over sail ships—especially in terms of speed, regularity, and maneuverability—and maximized the number of passengers on board to compensate for the higher fueling, operating, insurance, and maintenance costs. Overloading, haste, and audacious schedules created new kinds of accidents.[53] What follows examines how one such accident helped to create a world. Like the *Hugh Lindsay* in the previous chapter, the *Jeddah* is a heuristic device which should not be overloaded with meanings it cannot withstand. But it does symptomatize and dramatize similar accidents and helps to chart landscapes and seascapes of hazard and how these were tamed.

In particular, the paper trail left by the *Jeddah* affair[54] enables a discussion of the accident's technopolitical causes and implications, its territorial reverberations, and the underlying epistemological mechanisms of these technical, territorial, and political developments. As disruptions of everyday routine, accidents are especially revealing moments which allow historical actors, and then historians, to understand how technical systems work. If the mechanics of quotidian existence are "black-boxed"[55] when infrastructures operate smoothly, breakdowns, malfunctions, and their ill effects focus attention on what usually occurs behind the scenes.

However, such accidents offer more than a window for passive understanding. They were also powerful real-time driving forces for enhancement and improvement of the systems within which they occurred. Any technological system is developed through a process of trial and error. Accidents, as moments when error becomes catastrophe, prompt attention and inquiry, and subsequently technical solutions, new maintenance and training protocols, and novel management schemes, all intended to prevent similar accidents from recurring. Thus, food poisonings and fires on board, resulting from letting pilgrims bring and cook their own food on deck (while steamer companies provided the fuel), prompted the introduction of kitchens, supply rooms, and new hygiene standards on steamers. Harbors and canals were broadened and deepened, and side wheels were replaced with rear propellers after the bottom and sides of ships collided with the floor and banks of the canal and docks. Crews were better recruited, trained, compensated, and commanded following mutinies or disastrous human errors. Pirate attacks yielded stronger ship armor and enhanced securitization, as well as negotiations and agreements with local leaders; stowaways led to more inspection.

In the hajj trade, safety measures and technological innovation were especially important to European steam companies which sought to corner the market with the aid of regulatory measures imposed by colonial officials. According to Low, "government standards for shipboard fittings, anchors, cables, nautical instruments, safety equipment, overall tonnage, and speed during monsoon conditions"–stipulated, for example, in India's 1895 Pilgrim Ships Act–were introduced to push Muslim-owned shipping companies like al-Saqqaf's out of the market. These companies were thus forced to specialize in the lower-end clientele and consolidate resources to purchase older ships from P&O and Lloyd's.[56]

Accident's footprint can be traced in spheres as diverse as infrastructure and diplomatic relations, as well as religious reform, which was also part of this dynamic. Take Muhammad ʿAli Effendi Saʿudi, an Egyptian civil servant and member of Muhammad ʿAbduh's circle, who performed the pilgrimage in 1904 and 1908 carrying photographic equipment, and who liked to methodically compare the two experiences. Sa'udi was one of the many steam pilgrims who sailed to Arabia from Egypt down the Red Sea, but some of his experiences typified the crossing from South and Southeast Asia.

Sa'udi recalled that on his first hajj, pilgrims had come across the badly burned body of a woman hidden behind some luggage next to the ship's boiler room, yet on his second pilgrimage (in which the Egyptian contingent was double the total number of pilgrims in the first hajj), inspection had already prevented such a possibility recurring. Similarly, he noted that previously, Egyptian pilgrims traveling third-class arrived at the quarantine station of al-Tor on the Egyptian coast of Sinai and had to hand over their clothes and wait in icy conditions while being disinfected with antiseptic soap. The diseases they caught in this process led to the introduction of metal cylinders in which clothes were quickly fumigated through the injection of sulfurized steam, as well as to other new protocols such as the distribution of warm flannel *gellabas*.

After leaving a-Tor, our protagonist immersed himself in Ibn Taymiya's book on the rituals of the hajj and engaged in a heated discussion with fellow passengers about the merits and faults of this fourteenth-century theologian. Rough seas and a bad spell of seasickness terminated the controversy, presumably sending the interlocutors to the toilets or the upper deck for a breath of fresh air. It forced Sa'udi to concentrate on the comforting fact that every seesaw movement of the steamer brought the *haramayn* closer. With this thought in mind, and while seeking solace in the example and image of the Prophet, Sa'udi relates how he started considering Muhammad's Farewell Pilgrimage. Sa'udi calculated that the Prophet, traveling on land, had entered a state of *Ihram* (purity) at dhu al-khalifa, near Medina. He checked the charts with the ship's captain to find the exact point at which the steamer would come level with this landmark, and when it did he started his purifications, followed by other passengers. This occurred much earlier than usual for the steam hajjies. Besides enabling them to come closer to the Prophet earlier, another advantage was that the earlier *Ihram* obviated the usual rush to the toilets, as each of the first- and second-class sections had one lavatory for both men and women, while the third-class pilgrims had to queue for hot seawater brought from the kitchens for their ablutions. On Sa'udi's first hajj, water was even sold for the inflated price of five piasters a bucket. Tempest, nausea, and lessons learned from earlier sprints to the toilet churned out new Islamic solutions.[57]

Disaster, in short, is the lifeblood of progress, both material and spiritual. But as the *Jeddah* affair, to which we now turn, shows, the reverse was also the case and certain paths to progress (again combining materiality and spirituality), especially those involving new notions of risk and *rizq* management, were the lifeblood of disaster. Unlike most other ships that carried pilgrims to Arabia in 1880, the *Jeddah* did not reach its eponymous port of call. Its troubles started early on. The ship experienced the rough weather of the summer monsoon, including strong winds that snapped her auxiliary sails, giant waves, and reduced visibility from the moment it left Penang for the open sea. At some point during the trip, the *Jeddah* sprang a heavy leak, and after the auxiliary engine broke down the crew was unable to pump out the water. When it reached the Gulf of Aden, the captain decided to abandon ship. The European crew and the nephew of the ship's owner, 'Umar al-Saqqaf, left nearly one thousand Muslims to their fate, with only six lifeboats. When they reached the port of Aden, they told stories of violent Muslims and a foundering ship near Cape Guardafui.

There is no indication that anybody doubted this account, and indeed, this was an unusual story in terms of the scope of its tragedy, but not in its details. Ulrike Freitag mentions "reports from the 1870s which describe scenes of chaos and violence during departure due to lack of space on the ships," and on other occasions rebellions on ships due to long waits, for example in quarantine stations.[58] One year before the *Jeddah* affair, a British travelogue related the account of a ship foundering off the port of Jeddah after three hundred fanatical Muslim pilgrims lashed its drunken British captain and officers to the masts, causing the ship to hit a reef and sink, killing everyone on board.[59] A few days before the news of the *Jeddah*, the English press printed telegrams from Aden about the wreck of the steamer *Queen Victoria* near Cape Guardafui.[60] In 1881, it was again reported that the crew of the British steamer *Matthew Curtis*, which went ashore nearby, was attacked by hundreds of savages armed with spears.[61] Stories about the treacherous vicinity of Cape Guardafui kept piling up in 1882, 1884, 1885, 1887, until 1900,[62] comprising an archive of reports that blended pilgrim fanaticism, savage coastal communities, and the punishments of nature (figure 19 illustrates this graphically). Nevertheless, the tragic news of one thousand souls lost at sea, telegraphed immediately across the world, was still received with shock.[63] This was one of the worst naval accidents in recorded history.

However, a few days later, the *Jeddah* appeared in Aden, dragged there by a French steamer. It transpired that, in contrast to the story told

FIGURE 19. Shipwreck of the Mei-Kong steamer, from Messageries Maritimes, near Cape Guardafui, Somalia. Illustration by Brun from *L'Illustration, Journal Universel* 70, no. 1795, July 21, 1877. Getty Images.

by the abandoning crew, the pilgrims refrained from violence and even kept the *Jeddah* afloat with buckets, pumps, and whatever else they had at their disposal. A new flurry of telegrams ensued. The news of the ship's rescue was received with a global sigh of relief, but even more so with probing questions about the conduct of the crew and calls for inquiry into the scandal of abandoning ship.

TECHNICAL FORENSICS

The investigation was quite revealing. It turns out that while the *Jeddah* was taking in some seawater as a result of the storm, most of the water was actually coming from within: the ship was drowning in its own water supply. The *Jeddah* was an eight-year-old iron vessel, built in 1872 in Scotland. It belonged to a new type of screw-propelled steamer in which a two-hundred-horsepower engine generated steam in several boilers and also pumped water into a compartment directly underneath to compensate for the weight loss resulting from burning the ship's coal supply. This was the water ballast system examined in the previous chapter. The *Jeddah* left Penang with defective boiler fastenings. As a consequence of the rolling of the ship and the heavy seas, these fastenings gave way and caused a leak by breaking the pipes which connected to the ship's bottom. The leak was intensified by the large quantities of water stored below deck, and the need to empty the boilers into the ship's bilge several times, instead of into the sea, when repairs were being executed. The forensic examination of the crippled ship at Aden and the interrogation of her deserting engineers and officers revealed ample technical details about those broken water pipes. Wear-and-tear calculations revealed the limits of the system's components, especially in the rough conditions of the Indian Ocean (and it is perhaps not surprising that the failures occurred in small connective parts like fastenings and rubber washers). The accident was an object lesson in maintenance and repair. The experience also revealed that, given their astonishment that the ship did not go down, the abandoning crew had underestimated the ship's endurance. What the court and public opinion saw as a failure of moral fiber was also symptomatic of a general lack of familiarity with the capacities and limits of the various materials and structures that comprised the water ballast system.

The decade between the 1870s and 1880s was a watershed in the development of water ballast technology in oceangoing steamers, with widespread conversion from wood to steel, and the design of watertight bulkheads. Widely publicized spontaneous experiments in floating laboratories like the *Jeddah* generated important insights that informed the design and improved compartmentalization of cargo holds for ballast, port and starboard ballast tanks, fore and aft tanks, double-bottom tanks, and other spaces. (We will see below exactly how this knowledge traveled.) Improved fastening of boilers, more durable pipes, and recommendations for improving engineering and safety standards emerged out of the experiences of such accidents and from inquiries into how they occurred.[64] The 1880s also saw the development of early oil tankers, which applied similar lessons, primarily regarding the carrying of liquid cargo, to the shipping of petroleum.[65] On the technical level, the *Jeddah* affair encapsulated and revealed steam technology, but also played an active part in its development and enhancement, and later in enabling the replacement of coal with oil.

REMAKING TERRITORY

We already know that a steamship's technical composition is inseparable from the environment through which it sails. Maritime accidents affected these environments as well, and the *Jeddah* affair is again a good case in point. The ship was abandoned off Cape Guardafui, at the easternmost point of Africa, a spot leading up to Bab-el-Mandeb, at the entrance to the Red Sea—a notorious strait which has proved fatal to many a vessel and whose very name ("the gate of anguish") had a terrorizing effect. Passengers of all stripes and ethnicities were deeply fearful when approaching the strait. For many, European navigational innovations, steamers' maneuverability, and even the British flag were a source of true comfort. Here is Indian Islamic reformer and frequent steam passenger Sayyid Ahmed Khan: "Sailing vessels or non-British ships might be at risk in Bab-el-Mandeb. Our ship, however, did glide through the strait at night. Europeans have really taken navigation to new heights. They have such good instruments . . . They can run a ship at any angle they wish . . . They can clearly turn the ship full circle with ease."[66]

Alongside strong currents that vary in direction according to the wind, and the pirates of the nearby Island of Socotra, the principal danger in these waters was reduced visibility owing to sand and dust storms during the southwest monsoon. "As a natural consequence," wrote an anonymous writer to the *Daily Times* as the news of the *Jeddah* accident reached London,

> all distant objects become invisible even at noonday. The probability is that during a storm of this character the *Jeddah* ran upon one of the reefs of coral and lava immediately off Guardafui, without ever seeing the Cape itself. . . . The subject of erecting a lighthouse on Cape Guardafui, or on the neighbouring Cape Ras-Hafun, has been much agitated lately, and a slight increase on the dues of vessels passing the Suez Canal to meet the expense is talked of. The difficulties in the way, however, are twofold. In the first place, Guardafui is situated in a wild, lawless country, and it would, in the event of a lighthouse being resolved upon, be necessary to establish a fortification, with a garrison to protect it. In the next place, in the opinion of navigators, a lighthouse in such a position would only be an additional element of danger, in tempting ships to take a course which they should always avoid.[67]

The need for a lighthouse at Guardafui was identified about a decade before the *Jeddah* affair, owing to the increase in steamer traffic and accidents resulting from the opening of the Suez Canal. These accidents pushed the Glasgow Chamber of Commerce to call upon the Board of Trade to erect a lighthouse at or near Cape Guardafui.[68]

Although it turned out that the *Jeddah* did not hit a reef, such was the fate of two other steamers only a few weeks earlier, and of many others before. But even without grazing a rock, the *Jeddah* affair rekindled the lighthouse debate, revealing some of the technopolitical challenges of building such a structure in a foreign land. In addition to construction complications characterizing lighthouse building in general, the project required covert and overt British diplomacy with the Egyptian Khedives, the Direction Générale des Phares (the Ottoman Lighthouse Administration, established in 1855 after its French director experienced an accident at sea),[69] Socotran chiefs, the governing bodies of the Suez Canal, various European chambers of commerce, and even the European Red Sea lighthouse keepers, who tended to abscond frequently from their "very tiring" posts.[70] The eventual construction of this lighthouse, which took more than two decades to complete, was no stand-alone matter in other respects. Over the second half of the nineteenth century, changes in building techniques, reflective materials, and modes of illumination allowed the construction of ever more visible lighthouses. Whereas during the 1860s Red Sea lighthouses were visible, on average, from ten to twenty miles,[71] in the early twentieth century, when the Cape Guardafui lighthouse was erected, neighboring lighthouses at Mocha, Abu Ail, Zebayer and Jebel Teir, prided themselves on enjoying closer to thirty miles' visibility.[72]

Changes in visibility determined which courses the steamships could sail. A complementary way to promote safer passage was suffering accidents on alternative routes. The *Jeddah* affair, for example, helped promote a route through a channel along the north side of Socotra. While merely consulting a map suggested that sailing on the south side meant saving about twenty miles, the strong current was found "to set a vessel sixty miles out of her course in twenty-four hours, and where the land is obscured by perpetual sandstorms, is accompanied by risks which no prudent navigator would care to incur."[73]

In many respects, the Red Sea was a new environment. Before the nineteenth century, Ottoman sources did not name it as such, and its emergence as a separate and unified entity was tied to British cartography and steam navigation from the 1830s, and to attempts to curtail its dangers. As Alexis Wick has shown, "the scientific invention of the Red Sea" as a knowable object, for example in British navigation guides like the bestseller *Sailing Directions for the Red Sea,* was directly connected to adding the adjective "dangerous" to this sea's reefs, currents, and inhabitants. "Danger," British cartography, steam navigation, and "the Red Sea" were mutually constituting frameworks.[74] As we will see (and as the quote ending the previous paragraph suggests), over the following decades, what was designated "dangerous" soon became governed by "risk" and "security." Risk-prevention schemes like lighthouse erec-

tion became the norm, criminalizing and depoliticizing existing economies and ethics connected to maritime misfortune (as on the shores of
the British Isles, enjoying the spoils of salvaged goods around the Red
Sea was previously considered legitimate). Lighthouses came to the Red
Sea with new trends in cartography and navigation predicated on point-
to-point navigation. If new kinds of fuel, navigation, ethics, and politics
were required to build lighthouses, such structures, in turn, allowed
these new materials and epistemologies to flow safely. In the "List of
strandings which occurred on or near Cape Gurdaphui" between
1872 and 1886—a document produced to justify building a lighthouse
there—coal was the most widely represented type of cargo.[75]

Accidents also shaped social landscapes. While the *Jeddah* was saved
by a French steamer and by her own passengers, other accidents
prompted the cultivation of political alliances with local tribes. Take the
similar instance of the SS *Consolation,* shipwrecked near Cape Guardafui
in 1883. The steamer's crew was also saved by a French steamer and
taken to shore, where they enjoyed the hospitality of Yusuf ʿAli of Alula,
"chief" of a local tribe. A monetary reward secured by the British Political Resident in Aden was used to entice ten neighboring tribal leaders to
sign a treaty with the British government, according to which they would
bestow similar hospitality and protection on future British shipwrecked
steamers, and accompany passengers and crew members safely to Aden,
in return for an annual stipend. The treaty deliberately set out to translate the language of hospitality, goodwill, and friendship into that of the
contract, security, and British salvage law. It also helped integrate coastal
communities, as "tribes" ruled by powerful "chiefs," into the political
realm and political imaginary of British Aden.[76] Local values and ethics
were relied on and encouraged when they proved useful, or instrumen-
talized and taken over by monetary calculations when they did not, as
we will see below. The co-optation of local forces and actors comprised
part of protocols of securitization that were also part of risk management. Indeed, risk and accident were basket categories that collapsed the
monsoon and the fanaticism of pilgrims.

MILIEUS OF CALCULATION: RISK AND/AS TECHNOLOGY

The *Jeddah* left Singapore not only overloaded but also "heavily insured,"
and it was even suggested that the captain, together with owner Muhammad al-Saqqaf, "wished to collect on her."[77] I could not find evidence

for this conspiracy theory, but its evocation allows us to add African slaves and the "coffin ships" that transported Irish immigrants across the Atlantic—often as part of insurance scams—to our trafficking repertoire that until now consisted of pilgrims in various slave-like conditions. These chapters in the history of insurance also connect overloading and overinsuring.[78] Insurance was a key infrastructure that historically allowed converting African people into chattel as insurable cargo during most of the grim period of the Atlantic slave trade. Insurance's role in commodifying human beings was significantly transformed over the nineteenth century. A century before, the captain of the notorious slave ship *Zong* was able to have 133 slaves thrown overboard when his ship ran into trouble. The slaves were insured as cargo and massacring them was a profitable way of cashing in on the policy.[79] Yet in the intervening decades, lawyers representing insurance companies which were reluctant to pay up in disputes concerning the deaths of slaves or revolt generated some of the most effective arguments for self-ownership, which abolitionists later adopted. Arguments about the intrinsic value of human agency made people uninsurable as inert cargo.[80] During the nineteenth century, slave toil was replaced by various forms of indentured labor. Pilgrims like those sailing on the *Jeddah* regularly paid off their debts to shipping companies in this way. In such an environment, insurance no longer licensed the homicidal objectification emblematized by the *Zong*. Probabilistic frameworks yielded new thinking about responsibility and freedom. As we will see, as they developed new technologies for statistically abstracting people into "populations," insurance companies promoted a shift best articulated in Foucaultian biopolitical terms, from the power to take life to the power to "let die."

Formal abolition did not mark the end of subjugation but rather its transformation. It exploited new kinds of fuel and fueled new forms of exploitation, enabling both to go global even beyond the reach of slavery. Consider a vast compensation package paid to West Indian slave owners after the 1833 Slavery Abolition Act. As Kris Manjapra shows, a significant portion of these reparations, "the actuarial pricing of black slave life converted into money and sent to the East Indies," made a "swing to the east," animating a racial regime of indentured plantation labor in the lands around the Indian Ocean. This capital was also invested in steamer and railway lines and in other infrastructures of human trafficking in this environment (where mechanical devices were beginning to be seen as "slaves," as they were called in the British Isles), and in the insurance companies that underwrote them.[81]

Through monetary compensation, the scandal of transforming humans into chattel was given a new lease on life in the Indian Ocean world, where individual bodies were interpolated into large data sets. During the decade or so following the 1869 inauguration of the Suez Canal, the intensification of steamer traffic, and especially the increased number of steamer accidents on the new east-west route, allowed insurance companies for the first time to methodically take stock of, and develop annual "risk books" and actuary databases for, their gains and losses, and calibrate their premiums accordingly. As the chairman of the British Association of Average Adjusters (experts in maritime insurance) described it in 1873, "the law of insurance and perhaps even in a higher degree the law of general average is every year more and more rounding itself into a science."[82]

For example, Lloyd's, the major British insurance market that pooled underwriters operating in this environment, ascertained that "there would seem to be much greater risk and difficulty in making the Red Sea from the Indian Ocean than in issuing from it, as so many losses have taken place in this neighbourhood."[83] In other words, it was riskier to sail to Arabia from India than the other way around, and although steamers had overcome the wind regimes of the Red Sea, they still remained susceptible to the risk environments that inherited them. This pattern also suggests that higher insurance premiums on the *Jeddah*'s route were not unusual. Lloyd's statistical committee also discovered that after the 1860s adoption of marine steam engines, collisions at sea increased, as unlike sail ships, steamers steered narrow sea lanes. Coal transportation similarly emerged as a "bad risk," as colliers were much more susceptible to shipwreck: in 1866 half of all maritime shipwrecks happened in this class of vessel.[84]

Such banal calculations were more powerful than the conspiracy theory of insurance scam. As far as the *Jeddah* was concerned, there is little corroboration for the scam hypothesis; however, insurance was known to generally push captains to take risks they would otherwise have avoided.[85] As François Ewald explains, "by objectivizing certain events as risks, insurance can invert their meanings: it can make what was previously an obstacle into a possibility. Insurance assigns a new mode of existence to previously dreaded events; it creates value."[86] In 1884, a French advocate of insurance depicted it more starkly, as risk management being a partial license for risk taking: "One of the first and most salutary effects of insurance is to eliminate from human affairs the fear that paralyses all activity and numbs the soul . . . Delivered from fear, man is king of creation; he can dare to venture; the ocean itself obeys him, and he entrusts his fortune to it."[87]

Insurance was what transformed the *Jeddah* affair into an "accident" and politicized it.[88] Above I put forth a seemingly apolitical accident-centered explanation for technical development, suggesting that technologies and infrastructures develop through trial and error, with errors prompting innovation designed to prevent their recurrence. But in a market of smallholders, with shipowners and captains usually in command of only one or two ships, disaster at sea would be more likely to cause bankruptcy and inaction than generate innovation. Insurance allowed shipowners large and small to offload their risks onto larger corporations. These bodies have by and large remained under the scholarly radar: unlike the familiar risk-sharing frameworks, namely colonizing corporations such as the various East India companies, the importance of insurance companies and insurance in general in European imperialism in the Indian Ocean and Red Sea remains understudied.

The march of insurance was part of a larger transformation towards quantification. Neither the Select Committee on Steamboats (1817) nor the one on Calamities by Steam Navigation (1831) attempted a systematic quantification of such calamities.[89] A third committee, formed in 1839 to ascertain the number and nature of such disasters, indicated that, "in the absence of any official record or registry of disasters peculiar to steam vessels, we could only have recourse to the knowledge and recollection of individuals, of accounts given by newspapers . . . nor was it to be expected that the owners of steam vessels would, very generally, volunteer a disclosure of the losses they have sustained."[90] The year 1839 can be seen as a heuristic watershed, when "trust in numbers" rather than impressionistic storytelling became an organizing principle in addressing steamer accidents.[91] In that year, after the almost-failed occupation of Aden, statistics on the accident of dead calms were introduced to prevent such mishaps from recurring, recruiting the weather as a quantifiable actor in support of the introduction of auxiliary steam engines in vessels crossing the Indian Ocean. The first large data set calculating hours of "light airs," "foul winds," or "dead calm"—collected from a hundred ship logs—revealed the importance of a more dependable driving force than the wind (see figure 20). Its compiler, Henry Wise, gained his fortune and fame (or notoriety) that year by his involvement in coolie-aided coal mining in Labuan near Borneo, a key bridgehead for British steamers sailing to the east.[92]

Especially from mid-century and on, maritime insurance companies played an important role in ensuring that peril at sea would become a dynamic engine of innovation. The late 1850s and early 1860s saw the

GENERAL ABSTRACT OF VOYAGES FROM INDIA AND CHINA
TOWARDS ENGLAND.

Page.	Year.	SHIPS' NAMES.	Days between the Trades.	Hours calm and light Airs.	Hours fair Wind.	Hours foul Wind.	Total Hours during the Voyage.	Total Dist. per Log in Miles.
58	1792	Taunton Castle	7	835	1452	377	2664	13159
59	1794	Lord Thurlow	15	1151	1564	357	3072	13327
60	1805	Ceres	15	1774	1487	267	3528	13419
61	1808	Walmer Castle	14	1603	1630	341	3576	13115
62	1809	Winchelsea	14	1714	1451	387	3552	13962
63	1812	Batavia	10	1834	1446	488	3768	14988
64	1813	James Sibbald	7	1437	1722	393	3552	13476
65	1817	William Pitt	13	1234	1505	381	3120	14016
66	1819	Castle Huntly	7	1073	1332	403	2808	12682
67	—	Dunira	10	493	1821	458	2772	14577
68	—	Asia	11	1504	1540	364	3408	15082
69	1820	Windsor	5	637	1804	247	2688	14188
70	—	William Pitt	7	1926	1948	330	4104	16062
71	1821	Asia	2	705	1353	294	2352	12540
72	—	Bombay	5	737	1240	375	2352	12828
73	1822	Marquis of Wellington	9	984	1522	251	2760	13022
74	—	Windsor	3	729	1639	272	2640	14247
75	—	Asia	6	1063	1648	385	3096	14554
76	1825	Castle Huntly	3	747	1384	317	2448	12136
77	—	Asia	9	752	1589	299	2640	12988
78	—	Rose	4	1295	1765	304	3360	14309
79	1830	Buckinghamshire	8	1343	1560	361	3264	14409
80	1831	Castle Huntly	5	1061	1673	386	3120	14232
81	—	Thames	5	625	1512	239	2376	12815
82	—	Repulse	2	636	1534	278	2448	13180
83	1832	Duke of York	4	566	1590	424	2580	14317
84	—	Orwell	7	618	1395	339	2352	12827
85	—	Duchess of Athol	2	470	1461	253	2184	13120
86	1833	Edinburgh	11	855	1402	263	2520	11848
87	—	William Fairlie	8	758	1452	190	2400	13216
88	—	Earl of Balcarras	7	601	1546	205	2352	12479
89	—	Reliance	6	493	1577	210	2280	12594
90	—	Sir David Scott	6	630	1655	211	2496	12479
91	—	Scaleby Castle	4	616	1536	162	2304	12488
92	1834	Waterloo	6	558	1577	265	2400	12905
93	—	Farquharson	7	714	1449	261	2424	13472
94	—	Lady Melville	6	496	1574	186	2256	11986
95	—	Herefordshire	3	498	1436	226	2160	12240
96	—	Marquis Huntly	5	519	1694	235	2448	12646
97	—	Warren Hastings	5	639	1687	266	2592	13896
98	—	Vansittart	4	648	1636	285	2568	13488
99	—	Prince Regent	11	517	1566	149	2232	12919
100	—	Buckinghamshire	3	608	1634	254	2496	13892
		43)	301	38696	66988	12938	118412	576125
		Average Hours and Distance.		900	1558	301	2754	13398
			d.	d. h.	d. h.	d. h.	d. h.	
		● Days and Hours ...	7	37 12	64 22	12 13	114 18	

FIGURE 20. The whims of the wind quantified. Henry Wise, *An Analysis of One Hundred Voyages to and from India, China, &c,* 1839.

emergence of steam boiler insurance with associations for the prevention of boiler explosion formed to provide insurance and technical
expertise, inspection services and standardization.[93] Insurance companies, which concentrated more and more of the risks they offloaded
from shipowners for a premium, financed scientific expeditions, and
acted as powerful lobby groups pressuring governments to engage in
agreements with foreign rulers, fight piracy, map uncharted waters, and
invest in digging canals and building lighthouses, in order to reduce or
prevent the materialization of the risks they had taken upon themselves.

Below are three examples, taken from among many such instances,
which are relevant to these technical and environmental discussions. In
1877, Benjamin Martel, chief surveyor to Lloyd's, spoke at the British
Institution of Naval Architects, where he surveyed the history of water
ballasting systems—which had by then been in place for two-and-a-half
decades—for the first time.[94] Only an insurer would have access to the
cumulative knowledge, technical charts, and various consequences in
terms of life and property in the diverse environments and designs he
surveyed, and then provide feedback to the designers themselves. In
March 1880, a few months before the *Jeddah* affair, a discussion in the
British Parliament suggested that the construction of a system of lighthouses, including at Cape Guardafui, would reduce insurance premiums.[95] Another report from August 10, 1880, about the same time that
the *Jeddah* was abandoned, stated that "the occurrence of several disasters, such as the grounding of the *S.S. Venetia* and the wreck of the *Duke
of Lancaster,* off the Island of 'Jibb-el-Zoogur' in the lower Red Sea,
prompted the British government to concur with Lloyd's appraisal, that
erecting a lighthouse there is an absolute necessity." It was recommended
that the Egyptian government be urged to erect and maintain, "as part
of the scheme for which they receive heavy light-dues, a light-house on
Aboo Eyle Islet, eastward of Jibb-el-Zoogur." In relation to these events,
the Board of Trade in London passed a memorandum on the Lighting of
the Red Sea and Neighbouring Waters, which was warmly received by
officials of the government in India.[96] During the following decades,
Lloyd's extended its reach inside the Ottoman Empire.[97]

Insurance corporations like Lloyd's acted as "centres of calculation,"
as Bruno Latour calls them. Abstracting the specificity of fire, epidemic,
mutiny, foundering, or spoiled cargo into "risk" which materialized as
"accident," and evaluating them with actuary tables, insurers extrapo-
lated from large data sets premium rates that made their involvement
in a voyage worthwhile (see figure 21). These premiums, calculated in

FIGURE 21. Centers of calculation: "Lloyd's of London—underwriters examine the lists of wrecks and collisions in the Casualty Book in the mid-1880s," *The Graphic*, August 7, 1886.

London, informed the adoption of certain navigational courses and technologies over others in remote theaters like the Red Sea.

Even the adoption of the Suez Canal had less to do with the reduced distance and more with the ability to merge the categories of maritime risk at sea and in harbor.[98] In turn, these changes set in motion new material, technical, and social processes, such as furnishing ships with beams and lifeboats, widening canals, and requiring steamers to use canal company pilots for safer passage through the narrow artificial waterway.[99] Calculation and the messy world it sought to tame were mutually constitutive.

Increases in the size of ships and developments in maritime technology were associated with the development of risk reduction and distribution as early as the thirteenth century—a link forged in a world-system before European hegemony and considered a key enabler for the emergence of capitalism.[100] Especially in the Genovese trade with the Levant, the construction of large ships was made possible by subdividing ownership into shares held by each of the mariners on board.[101] The emergence of joint-stock enterprises, especially from the seventeenth century, allowed European merchants who did not enjoy the support of their crown for overseas commerce (like the English) to share the risk of maritime trade, and enabled them to raise more funds as needed, and thus maintain the high payment standards overseas. This form of capital pooling allowed for the construction and maintenance of large fleets capable of long-distance travel and trade.

Risk distribution was a political principle. European corporations monopolized foreign trade as the sole representatives of their own governments, in commercial as well as diplomatic terms. By the seventeenth century, it was common practice for stockholders to recommend their country's ambassador to the Ottoman sultan's court.[102] As we have seen, insurance companies, which took this principle to the next level of quantifiable risk management, were political entities too, offering avenues whereby accidents pressured governments into action.

As the history of the corporation reminds us, the ability to distribute risk to shareholders, or offload it onto an insurance company or other entity, was a powerful incentive and enabler of risk. Yet "risk management" was a wide umbrella bringing together various practices and collectivities. Formally beginning with the Britannia Steam Ship Insurance Association (est. 1855) and the British and Foreign Marine Insurance Company (est. 1863), and in earnest from the 1870s, many steamers were insured via "clubs" of shipowners that offered separate maritime risk pooling from that of the insurance companies. Alongside insurance firms and clubs, self-insurance by large shipping companies was common. The EIC tended to underwrite its own risks and the P&O operated its own insurance fund.[103]

The fate of the *Jeddah* again offers an illustration of the compound nature of risk management, and also of the incentives it offered for risk taking, even beyond the £30,000 in insurance money that might have incentivized its owners to sail overloaded during the monsoon. After its crew abandoned ship, the *Jeddah* was salvaged by the French steamer *Antenor*. The Vice Admiralty Court ruled that its officers and owners would be paid £6,000 in compensation. In calculating this sum, the court considered several factors, including the degree of risk taken by the crew of the *Antenor* in saving the *Jeddah*. To determine this risk, the court used the decision to abandon ship as a yardstick.[104] Risk calculation incentivized behaviors ranging from the reproachable to the laudable, from recklessly sailing in rough weather to aiding a vessel in distress.

The salvage jurisdiction of the English Court of Admiralty grew in importance with the advent of steam.[105] Salvage crews and shipowners were handsomely compensated "to encourage people to assume the risk of saving other people and property from loss at sea . . . The purpose of recovering vessels or cargo at peril or lost at sea is to bring valuable property back into the stream of commerce. The laws of salvage are generally designed to encourage individuals to take risks in order to achieve this goal."[106] Before the end of the nineteenth century, salvage became an industry which relied on an appropriate "business climate." Near Aden, it was the Perim Coal Company connected to Lloyd's that held the lion's share of this sector.[107] As with benevolent tribes on the coasts of the Red Sea, mariners were also offered a new quantitative epistemology to frame their less honorable actions as well as, and maybe instead of, their more ethical behavior.

As well as tightening their grasp on the material and human environments in which they engaged, insurance companies paid close attention to and helped shape these legal spheres and adjusted their risk estimations accordingly. Establishing legal and bureaucratic standards facilitated salvage operations. The salvage of the steamship *Medina* which, like the *Jeddah* five years later, left Penang with more than five hundred pilgrims and ran into trouble near Aden, commenced only after lengthy haggling at sea about the salvager's fee. Such widely circulated precedents expedited the process and established a stable and familiar legal and economic environment which mitigated much of the turbulence of engagement at sea.[108] Along with stabilizing the ship, the sea and the weather and, as part of the process, risk calculation and standardization stabilized the legal and political environments through which a ship passed. Especially in a multiethnic setting, where human life was valued in different ways, salvage and insurance quickly became a lingua franca. Is it any wonder that it soon supplanted the various incommensurable ethical dialects along its course?

"Risk," "accident," and the insurance industry were political in other profound ways. First, they were attached to the political philosophy of liberalism and individualism, which assumed that every man was free to buy and sell his own risk. Slaves' risks and futures, by contrast, were assumed by others. A free man was the master of his own personal destiny and endowed with a moral responsibility to attend to the future. This meant taking risks, but also offloading them onto new financial corporations. As Jonathan Levy puts it, "Liberal notions of selfhood had long emphasized the need for self-mastery, even in the face of uncertainty. But only in the nineteenth century did self-ownership come to mean mastery over a personal financial 'risk.'"[109]

Key presuppositions animating the insurance industry were secular: risk management was a defiance "against the gods" (as one of the best-known histories of insurance is titled), based on a worldview promoting man's control over uncertainty. It was hailed as bringing about emancipation of action from fear, blocking the "transposition on to the earthly plane of the religious faith that inspires the believer."[110] This secularism was also antiethical, as risk management divorced ownership from responsibility and responsibility from blame. "It proposed a mode of thinking completely foreign to the moral, moralizing mode which underpinned and was supposed to validate the juridical notion and practice of responsibility, and yet it did this without entering into conflict with juridical practice."[111] The politics of risk also had a distinct class bias: while risk benefited "those who enjoyed a certain comfort," it remained "quite fruitless, not to say inaccessible, for the poorer classes."[112] Finally, the insurance industry and related sectors had a particular disposition vis-à-vis the future, laying the foundations for what Arjun Appadurai called our current neoliberal "politics of probability" in which a limited calculating mentality replaces acting on hope, possibility, and values.[113]

RISK AVERSION

This set of presuppositions regarding self-ownership, individualism, fear, God, responsibility, class, and the future accounts for the strange career of insurance in the Ottoman and Indian Ocean worlds—spheres which might reveal something about the unpredictable dynamics of the emergence of our global "risk society."[114] As Jonathan Levy has shown, until the nineteenth century "risk" was a specialized term: it was the commodity exchanged in a marine insurance contract. Risk's ability to move inland, promote far-reaching changes in how people lived their lives, and animate a comprehensive philosophy depended on new energy sources, technical infrastructures, and geopolitical connections. In the Ottoman case, it was paradoxically a strong aversion to much of what risk represented that facilitated the penetration of Western insurance companies. While Ottoman merchants tended to insure their merchandise at sea before the nineteenth century, relying mainly on British and other European insurers, insurance was first regulated formally in the 1863 Ottoman Maritime Commercial Law. From the mid-1840s onwards, European companies opened offices in the port cities of Istanbul, Izmir, Beirut, Trabzon, and Salonica. Maritime transportation insurance was followed by fire, life, and accident insurance on land.[115] According to a report presented to the Ottoman government in 1887, by that year there were thirty-six foreign unlicensed insurance companies operating in the Ottoman Empire. The government prepared a regulation in the same year to order the insurance business. In the 1890s, insurance started spreading from port cities to the empire's interior.[116] Religious authorities were by and large antagonistic to allowing insurance inland (with a few prominent exceptions, such as Muhammad ʿAbduh). Yet several *fatwas* stated that if an insurance contract was signed with a foreign company which paid indemnities of its own volition, the contract was permissible.[117] By the First World War, over one hundred international insurance companies were registered in the empire.[118]

These fatwas and the world of the *Jeddah* and the al-Saqqaf are again instructive for the march of insurance in non-Western settings. Let us examine one more closely: In 1905, another member of this family, Singapore Mufti ʿAlawi b. Ahmad al-Saqqaf responded to a question from Aden about the permissibility of a Muslim purchasing maritime merchandise insurance. Being a Shafiʿi himself, al-Saqqaf pointed out that according to the Hanafi school of Islamic jurisprudence—the school dominant in the Ottoman Empire—Aden could be seen as *dar al-harb* (the abode of war), where transactions that are prohibited in *dar al-Islam* (the abode of Islam) would be permissible.[119] Al-Saqqaf was extending to Aden opinions that legal scholars expressed about British India, like nineteenth-century anti-imperialist Mufties citing the

famous 1803 pronouncement of Hanafi scholar Shah Abdul Aziz Dih-
lawi, who declared British-controlled India to be *dar al-harb*.[120] Cement-
ing the *shar'i* definition of Aden as *dar al-harb* in legal and commercial
infrastructures is a further indication of the fruitful uses of the Indianization of Aden. Removing it from the abode of Islam and the Ottoman
Empire (paradoxically by relying on Islamic Hanafi jurisprudence) enabled steamers plying its waters to be insured, hence its growth and the
growth of the maritime hajj. Reprinted in *al-Manar*, and endorsed by its
editor Muhammad Rashid Rida, this *fatwa* also prefigured the reform-
ist principle of *la-madhhabiyyah*, the licence to abandon the protocols
of interpretation of a particular legal school, choosing from several
other legal options instead. Madhhabs were not only simple doctrinal
schools united by a legal interpretation, but equally a set of intergenerational relationships and scholarly genealogies whose desolation we wit-
ness here. A new attitude to the future as a realm of unhindered growth
and speculation meant reconfiguring the past.

This transformation did not go unchallenged, however. A year later,
the future Mufti of Egypt, arch-traditionalist and Rida's nemesis Muhammad Bakhit, issued his own opposing view on the permissibility of
insurance. Bakhit's stance is based on insistence on coherence between
one's disposition about futurity and the retrospective interpersonal connection to the community's sources of authority. In this way, intergenerational responsibility could become a prospective principle too.
According to Junaid Quadri's analysis of the heated exchange that followed, Bakhit's was a defense and reinvigoration of *madhhab*, tradition-
alism, and even taqlid, the principle of imitation of former generations
and their rulings.[121]

Notably, the word *risk* diverged from the Arabic word *rizq* in the
seventeenth century, roughly when European risk-sharing entities,
namely corporations, developed out of and departed from Islamic
schemes of collective pooling of resources.[122] While joint-venture formulae were developed in the Ottoman world, and the Islamic *mudaraba*
has even been found to be the inspiration for the European *commenda*,
the legal arrangement that preceded the joint-stock corporation,[123] these
partnerships were generally not considered legal entities, and individual
rather than corporate liability remained the underlying principle in
commercial arrangements covered by Islamic law.[124] *Rizq* thus represented an aversion to risk's amorality. More specifically, insurance
transactions in which future uncertainty was the commodity exchanged
were usually seen as a form of prohibited gamble. Legal maneuvres like

al-Saqqaf's facilitated risk's march by relying on the very mechanisms that chastised insurance as un-Islamic. They laid the foundations for subsequent Islamic movement in the post-Ottoman world to effectively combine piety and neoliberalism.[125] But they also bolstered the strong risk aversion encapsulated by the notion of *rizq*.

Unlike risk, *rizq* animated quite different politics, demonstrated by attending, again, to the *Jeddah* affair, now through the eyes of the deserted pilgrims or as close as possible to their perspective, which has not been preserved or inquired after in court. The notion of *rizq* has been closely associated with the path of God, in the metaphorical sense of pious conduct as well as in the literal sense of the pilgrimage to Mecca. As *surat al-hajj* (Quran 22:58) states: *wa-alladhina hajaru fi sabili Allahi thuma qutilu aw matu* **layarzuqanahum** *Allahu* **rizqan** *hasanan wa-'inna Allaha lahuwa khayru al-raziqina* (And those who emigrated for the cause of Allah and then were killed or died—Allah will surely provide for them a good provision. And indeed, it is Allah who is the best of providers). If risk offloaded possible future consequences on financial institutions, *rizq* did so on God.

Like the hajj, *rizq*-seeking intensified with steam and bolstered it in its turn. Steam navigation made travel safer, quicker, cheaper, and more regular; but paradoxically, in experiential terms, for many people it probably made it scarier. This was precisely because steam engines allowed ships to sail in rough seas and bad weather, with large numbers of people on board. The opening out of the pilgrimage meant that Muslims of lower social standing, many of whom were unaccustomed to travel, took to the sea for the first time. More often than not, they ventured on the passage to Mecca on board old steamers, as second- and third-hand ships usually found their way into the "hajj trade." While most of these pilgrims eventually reached Arabia safely, many of them experienced the ordeals of a rough passage. Risk and *rizq* thus formed a synergy not unlike the technological sublime and technological terror, or the hajj and steam navigation and the dhow and the steamer. *Rizq*-seeking was a major reason for the intensification of the hajj, and for the willingness of hajjies to endure miserable conditions of passage, and later of debt repayment. Europeans were well aware of the Muslim willingness to die on the way to Mecca (during this period on average about 20 percent of pilgrims did not survive the journey).[126] While they criticized it as fanaticism, this state of mind fueled both steam navigation and the insurance industry. Divergent as they were (and perhaps precisely as a result of this), *rizq* and risk were co-constitutive.

Like risk, *rizq* was an umbrella for heterogeneous Islamic dispositions to danger (even though some of their proponents were not Muslim). Diverse as they may have been, accounts of *rizq* differed from risk not only qualitatively but also quantitatively: first, we have far fewer such tales compared to the large data sets of risk management; also, unlike the abstract mathematical character of the former, accounts of *rizq* had to do with storytelling, not calculation. My account of this set of positions is similarly impressionistic: we do not have the pilgrims' perspective of the *Jeddah's* salvation; what follows uses several approximations.

Lord Jim can be seen as a first stab. Joseph Conrad may be faulted for not recounting the perspective of the pilgrims in the novel.[127] But reading *Lord Jim* together with the novella *Typhoon*, another sea yarn he began writing in 1899, suggests this silence might have been intended. *Typhoon* dealt with the inability of European steamer captains to acknowledge the perspectives of the Chinese indentured laborers they transport during a storm (and a familiar model for our Singapore *hajj-coolies*); indeed, this was a text about the British very incapacity to regard these people as passengers ("Never heard a lot of coolies spoken of as passengers before").[128] In *Lord Jim*, Conrad briefly describes the natives' response to the ship's troubles as informed by an incapacity to fear death and a complete disassociation of fear from the possibility of abandonment. As he states through the words of the steamer's Malay helmsman during the trial, "it never came to his mind then that the white men were about to leave the ship through fear of death." While Europeans—Conrad included—usually regarded the absence of fear as a sign of a flawed individualism and interiority, locating the heart of the matter in diverging attitudes to danger and fear is nonetheless insightful.

The flaws of *Lord Jim* may speak to the limitations of the novel form more generally, and to its debt to bourgeois notions of risk, futurity, and speculation. Yet this literary legacy can be pluralized. Elizabeth Holt suggested that nineteenth-century early Arabic novels written in Syria and Egypt may be connected to "older tales of risk at sea, the likes of *Sindbad* and sometimes *A Thousand and One Nights*."[129] One such novel, *Fatat Misr* (Girl of Egypt), written in 1905 by Ya'qub Sarruf, a Protestant Syrian émigré to Egypt and coeditor of the popular scientific magazine *al-muqtataf*, was serialized in this periodical six years after the publication of *Lord Jim*. Beyond offering its readers a cautionary tale about futures markets and financial speculation,[130] one scene closely mirrored and inverted the world of *Lord Jim*.[131] The setting is hauntingly familiar: a loaded steamer springs a heavy leak near the Island of

Socotra after hitting a submerged rock (ostensibly the result of limited visibility due to the lack of a lighthouse and the ship's proximity to the notoriously strong underwater current there). Here, too, when the crew fails to pump out the incoming water, the captain decides to abandon ship. Unlike in Conrad's account, however, different groups of passengers handle "the fear of death" in their own ways, including prayers and the comfort of bodily contact with loved ones. The evacuation is tense but orderly enough as the women and children leave first while the men have to swim for their lives. Notably, when one of the women reaches shore she begins to calculate the statistical chances of her brother's survival. Despite the ratio of 100:1 against, she does not give up hope or abandon her lookout post, and indeed, after much anxiety, he survives. It was not calculation and probability *per se*, in other words, but acting on them as if nothing else mattered that made risk management rhyme with ship abandonment.

As the travelogue of Shibli al-Nu'mani reveals, Christians and Muslims could see eye to eye when facing trouble at sea, and could share calming techniques:

> The next day after the steamer took us from Aden [en route to Port Said], a dangerous accident caused me anxiety and confusion for a while. When I woke up on the morning of May 10 [1892], a friend informed me that the engine had broken down, and I saw the captain and officers concerned and anxious, trying to fix it. I hurried to Mr. Arnold[132] and found him immersed in his studies, reading a book in relaxation. I said to him, "Have you heard?" He replied, "Yes, the engine has broken down." I said, 'Aren't you concerned? How are you able to read a book?" He said, "The steamer's malfunction actually makes this time extremely valuable and it would be stupid to waste it." And this filled me with relaxation. Eight hours later, the engine was fixed.[133]

Maritime fears often pushed passengers—from Shibli to Sayyid Ahmed Khan and Sa'udi—to focus on their priorities and seek solace in God, scripture, the Prophet, and other comforting religious figures, sometimes finding there sources of creativity and problem-solving.

Beyond these middle-class narratives, popular North African hajj songs often expressed admiration for *babur* passengers (from the French word *vapeur*, or steamboat) facing their fears and eventually persevering through maritime hardship en route to the port of Jeddah, as shown by Samir Ben-Layashi, who analyzed a corpus of these works from the late nineteenth century.[134] Other accounts of miraculous rescues at sea during the hajj were quite a developed subgenre of the *al-rihla al-*

higaziya travelogue genre, called *'a'zam al-karamat*. Our acquaintance from chapter 4, Fadl ibn 'Alawi, theorized paranormal events like *'aja'ib* and *karamat* as instances in which quotidian calculative rationality is suspended by God.[135] Sufi devotion, in his book, is the suspension of calculated disbelief. Historian Nile Green, who has studied this genre, referred to it as "supernatural insurance."[136] The term is useful in the sense that it suggests that both capitalism and Islam are predicated on metaphysics, whether by means of the invisible hand of the market or the hands of saints pulling a steamer in distress from the waves. Yet this is also where the similarity ends. Many of these stories speak of great fear in the face of mortal danger, but also of fear as an ethical force directing man to the righteous path.[137] Such stories also speak of communities joining hands in the face of peril, as the passengers on the *Jeddah* did as they kept the ship afloat after her crew had abandoned them. Against the ethos of risk predicated on the self-owning individual who ignores fear and divine forces up to the point where he no longer can, to the point of abandonment, they seem to have marshalled a *rizq*-based ethos of collective responsibility and piety.

The introduction of insurance into the Ottoman and Red Sea worlds via foreign companies is indicative of how, despite fierce attempts at securing a degree of independence, the Ottomans were annexed into the global economy under terms they did not determine. At the same time, the Ottoman will and ability to reject key epistemologies of risk, especially those which had to do with eroding forms of community, solidarity, and piety, are indicative not only of resilience, but also of how steam power was simultaneously connective and differential. As we saw with the racialized and thermodynamic bodies, and for carbon democracy and carbon autocracy, risk and *rizq* together were the zeitgeist of coal. *Rizq*, though, was a hidden partner, written under erasure as the "constitutive outside" of risk, insurance, and a secular nonethical epistemology.

However, if the age of coal persists into our own age, and if the ethics deposited in this fossil fuel wait as potentials, however inert and dormant, this Janus-faced phenomenon might have important implications. First, as a historical structure it affords a more complex and hybrid account of the advent of key economic, political, and religious institutions and transformations—materializing around the sector of passenger transportation in the Indian Ocean and Red Sea. On a methodological level, it reveals how contingency was a historical force that yielded not only disarray and unpredictability, but also the mechanisms to curtail them.

Such an explanatory scheme specifically enables us to examine the liberalism of risk together with the collectivism of *rizq*, thereby revealing worthwhile attitudes that have been abandoned. In turning their backs on the *Jeddah* and its Muslim passengers, it did not suffice for the deserting European officers to blame the rough weather or the leak. To justify their amoral decision publicly—and perhaps also to themselves—they described the pilgrims' collectivity as malicious; and in so doing, they tapped a gushing fountain of similar stories which made their account all the more compelling. Yet as their own account was revealed as spurious, it now allows for reading this broader genre, predicated on expressing natural calamity in terms of human agency and intercommunal violence, for what it was: a construct. Such accounts drew on the tendency, discussed in the previous chapter, to regard Muslims as a force of nature, and as parts of the environments they inhabited or traversed. The other side of this coin was a new, coal-fired politics of abandonment. Risk might have effectively waived fear and set ships in motion, but when things went awry this scheme could neither contain real dread nor direct mortal fear towards collective action or creative solutions.

Mine is neither the only nor the first attempt to salvage alternative forms of togetherness from accounts of shipwreck seen as microcosms of human cooperation.[138] Clearly, multiple forms of actual politics teem under the hood of actuary politics, offering a rich repertoire of alternative forms of collectivity and action. If shipwreck emblematizes the modern condition (a proposition Conrad himself entertained),[139] actual steam navigation disasters (and their fictionalization) reveal broader crises and the ecologies that cause them. The *Jeddah* affair is an especially telling case, as it encapsulates Western insurance and *rizq* management and mismanagement by both the al-Saqqafs and their passengers, and allows discussing freedom and multiple degrees of unfreedom including continuities with slavery that teem underneath our modern, liberal risk society. Bonded people, be they slaves whose manumission financed plantations, steamer and railway lines, and insurance companies; indentured laborers debt-bonded to steamship companies; or pious pilgrims whose bondage to God and *rizq*-seeking drove towards Mecca, were all enablers of steam navigation. Some of these intersecting and mutually animating forms of bondage were among the most abominable forms of human exploitation. Yet even enslavement was a form of obligation, and even manumission, under certain conditions, could be a form of abandonment that was far worse. This chapter has shown that like "carbon democracy," finance capitalism and liberalism are indebted to forms of bondage often relegated to capitalism's past and that this is the case even if we examine these systems through the technologies and fuels that allow them to function. As a result of this erasure, the ethical systems developed by bonded people have no place in our current "politics of abandonment." (This term is derived from Elizabeth Povinelli's account of late liberal economies of abandonment, as well as from the etymology of the word for freedom in Arabic, *huriyyah,* previously denoting the manumission of slaves.)[140]

As all historiographical pursuits of a usable past are written with an eye to the future, contrasting different approaches to futurity itself helps to kindle a political conversation from a (re)new(ed) position. The future we eventually opted for is one devoid of fear. As the lessons of the *Jeddah* suggest, risk and freedom from fear pin us to a course in which the choice is either acceleration or abandonment. Instead, *rizq* made fear an invitation to ethics, one based on coming together. This is not an escape route: as a protocol of handling fear, *rizq* teaches one how to stay with the trouble, whether actual shipwreck or the hyper-object of the Anthropocene. If above we prodded this avenue in a fluid and rather sketchy manner, the next chapter tries to ground and even underground this possibility more systematically.

CHAPTER 6

Fossil

COAL'S AMBER

We tend to think of oil and coal today almost exclusively in terms of
(fossil) fuel and practically only as sources of thermal energy. There are
many problems with such tunnel vision, beginning with the fact that
this terminology is unstable since, like most other words (including
energy itself), *fossil fuels* has meant different things at different places
and times. During the nineteenth century, the English term was used in
the singular, to make a distinction between mineral coal and charcoal,
the latter being the usual referent of the word *coal*.[1] In the early twentieth century, *fossil fuels* (now in the plural) came to denote the common
properties and uses that bind coal and oil together. Later that century,
and still in the plural, the notion became a reference point for different
energy sources (including natural gas) produced by different means and
put to different uses, but united in their detrimental effect on the environment, and contrasted with "sustainable" motive powers.

More importantly, the concept of fossil fuels obscures other histories
of oil and coal. Petroleum became a commodity more than half a century
prior to its twentieth-century emergence as a fuel and only after new uses
were sought for gasoline which, up to that point, had only been the useless by-product of refining crude oil into kerosene for illumination. Oil's
adoption of a commodity form resulted from its being modeled on coal.
In the 1850s, coal-oil started dominating the US market for lamp oil and

194

lubricants, a market which had previously been based on organic matter like whale oil. After coal opened it up to other mineral, nonorganic substances, other synthetic products came into their own, especially petroleum, which unlike coal flowed naturally in liquid form.[2] Further, coal-fired steam engines with their fast-moving metal parts were central in generating the need for petroleum-based lubricants.

Coal and oil are also connected by their common prehistory as dead organisms that have undergone anaerobic decomposition. As such, they share a hydrocarbon base and a range of potentials associated with how heat is released when they burn. The prehistoric catastrophe that created them is echoed in the posthistoric apocalypse that is said to result from their combustion. The obfuscations of fossil fuels and the energy released by their incineration are joined by another loose term, the *Anthropocene*, thus bundling together the results of various regimes of extraction and use on the levels of the stratosphere and the underground, industrial modernity, and geological time. A powerful critique of the term *Anthropocene*—emerging from a trend sometimes called *posthumanist*—claims that "the human" is far too narrow and incoherent, a poorly insulated category that is useless for serious analytical work; there is no room for anthropos-centric methodologies in the convolutions of what is perhaps better called the *Chthulucene*, evoking the octopus-like deity as an emblem for the complex and multivalent conundrum of our times. What we need instead is to develop appropriate modes of tentacular thinking attuned to a multiplicity of creatures and things with which we are embroiled.[3] By contrast, according to another major critique of *Anthropocene*, human agency itself is not the problem. For some of its proponents, it is even the beginning of the solution: what is needed in their opinion is "a new anthropocentrism" that takes full responsibility for anthropogenic climate change, or another "theory for a warming world" that begins with disentangling nature and society (the very domains posthumanists insist on entangling).[4] Rather, the problem is that humanity is far too wide, and pointing a finger in its vague and general direction removes power and politics from the picture altogether. What is often called for in this line of argumentation is a class-first perspective to better understand the *Capitalocene*, and/or an analytical approach and terminology attuned to the roles of empire, race, Christianity, and the West in anthropogenic climate change.[5] Controversy abounds between, and no less within, both camps.

Like the fossil fuels that now energize it, the *Anthropocene* is a shifting term, hard to pinpoint and periodize. Yet language's efficacy, as troubled as it may be, only rhymes with accuracy. We have come to define our age as a geological epoch, one marked by fossil fuels that have been turned into capital with detrimental effects. All these themes—geological epochal thinking touched with catastrophism, fossil fuels (and the relations between coal and oil), and capital—are the main objects of analysis of this final chapter, which examines a case in which they were tangled together in unfamiliar ways. In what follows, I try to retrace the junctions where they initially intersected, in Europe and the Ottoman Empire, beginning with mineral coal, which was the first to lump them together. To achieve this, I will attend to the *fossil* in fossil fuels. While much attention is given to how coal and then oil were transformed into fuels and used as energy sources, it is their complex and possibly open-ended legacy as fossils which I seek to situate as the key contact point between Anthropocene and Capitalocene (and perhaps Chthulucene too).

FOSSILIZED KNOWLEDGE

During the nineteenth century, the parts of natural history that later developed into the semi-independent sciences of evolutionary biology, geology, ecology, and physics (especially thermodynamics) shared several important features, key among which was the prominence of fossils for their development. Whether as fuel animating a steam engine, an earth sample indicative of underground processes and morphologies, or the petrified traces of prehistoric flora and fauna that cast light on progeny living in the present, fossils and the knowledge they afforded were one of the main engines of scientific advance during the late eighteenth and early nineteenth centuries.

The fossil was a unique medium: part organic, part inorganic; part stone, part creature; part lowly material, part objet d'art; part fuel, part treasure. This hybridity allowed the multidirectional flow of knowledge and methodologies between West and East and between the human, social, and natural sciences through carbonized conduits. An example of this epistemological flow, one also worth explicating for what follows, was the application of statistics to the natural world. In contrast to the early-nineteenth-century catastrophist geological paradigm, which postulated that the history of the earth was shaped by abrupt, major disasters akin to the Biblical Deluge, around the 1820s European naturalists began to posit more gradualist explanations, which toward the end of the century would cohere into the theory of uniformitarianism according to which "nature does not make leaps" (which was also the foundation for Darwin's evolutionary thinking). These explanations were informed by methods for collecting data for political use by the "statist," or "statesman," and thus appropriated the term *statistics*, applying it to the tabulation of quantitative data on petrified flora and fauna. State-collected information mainly focused on birth and death rates in a population of human subjects. Naturalists borrowed such proto-social-scientific tools to quantify "the birth and death of species." Geologists such as Charles Lyell analogized their fieldwork to that of itinerant census takers and started regarding preserved fauna as "statistical documents."[6] Governance of a human population was thus recruited to order an unruly natural field.

During the second half of the century, fossil knowledge was rehierarchized in a quickly industrializing Europe, with physics and physicists setting the tone and providing the engine to which geology and biology were now yoked. Thermodynamics and the science of energy grew out of the engineering challenges of using coal in steam engines. These imperialist epistemologies promoted and helped finance geological expeditions and surveys in search of fossil fuel. They set the standard for calculating the age of the earth. They also invigorated the reconceptualization of biology and ecology around the notion of an organism striving for internal and environmental energetic balance (according to the first law of thermodynamics) and constantly losing energy (according to the second law).[7] Human physiology, too, was now understood within the framework of "the human motor." Consequently, human society and economy could be seen as thermodynamic.[8] These fossilized bridges now allowed methods and concepts to travel in the other direction, towards the social sciences.

However, as we have already seen, their contact points and hierarchies notwithstanding, fossil knowledge was animated and simultaneously informed by the humanities and social sciences long before the emergence of thermodynamics or statistics. This is especially evident in the case of geology, "the most historical of all sciences."[9] In the seventeenth century, naturalist Robert Hooke regarded fossils as "nature's documents" and the bedrock as "nature's archive." Buried fossilized remains were seen as "medals" or "coins" akin to the Roman coins found by antiquarians in northern Europe, demonstrating that the Roman Empire had extended far beyond Rome.[10]

In the nineteenth century, fossils helped dissect deep time into different eons and eras, epochs and periods, each characterized by the different creatures living in them and dated by their extinction. Our current controversies about the Anthropocene and about the present human condition vis-à-vis its past and future have their roots not only in the burning of fuels and their emissions into the stratosphere, but also in such fossil-based historical taxonomies. This explains the urgency of focusing on the connection between natural and human history.[11] The fact that the natural sciences were and are still informed by politics, aesthetics, and religion does not make them any less "objective." Rather, it opens up a window through which humanists can probe scientific objectivity and suggest how it may be used and informed.

In what follows, I explore the convergence of the natural and human sciences and examine how during the nineteenth century Ottoman coal was transformed from a humble black or "burning" stone into a buried treasure deposited by God for the benefit of the entire community of the faithful, and thereby infused with ethical meaning. This alchemical transformation involved a classical literary aesthetic that praised knowledge of all sorts by equating it with precious stones. From the 1830s, this convention started animating translations of modern geology books into Arabic and then Ottoman Turkish. Nineteenth-century interest in geology and mineralogy had everything to do with the new importance of mineral coal, previously a lowly material that was now elevated to the status of a coveted resource. This was true in many places. In Europe, transforming fossils and then fossil fuels into treasure meant economizing them in ways that ostensibly liberated these substances from important material and ethical constraints. In the Ottoman Empire, however, defining fossil fuels as treasure helped to include them under the umbrella of the sultan's Privy Treasury. The process resulted in the management of these substances according to the established Shar'i stipulations for the conduct and dispensation of buried treasure. In other words, through protocols of translation, institutional and budgetary changes, inter- and intraimperial struggles, and acts of governance, coal, and the science of geology were included within a particular theological and ethical framework just as coal in western Europe was being secularized in ways that masked its connections to these domains. Both processes have ongoing consequences, and the former may offer valuable insights for the present environmental conundrums that were created by the latter.

ARABIZED AND TURKIFIED STRATIGRAPHIES OF KNOWLEDGE

The mounting and global importance of fossils and the subterranean world during the nineteenth century is evident in the fact that during this period books on these matters were translated into languages such as Arabic and then Ottoman Turkish early on and in growing numbers. Tracing the career of this literature-in-translation documents Ottoman and post-Ottoman Middle Eastern attitudes toward fossil fuels and toward science at large. It also fleshes out the dependence of scientific protocols on literary hermeneutic conventions, in Ottoman and European settings alike. The first modern geology book translated into Arabic in 1833 was Cyprien-Prosper Barard's *Minéralogie Populaire*. As its name suggests, the book offered a simplified popular science, one explicitly contrasted to the image of natural science as "that which is mathematically exact."[12] A member of the circle of Georges Cuvier, the main proponent of geological catastrophism in France, Barard was a mineralogist and economic geologist who in 1813 was appointed director of the copper mines of Servoz, where he was the first to use anthracite coal as a fuel. His single volume dealt with sparkling gems and dirty coal. Its translator, Rifaʿa Rafiʿ al-Tahtawi, was an al-Azhar graduate turned translator. Tahtawi is famous for importing European political terminology into the Middle East, and is hailed as the grandfather of Arab territorial nationalism due to his role in Arabizing notions such as *patrie* and *patriotism*, which he rendered into *watan* and *wataniyyah*. Less attention was given to how he imported and Arabized European subterranean concepts.

Tahtawi was no geologist. His first book was a *rihla* or travelogue he published in 1831 about his sojourn in France between 1826 and 1830 as an imam of an Egyptian student scientific mission. It was titled *Takhlis al-ibriz fi talkhis bariz aw al-diwan al-nafis bi iwan bariz*, which is usually translated into English as "The Quintessence of Paris." But literally, the title reads, "The extrication of gold in the summarization of Paris; or the treasured *diwan* in Paris's *iwan*." A *diwan* is a literary salon or a volume of collected texts, usually poetry. An *iwan* is a palace where the entrance is typically decorated with calligraphic and geometric designs (like the one depicted in figure 27, towards the end of this chapter). Tahtawi's heading thus anchors his *rihla* in a longstanding tradition whereby books are treasured as precious physical objects, and textuality is visibly heterotopic and seen as a gateway to distant locations. Published two years before the translation of Barard's book, Tahtawi's travelogue has nothing to do with geology, it simply serves here as an example of an established style of embellishment. It also helps me to tease out the context for the flow of European scientific knowledge to Egypt via such student missions. Yet the application of this mode of textuality to modern geological literature would have unexpected implications.

The title Tahtawi gave Barard's book was *Ta'rib kitab al-mu'allim Firad fi al-ma'adin al-nafi'ah li-tadbir ma'ayish al-khala'iq* (the Arabization of the book by the scholar Firad on the minerals/mines useful for the living of created beings). Unlike translation, the notion of Arabization meant adapting a text to its target language both lexically and culturally (thus, the *ta'rib* of Shakespearian sonnets written in iambic pentameter entailed rendering them into classical Arabic meters). As it developed from Tahtawi onwards, the process usually entailed at least two people—a speaker of French (usually) and an editor or *musahih*, typically an Azharite shaykh specializing in classical Arabic—working on a single foreign text. After Tahtawi's appointment in 1835 as director of Egypt's School of Translation, this protocol became all the more entrenched. This collaboration involved the blending of the physical and metaphysical aspects of *'ilm* (knowledge, as in the word *scholar* or *mu'allim*)—knowledge both natural and spiritual.[13] We may contrast these notions of *'ilm* and Arabization to the now canonic "sociology of translation" and the philosophy of reference that Bruno Latour developed in *Pandora's Hope,* which was so useful in debunking a vulgar dichotomy between representation and reality, the world of things and that of signs. There, too, the first case in point was how soil scientists translated findings from a stratified subsoil and bedrock into textual arguments published in scientific journals. Latour's pedologists gradually "translated" earth samples into color codes, then into numerical values, and finally into graphs in peer-reviewed articles. This translation necessarily entailed making two realms commensurate with each other by retaining something of the logic of each.[14] Arabization, by contrast, supposes a single overarching logic and refuses altogether the meta/physical divide, notwithstanding how slow and gradual the bridging of the realms may be.

Like Tahtawi's notions and practices of Arabization, authorship was also diffuse and hardly ever inhered in a single person. Rather than the genius of a scholar/author or translator, knowledge emanated from God via a progression of authority that could be mapped onto the physical composition of the world. Tahtawi's preface to Barard's text established the gold standard for handling modern geology in Arabic and later in Ottoman Turkish:

> Blessed is Allah, the precious stones [*ma'adin,* a word denoting also "mines" and derived from the same root as *Eden*] in Whose treasury are too many to number. And Whose buried treasures are boundless. May the gold of his gifts never end . . . [He is] the one who has laden the treasures (*kunuz*) of grace

and uprightness . . . Whose Prophet [PBUH] beautified the Arabs and Persians . . . Our noble Muhammad, the last of the emissaries and master of all . . . This yarn was translated from the French into the Arabic by the humble Rifa'a Badawi Rafi' al-Tahtawi, but truthfully it was woven by the pleasant hand of the ruler (*wali al-na'm*) [Mehmet Ali].[15]

Tahtawi proceeds to refute the saying according to which the *salaf*—the blessed ancestors, those first generations of Muslims whose example is followed from the beginning of Islam—have an inherent advantage over those who succeed them. Egypt's ruler is the living example that this is not the case, especially because of his efforts in spreading knowledge, and the arts and sciences (*funun wa 'ulum*).[16] This protocol of attribution of earthly knowledge followed a sequence beginning with God, passing through the Prophet, and reaching the translator via the patronage of the current ruler, while simultaneously likening the different strata of the earth to cohorts of men—prophets or emissaries, rulers, or pious forefathers—and became the established formula for Arabizing/Ottomanizing/Islamizing European geology. It also prepared the ground, and in this case the underground, for informing the matter of subterranean extraction with intergenerational awareness, with responsibility to forefathers and descendants.

The next geological book to be translated in Egypt, in 1841, *Geologie populaire à la portée de tout le monde appliquée à l'agriculture et à l'industrie* (Paris 1833) by Nérée Boubée, another French catastrophist, followed the same process. Boubée was the student of Louis Cordier, one of the savants who started his career accompanying Napoleon in Egypt during the turn of the nineteenth century. Boubée's writings were used as student textbooks in France, which is likely where the book's translator, Ahmad Afandi Fa'id (himself a student of Tahtawi), encountered them during his sojourn there as part of Mehmet Ali's student missions. Fa'id, who also added a lexicon of scientific terminology to the book, collaborated with another Azharite Shaykh, Ibrahim al-Dasuqi, who "corrected" the translation. The Arabic title, *al-Aqwal al-murdiyah fi 'ilm bunyat al-kurrah al-ardiyah* (The agreeable sayings on geology), has rendered the science of geology into an *'ilm,* that of "the construction of the earthly sphere."

From the 1830s through the early 1850s, various European scientific books Arabized in Egypt, on subjects ranging from foot disease to astronomy, were rearticulated as "buried treasures" (s. *kanz,* which is also the root for Geniza that likewise connects text and material wealth), underground relics (s. *tuhfa,* built with the root from which *museum* or *mathaf* would be derived), or other forms of precious metal and stone.[17] Gradually, similar conventions also began to characterize scientific translations into Ottoman Turkish, where written texts, periodicals, and the very imperial archive were already couched in language that described them as treasure troves.[18]

It is noteworthy that during the 1840s, Egypt, which had energy needs in inverse proportion to its available resources, was one of the main driving forces promoting coal exploration and the dissemination of relevant knowledge across the Ottoman Empire. In 1845, for example, Mehmet Ali Pasha suggested sending Egyptian engineers to develop coal mines in the Tauris Mountains of southern Anatolia, including İçel, Alaiye, and Marmaris. If discoveries were made, a portion of the coal would be allocated to the needs of Egypt. The Pasha tried focusing on an Islamic sensibility which suggested that instead of paying high amounts for European coal, it would be better to circulate the money among Muslim countries. The sultan was not convinced.[19]

The Turkification of Arabized French geological writings had surprising consequences. Take, for example, Mehmet Ali Fethi, who adapted Fa'id's Arabization of Boubée's book; the resulting *Ilm-i tabakatu'l arz* was the first geology book in Ottoman Turkish.[20] It was printed in six hundred copies by the royal printing house in Istanbul in 1853. Like his Egyptian predecessors, Fethi became a translator unexpectedly and unintentionally after beginning to compose *El-Âsarü'l-aliyyetü fi hazaini'l-kütüb* (The noble relics in the treasury of manuscripts). This was a comprehensive catalogue of the books in the Istanbul libraries, a work whose title neatly complies with the aesthetic convention of rendering literary works into material treasures. The first volume (the only one he managed to complete) earned him a place at *Meclis-i Maarif* (Council of Education) and later at the *Encümen-i Daniş* (Privy Council), as well as 150 golden liras from Sultan Abdulmecit I. However, promotion to these prestigious boards did not come without a price. Fethi's "dear friend," probably a member of the Education Council, suggested that he take on translation work, and Fethi did so "to express his gratitude."

Little in Fethi's education had prepared him for translating a geology book, as he specialized in traditional Islamic texts and *tafsir*; however, his command of Arabic allowed him access to texts like those of Fa'id. During the years of translation, which stretched through most of the 1840s, the Ottoman Empire started breaking with its earlier policy of employing European mining experts in the Empire's main coal mines—

newly discovered during this decade—and began investing in building a local base of knowledge and expertise. Together with translation, Ottoman engineering student missions were sent to Europe, something the Egyptians and Persians had been doing for decades; the first mission aimed at training Ottoman mining engineers was launched in 1851 and dispatched to England and France.[21] These students had to demonstrate a sound command of Arabic, as did students seeking admission to the mining school that opened in the mid-1870s. Like them, and as a result of Egypt's pioneering role in introducing science to the empire, Fethi's knowledge of Arabic gave him access to this kind of scientific knowledge. Turkification of Arabized geology helped incorporate people like Fethi, members of Ottoman social strata rooted in traditional Islamic learning, in the new scientific circles.

These groups came with their own class politics, aesthetics, and ethics. We have seen how knowledge in general was understood as "treasure" as a matter of aesthetic convention, as neatly revealed in the title of Fethi's catalogue and in the titles it contained. (Catalogues like World-Cat today have over a thousand entries of Arabic books alone with the word *kanz* in their titles.) Yet in the field of geology this had concrete implications. The first pages of *Ilm-i tabakatu'l arz* (literally, the science of the earth strata) were dedicated to supportive statements written in Turkish, Arabic, and Persian by prominent Ottoman officials and *ulema*. These included a statement equating stratigraphy, as well as the book that rendered this science from Arabic into Turkish, to a gem left unnoticed under the earth until the sultan had exposed it. Another supporter congratulated the translator and the sultan who sanctioned the project for their contribution to the "gems and glories of learning." As in Egypt, here too the book itself was repeatedly likened to a treasure trove. Supporter Ahmed Cevdet Pasha, for example, waxed poetic about the blessed travel between textual lines and treasures of meaning.[22]

Moreover, again and again the endorsements anchored the book within a gradualist and stratified cosmology in which Allah, "who created heaven, earth, and what is under the earth," is followed by the Prophet Muhammad, "who divulges what is hidden"; his family and friends; and then by the sultan, who sanctioned the project, and the book's translator, who "disclosed underground treasure troves and mines and opened the closed layers of the earth to the scholars who are eager to explore them." Finally, several of the book's endorsers, including the Grand Visir Ali Pasha and Foreign Minister Fuad Pasha, defined the community of those benefiting from the wealth of Sultanic-

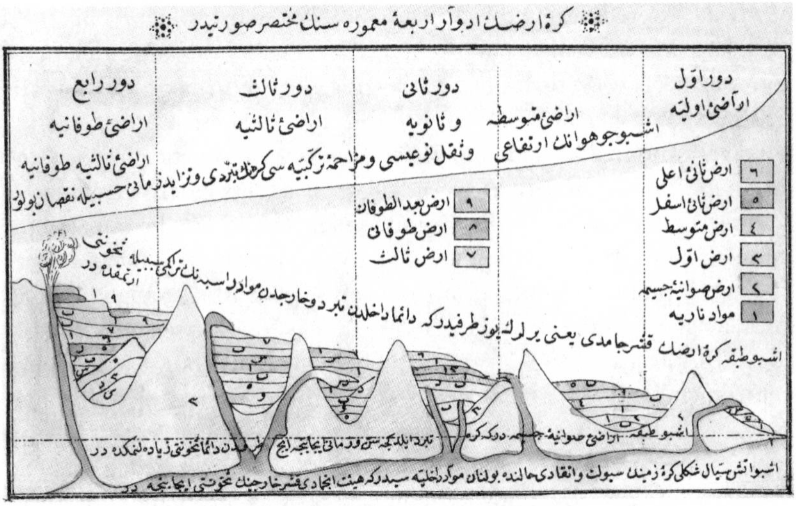

FIGURE 22. Ottomanized stratigraphy. *Ilm-i Tabakatu'l Arz.*

sanctioned patronage of geological knowledge as members of the three monotheistic religions in the empire. Indeed, according to Education Council member Subhi Beyefendi's endorsement, the true meaning of *tabakat* is like strata in the world's population, and he therefore praises God for creating men of different social, religious, and linguistic strata.[23]

The combination of scriptural and geological knowledge, especially as it pertained to human and earthly strata, was visualized by the book's geological maps (see figure 22). In drawing these, Boubée was inspired by the work of Élie de Beaumont, geologist and director of the École des mines. Even more than Barard and Boubée, de Beaumont was a key promoter of geological catastrophism, the belief that catastrophes were principal agents in the history of the earth. As in France, in early-nineteenth-century Britain, catastrophism was developed in order to reconcile geological science with religious traditions of the biblical Deluge. De Beaumont, who studied the geological map of Great Britain, fused together French and British catastrophism which, in turn, was Islamized by Fa'id and Fethi, whose maps speak of Delugian (*tawfani*) and post-Delugian (*ma ba'ad al-tawfan*) earth strata.

Especially in an Ottoman context, the notion of *tabakat,* or strata, was useful for mapping the layers of the earth onto the familiar protocol of knowledge accumulation by way of transmission between *tabakat* of scholars, often recorded in the genre of *kutub al-tabakat* (books of *tabakat*). Indeed, the passage of knowledge between specific members within cohorts of scholars was the definitive scheme for the paradigmatic Islamic "science," ʿilm al-hadith, arguably the prototype upon which all ʿilm was predicated. Unlike other Islamic rulers who gained legitimacy by claiming prophetic lineage (via what we call genealogy), the Ottomans depended on other channels, such as the system of intisap or the patron-protégé relationship, to cement social hierarchies, and this was a key avenue for the patronization of scientific activity in the empire.[24] How were these knowledge architectures mapped onto stratigraphy and treasure?

BURIED TREASURE

The notion of buried treasure with which Ottomans understood geology, as well as the ruler's role as patron of the earth sciences, had deep roots. In the biographies of the Prophet Muhammad (passed between generations of transmitters), it is stated that:

[ʿAbd al-Muttalib, the Prophet's paternal grandfather] continued digging until the top of the [Zamzam] well appeared to him. He praised God because he knew that he had been right [about its buried location]. When he continued digging he found in it two gazelles of gold. These were the two gazelles which the Jurhum had buried in the well when they left Mecca. He also found in it Qalʿi swords and armor . . . [He] made the swords into a door for the Kaʿabah, and pounded the two gazelles of gold into the door. This was the first ornamentation of the Kaʿabah. Then ʿAbd al-Muttalib began giving the water of Zamzam to the pilgrims.[25]

As Brannon Wheeler demonstrates, the contents of this treasure indicate that these were votive offerings made by pre-Islamic kings to the Kaʿaba and Zamzam. Moreover, various Islamic texts stress that the Prophet had Qalʿi swords (made from tin from a mine in Syria) similar to those found by his grandfather.

Many of these weapons came to be preserved in the Topkapi Museum in Istanbul and are mentioned as having been included in the treasuries of the Abbasid, Fatimid, Mamluk, and Ottoman dynasties. Such treasures connected these dynasts to their source of legitimation and to former custodians of the sanctuary at Mecca. The discovery and bestowal of such hidden treasures is closely associated with accumulating and passing on royal and prophetic authority.[26] Especially to rulers unable to claim a blood lineage to these sources of authority, like the Ottomans, such stratigraphic alternatives to genealogy were significant. During the nineteenth century, treasury items originally kept in chests were put on display by Sultan Abdulmecit (r. 1839–61). His successors Abdulaziz (r. 1861–76) and Abdulhamit II (r. 1876–1909) continued and expanded this practice.[27] More specifically, as Wheeler and others indicate, alongside guardianship of miscellaneous relics like the Prophet's hair or fingernails, custody of the underground realm positioned a ruler in a particular nexus between the holy cities Mecca, Medina, and Eden.[28]

Beyond offering legitimation by (under)grounding Islam in older religions and connecting a current sovereign to the authority of previous Islamic and pre-Islamic rulers, the notion of pre-Islamic buried treasures was developed into a special Shar'i tax category, *rikaz*, whereby a fifth rather than the standard 2½ percent of *zakat* was transferred to the state; this also came to incorporate minerals, including coal, which were seen as divine deposits. According to several *hadith*s, Allah's apostle said, "There is no *diya* for persons killed by animals or for the one who has been killed accidentally by falling into a well or for the one killed in a mine. And one-fifth of *rikaz* is to be given to the state."[29] The aforementioned category of *kanz* is also a particular case of the general category of *rikaz*.[30]

These stipulations found their way, with very minor variations, into article 107 of the 1858 Ottoman Land Law (*Arazi Kanunnamesi*).[31] Indeed, new ways of thinking about underground realms and the minerals deposited there informed conceptualizations of the topsoil.[32] Land (*arazi*) itself emerged as a source of private property during the 1850s, just as geology was reconceptualized from the science of stones (*ilm-i Cemadat*)[33] to the science of earth (*ilm-ul arz ve'l maadin*).[34] In some respects, the reforms (*tanzimat*) of that decade were attempts to respond to a painful process of territorial contraction. In the past, conquest and distribution of captured territory constituted one of the main engines of Ottoman dynastic legitimacy and sovereignty. This avenue was closed before the age of coal, when European empires started dominating battlefields. The emergence of new notions of geology, the activation of Ottoman coal mining during the 1840s, as well as tensions around land reclamation near ports (discussed in chapter 4) can be situated within a growing repertoire of different ways of creating new land through draining marshes and shifting populations, again often following the Egyptians' cue.[35] With horizontal expansion through warfare foreclosed, Istanbul was focusing on existing resources and on verticality for situating sovereignty above and below ground. By the end of this process, *arziyat* was established as the new term for geology, as *arz* (land) became a geological term and *arazi* a new term for property.

From around the same time, revenues from underground mineral repositories have been regularly channeled to causes that address a communal subject, often in places far removed from the mines, sometimes serving the entire *ummah* and particularly travelers (most notably pilgrims) and the poor. Donald Quataert has shown that during the last third of the nineteenth century, Ottoman coal mines were connected to Islamic charity. Coal revenue was used to support accident victims (a connection implicitly suggested already in the *hadith* above), mosque maintenance, educational sessions and institutions, support of *ulema* and Quran reciters during festivals, and various other activities.[36]

This connection between fossil revenue and charity was also forged by the institution of the *hazine-i hassa,* which was long associated with the ruler's generosity in the Ottoman Empire. The sultan's *hazine-i hassa* (often translated as "privy purse," but more accurately rendered into "privy treasury") was the direct administrator and main benefactor of mineral deposits in Ereğli and elsewhere in the empire,[37] even during periods of administration by the Ottoman naval ministry or foreign companies, from the establishment of Ottoman coal mining until the demise of the empire. Defining coal and other minerals as treasures hidden deep in the ground, which were revealed under the auspices of a benevolent sultan with specialized knowledge, was befitting for their placement in this inner treasury.

From the mid-seventeenth century, the Ottoman state and state bureaucracy emerged as a domain separate and independent from the ruler, and vice versa, by contrasting the ruler's *hazine-i hassa* and the "public coffers" (*beytulmal*); at other times, the former treasury was contrasted with the *hazine-i devlet-i aliye* or *hazine-i amire,* the "imperial treasury," associated with the state's regular budget and the scribal know-how required to manage it. These were often couched in such terms as the inner (*hazine-i enderun*) and outer treasuries, respectively.[38] Through public mass gift-giving ceremonies and special bonuses paid from the imperial coffers to deserving officials, Ottoman sultans—especially Abdulhamit II—used the privy treasury to forge personal loyalty to themselves rather than to the state.[39]

The connection of coal, treasure, and divine deposit was not an Ottoman invention; it was common everywhere from China to Europe.[40] British and French geologists often considered coal and other minerals as a gift from nature, or from God, to mankind.[41] For example, English geologist William Buckland, who like de Beaumont sought to reconcile scriptural accounts about the flood with geological finds, saw providence in the depth of England's coal mines: "we may fairly assume that . . . an ulterior prospective view to the future uses of Man, formed part of the design."[42] Christian notions of hidden treasure are rooted in Roman law about the treasure trove,[43] and in this respect Robert Hooke's above-mentioned analogy between the spread of fossils and Roman coins was right on the money. For example, in the parable of the hidden treasure (Gospel of Matthew, chapter 13), Jesus likens the Kingdom of Heaven to a trove located in an open field. According to Roman law, a finder might collect funds to purchase the field and claim the treasure found there. Peter Brown has shown that during the fourth century, the Christian gift to churches connected heaven and earth to one another, as well as assuring many divine returns on monetary gifts in this world, especially to the poor.[44]

When geology was overtaken by thermodynamics in the second part of the nineteenth century, the Christian metaphysical attitude was retained and even emboldened. Seen by the Scottish Presbyterian physicists as a "gift of grace," fossil fuels were framed as an obligation: "once the gift of grace had been accepted, man had a moral duty to direct, and not to waste, the natural gifts."[45] But it was also modified significantly and pushed toward the abstract.[46] Waste, for example, was increasingly seen through the lens of time-thrift, connecting the deep time of fossils to the quotidian work-time of humans (now increasingly seen as "human-motors"). Such ideas about the value of time (Arabic: *qimat al-waqt,* the claim that "time is a fleeting treasure") reached the Ottoman Empire and were eventually debated there, and elsewhere in the Islamic world, by the century's end. But they were inflected by existing interpretive frameworks, as well as by the global geopolitical balance of power and the colonial condition.[47] Islamic reformers in India were taking issue with Christian ideas such as the assumption that financing a madrasa in this world would cause the building of a pearl palace in heaven.[48] Similarly, in Egypt, founder of the Muslim Brotherhood Hassan al-Banna rejected the equation of time and money or gold, insisting that the former cannot and should not be reduced to the latter.[49]

Yet transformations in Europe, namely the inclusion of supposedly lowly substances like coal in the trove of divine buried treasure, nevertheless had a significant impact, albeit an inflected one, in the Ottoman Empire. At first, coal in these territories was almost antithetical to treasure: the first coal repository in Ottoman lands was discovered accidentally in Bosnia in 1731 when Comte de Bonneval (Humbaracı Ahmet Paşa) was digging there for Balkan gold. The disappointed workers soon abandoned the mine.[50] A century later, the discovery of a larger repository, in Ereğli in 1840, was reported in the first issue of *Ceride-i Havadis,* one of the earliest newspapers of the Empire, as a discovery of "burning stones," known to be located in many parts of the Empire but not used before their application to steam navigation had been realized.[51] The story of this discovery was later reframed as one of stumbling upon "black stones" previously deemed useless.[52] In 1841, just after this important discovery, Granville Withers, who owned a coal mine in Belgium and was involved in Ottoman mining, complained that Ottoman ignorance about coal mining would surely keep "this important treasure in a state of unproductiveness."[53]

Yet during the following decades, coal became less of a stone (just as geology was no longer the science of stones) and more of a treasure, until the two could finally be fused together, sometimes literally. For example, due to deforestation and the shift to scientific forestry in the 1840s, the *Darphane-i Amire* (Ottoman coin factory), which had previously relied on charcoal for smelting, started using mineral coal. Alongside a growing interest in European fossil knowledge, acquired through scientific translation, the Ottomans also sought to attract foreign awareness to, and investment in, their domestic mining sector, for example by exhibiting minerals and nonagricultural raw materials in international exhibitions during the 1850s and 1860s.[54] Such a comparative disposition was key to the inclusion of coal in the trove of Ottoman buried treasure, and in the *hazine-i hassa.* Consider an 1889 report written by an officer from the technical commission of the Ministry of Public Works:

> General wealth (*servet-i umumiye*) depends, on the one hand, on the cultivation of fertile lands with modern methods and, [on the other hand, on] the exploitation of the various mines which can be seen as "buried treasures" (*hazine-i medfûne*). If they did not have coal and iron mines in their countries, English, French, and German people who are considered the most developed nations could not reach this high degree even if they progressed in civilization. The progress of humankind is a result of industry rather than agriculture, and mining constitutes the backbone of the industry. Considering the miracles of coal and iron, mines are more valuable than things like copper, silver, gold etc. and will obviously provide more benefits. There is no need to prove what a great wealth will emerge in the imperial lands where every corner can be considered as a natural treasure because of the abundance of mineral resources.[55]

During the Hamidian period, foreign attention to the local mining sector was turning into unwanted intrusion and exploitation. Eventually, it prompted a defensive policy of including new mines and revenues from existing ones under the umbrella of the *hazine-i hassa*.[56]

If the comparative aspect of Ottoman and European mining sectors is important, contrasts are also significant. As we will see in the next section, in western Europe and North America, coal was gradually monetized and secularized through the Presbyterian/thermodynamic mediation of "energy." While from the 1840s energy was predicated on the interchangeability of work, heat, and motion, the former aspect—a system's ability to perform work—became the central platform for developing thermodynamics and for putting it into global action.[57] Alongside the centrality of work, abolition campaigns during this period further entrenched this articulation of energy in thermodynamic terms of waged work, seen through the notion of time-is-money. Yet like monetized time, thermodynamics, and energy, abolition was slower to arrive at the Ottoman Empire, and when it did, it encountered traditions of bondage that were hard to reconcile with Atlantic slavery, or with quantifying freedom with time and money. For example, many Ottoman slaves were treasured and often even named after precious stones and metals (Lu'lu, Mirjan, Almasi, Firoz, etc.).[58] Their attachment to present and former masters, as well as their manumission, did not always conform to market models that helped commodify and quantify work in western Europe and North America. Such particularities focused attention on the materiality and specificity of treasures, including fossil fuels, hampering their abstraction. The sections that follow examine these existing legacies and some Ottoman inflections of the statistical and thermodynamic abstractions that fossil knowledge underwent.

INTO THE META/PHYSICAL LIMBO

The approximate mid-century fault line that separated geological catastrophism and uniformitarianism also represented other scientific transformations, including the emergence of thermodynamics and the not-unrelated increasing impingement of quantitative models on domains of natural history previously based on fieldwork. We have already seen the burgeoning of geostatistics, which similarly symptomatized a larger innovation—the mathematization of subterranean cartography. Maps of the subsoil, like the one that inspired de Beaumont and then Boubée et al., were pioneered in Europe by the British coal-mining inspector William Smith in 1815. Smith used the sequence of fossils, which he took as a marker of geological strata, to predict the presence of coal in a given subsoil. The maps he and his successors created constructed an image of what was underground as invisible yet predictable, a realm composed of continuous mineral strata. Although, based on the partial view they had, mine owners feared the depletion of coal and were reluctant to invest in seeking out new mines, geological maps boosted the mining sector by enabling less-risky investments and conducting probes in favorable areas.

Geological maps were optimistic instruments, supporting "statistically probable reserves" and "potential discoveries" in England and quickly also elsewhere.[59] They replaced the empiricism of observable patchy facts-on-the-ground with an unbroken and reassuring image in the British Isles, western Europe, and further afield too. George Bellas Greenough, whose 1820 geological map (parts of which were plagiarized from Smith's) had inspired de Beaumont, went on to produce the 1854 *General Sketch of the Physical and Geological Features of British India,* a country he never visited. De Beaumont himself took the geological map to France, and its optimism later traveled in translation to Cairo and Istanbul, as we have seen. This globalization of geological prospecting in the second half of the century further amplified the optimism about the abundance of coal worldwide.[60]

William Stanley Jevons's *The Coal Question* (1865) targeted precisely this geological optimism animated by probabilistic and global perspectives. Applying mathematical calculations that were more rigorous than the geologists' statistics, he predicted the foreseeable depletion of British coal. Jevons's new statistical quantitative methods represented a connection between neoclassical economics and thermodynamics, thus creating new terrain for the impingement of abstract mathematical physics on geology. (This happened alongside direct inspiration from physicists like William Thompson, Lord Kelvin, whose 1860s calculations of the age of the earth were adopted by geologists.)

Jevons was also among the first to draw attention to the importance of the global context for the British coal economy. In a British trading system that was focused on carrying homebound cargoes, coal was a readily available item that at the very least paid a shipowner's expenses and was increasingly finding customers around the world. Shipping coal away from the British Isles to places like the Ottoman Empire was crucial for keeping the empire's mines open and factories working. For this reason, coal from Britain's "daughters" (i.e., its formal, semi-, and ex-colonies) would do the metropole no good.

Margaret Schabas, who recounted the transformation of economics from a natural to a social science and traced the leap of its adherence from physical nature to human agency, identifies *The Coal Question* as "the point at which economics turned away from natural resources." Putting land, labor, and capital on a level playing field as perfect substitutes—a trick he borrowed from the first law of thermodynamics, which does the same for heat, labor, and motion as commensurable manifestations of "energy"—Jevons ushered in the neoclassical moment. Coal energy allowed him to depart from David Ricardo's classical economics, from a conception of capital as personal wealth, and from the latter's attempt to ground value in the price of gold.[61]

In 1864, just as Jevons was completing the book in which he used statistics to articulate British anxieties about what today we would call "peak coal," on the other side of the channel another pioneer of statistical representation gave shape to French anxieties about contemporary British coalonialism. Charles Joseph Minard's "thematic cartography" of British coal exports (figure 23), an expansion of a map he created in 1850, sought to capture the trajectory of the decade's fossil economy (represented by the lines' direction) and its volume (represented by their width). Trying to depict the flow of coal and accurately represent its trajectories, Minard was forced to make some hard choices. Lest the maritime routes of coal export be mistaken for overland passage, he was compelled to significantly expand the Straits of Gibraltar in the earlier 1850 map. The 1864 map depicted much-increased coal production and trade—an upsurge of nearly one hundred percent in less than a decade and a half—resulting in a further widening of the straits. The naïve mimetic realism of familiar two-dimensional maps was abandoned in favor of what we might call "statistical realism."

Examining Minard's 1864 work in 1878, French thematic cartographer, scientist, and pioneer of photography Étienne-Jules Marey defined the resulting shape of British coal export as a *poulpe immense,* a giant octopus/squid.[62] This was not an unfamiliar depiction of the British fossil economy. In the early 1840s, naturalist William Gourlie claimed that coal deposits "enabled [Glasgow] to stretch a hundred arms to the most distant corners of the earth."[63] After Minard, this image became all the more appropriate to both imperialists and their critics in Europe and its colonies. After the 1882 British occupation of Egypt, for example, an American cartoonist depicted John Bull as "The Devilfish in Egyptian Waters" (figure 24). Choosing to include the major coaling depots of Malta and Gibraltar in the map (where they are represented as dispro-

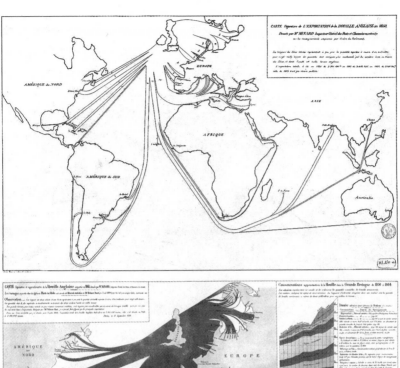

FIGURE 23. Charles Joseph Minard's maps of British coal exports, 1850/1864.

portionately large) makes it clear that coal was still sustaining this monster, just as it was extending its final tentacle towards the Suez Canal
and the coaling station of Port Said. Similarly, in the introduction to his
1899 *Ta'rikh al-afghan* (History of the Afghans), anti-imperialist
Islamic reformer Jamal al-Din al-Afghani, the intellectual grandfather
on whose shoulders the aforementioned *manarists* stand, who had
ample opportunities to see the squid metaphor in Europe and elsewhere,

FIGURE 24. John Bull as "The Devilfish in Egyptian Waters." Granger Collection—All Rights Reserved.

claimed that the British Empire was "a monster squid which had swallowed twenty million people, and drunk up the waters of the Ganges and the Indus, but was still insatiable and ready to devour the rest of the world and to consume the waters of the Nile and Oxus."[64]

Minard's carbon squid resulted from its maker's attempt "to convey promptly to the eye the relation not given quickly by numbers requiring mental calculation"[65] or, in other words, to make the unseen visible. It reared its uncanny head in a particularly fraught moment, and its appearance was one occurrence among many similar metaphysical encounters. During the nineteenth century, the familiar physical world was expanded beyond recognition on many fronts and in numerous ways with a totally disorienting effect. It is worth remembering that

FIGURE 25. "Monster Soup," William Heath, 1828. Credit: Trustees of the British Museum.

"objectivity" is itself historical[66] and that the physical and metaphysical are never stable, clearly separable domains. They are better seen as entwined, moving targets animating one another. Especially in periods of accelerated scientific advancement, yesterday's metaphysics becomes today's uncontested ontological and physical reality, and vice versa. In this respect, Minard's map represented an intersection of several significant processes. These included scientific developments, like the emergence of statistics and the challenges of "statistical realism," which contrasted the contiguous underground of geological maps to the patchy one of fieldwork. Similar processes occurred in the aquatic realm, in scales ranging from the planetary to the microscopic, including submarine cartography and the deployment of gigantic underwater intercontinental telegraph cables—perceived as having shrunk the globe, creeping in the deep "where the blind white sea-snakes are"[67]—and explications of drinking water statistically revealed to be plagued by cholera and microscopically exposed as "monster soup" (see figure 25).

Monsters seemed to be creeping out of every crack in this broken reality, or from the porous divide between the meta and the physical. The sciences most closely related to fossils, geology, organic chemistry, and thermodynamics—among the most dynamic scientific domains of the period—were undergoing an especially heightened encounter with the monstrous. For geology, it was the "monsters of deep time."[68] In the 1840s, Justus von Liebig, the father of organic chemistry, compared Britain (the major importer of mineral fertilizer) to a vampire or blood-sucking squid: "Great Britain seizes from other countries their conditions of their own fertility . . . Vampire-like, it clings to the throat of Europe, one could even say of the whole world, sucking its best blood."[69] For thermodynamics the monstrous usually appeared in the form of Arabian genii—the most familiar articulation of slavish supernatural "energy" (equated with a system's ability to perform mechanical work)—alongside spirits like "Maxwell's demon."[70]

As these ideas—in particular, thermodynamics' focus on work—entered the social sciences, the process further enlivened the monsters. The best example is probably Marxism, rife as it was with ghostly aspects, romantic and gothic connotations, and theological ploys.[71] Perhaps the most famous among these was the definition of capital as "dead labour which, vampire-like, lives only by sucking living labour, and lives more, the more labour it sucks." It is often the case that monsters are the harbingers of "category crisis," in the words of Jeffrey Cohen, as they are forms suspended between other forms. Monsters are often called forth to de-*monstrate* the invisible, as the root *monstrare* (to show forth), *monstra* (to warn or show), or *monstrum* (that which reveals) all indicate.[72]

Clearly, then, the Ottomans were not alone in ascribing incorporeal qualities to the underground and its fossils. But as al-Afghani's appropriation of Minard's octopus suggests, when new epistemic values spread, they encountered existing formulae and forces which helped people make sense of them (as we have also seen repeatedly in each of the previous chapters). Consider what happened to thermodynamics' Arabian genii when they traveled to the Islamic world where the jinn were already bound to existing understandings. In a *fatwa* by al-Afghani's disciple, Muhammad Rashid Rida, prompted by an inquiry about the existence of the jinn, he argued for the insufficiency of existing positivism: "If we rejected everything we cannot see or perceive with the senses, we wouldn't have been propelled to discover great mysteries such as electricity, whose effects we can sense and which has been attributed to the jinn." In yet another *fatwa* concerning the existence of the jinn, Rida claimed that "we regard as a type of jinn these tiny beings that cannot be seen without a microscope, and the expression '*janna*' [to hide, conceal] supports it. The hadith according to which the plague is caused by the sting of the jinn offers further support." As with electricity, venturing with a microscope to see beyond the purview of the naked eye ties modern science to the world of jinn. And if receptiveness to what is beyond the empirical helped break new ground in the sciences, modern science in turn sustained the supernatural, helping to separate real jinn from spurious ones: "If we were told before the discovery of microbes that tuberculosis and the plague and other maladies are caused by small physical entities capable of quick reproduction inside the body, we would have regarded this claim as superstitious or imaginary."[73] Electromagnetics or microbiology did not disenchant the world. Rather, the opposite was often the case.

The same applies to energy itself. In 1870, Butrus al-Bustani, one of the key importers of European scientific and technical terms into Arabic, compiled a large Arabic dictionary, *The Ocean of Oceans* (*Muhit al-muhit*). He defined *taqah,* the word that would become the proper Arabic synonym for *energy,* as "the capacity to perform something; and it is said that it is the name for the human ability to do something with difficulty. This is a simile to the band [*tawq*] that encircles the thing in question." According to al-Bustani, himself a Syrian Christian, this definition draws on a reading of a verse from the Quran, "Load not on us that which we have not the capacity (*taqah*) to encompass."[74] As with the Presbyterian version, this model for energy is monotheistic and can be articulated in terms of human work, but differently so, as the totality of the force in question is manifest from the outside in, rather than from the inside out. The specific vortexes through which energy entered Arabic and then Turkish obviously reconfigured it in the process.

This mutual interanimation and contextualization of science with other metaphysics should be inflected in the plural. Thus, to contextualize al-Afghani's statement, we must realize that as Minard's octopus passed the Straits of Gibraltar heading east, it may have morphed into *Tinnin,* the dragon-like sea monster, creator of terrifying storms, who resides near the Straits in the tumultuous *majmaʿ al-bahrayin* (the meeting of the two seas).[75] It might have also been informed by Yajuj and Majuj, mentioned in Suras 18 and 21 of the Quran: as a famous hadith explains, every night these creatures try to dig under a wall built to imprison them underground, only to discover that the wall has been fortified by God. When Allah decides to let the wall collapse and free Yajuj and Majuj, they will appear in the northeast of the ancient world as portents of the end, then proceed south to drink up the waters of the Tigris and Euphrates.[76] In what follows, I explain some of the implications of these inflections of unseen subterranean and submarine domains that animated Ottoman fossils and fuels.

(UNDER)GROUNDING PAN-ISLAM

The institutional and epistemic connections between fossil fuels and Islamic charity mentioned above had various consequences, both above and below ground. During the first decade of the twentieth century, the Ereğli Company operating the main Ottoman coal mines supported *Darulaceze* (the poorhouse in Istanbul) with a free yearly supply of coal, also paying for its transportation.[77] The Ottomans used coal and steam power to promote other charities as well.[78] In 1908, a committee of mine owners and local administrators was formed in the Zonguldak region to devise guidelines for establishing facilities for mining-sector workers, specifically a hospital, a bazaar where goods could be sold at cost price, mosques and public baths around the mines, and a technical (mining) school. The committee also sought to create disability pensions and benefits for dependents. One piaster from every ton of coal extracted in the region was allocated to finance these endeavors.[79]

A commitment to charity was also prevalent among lower-ranking mine officials. In 1909, for example, the newly appointed supervisor of the coal workers around the Ereğli and Zonguldak mines officially committed 25 percent of his revenue to the Islamic school in Ereğli and 25 percent to Darulaceze.[80] After the hard winter of 1911, government-controlled mines and forests provided free firewood, charcoal, and lignite to the poor; these fuels were exempt from forestry and mining taxes between March and mid-May.[81]

Carbonized minerals and related support for the poor were allocated not only to domestic heating[82] or, in the case of Darulaceze, mass laundry services, but also for facilitating transport to Arabia by land and sea, for the hajj and other religious occasions. The allocation of *rikaz* monies to support wayfarers was well-known, but at the beginning of the twentieth century, it started to be conducted differently. On his birthday in 1900, Sultan Abdulhamit II announced one of the grandest infrastructural projects in Ottoman history, the Hijaz Railway, which he described as a "sacred line" and which others called "Pan-Islamic." Beyond any religious aspects, this was a reaction to the shift of the hajj to marine routes dominated by European powers, and the economic and political implications of this transition. In a sense, it was a response in kind to British coalonialism, one which used the terrestrial rather than naval infrastructure of steam. As

FIGURE 26. Groundbreaking, *Servet-i Funun* (the treasure of knowledge), 1902.

we have seen in chapter 4, the Ottoman "infrastructural turn" was a thing of its time. Yet it was also a thing of its place, and the particular articulations of "the technological sublime" of the railway and other grand infrastructural projects in Islamic and Ottoman parlance gave it an independent life. First, the terrestrial and subterrestrial aspects of the railway were repeatedly stressed, for example in a series of ground-breaking ceremonies-turned-photo-ops published in the scientific and literary periodical *Servet-i Funun* ("the treasure of knowledge," see figure 26). In 1902, a special issue commemorating the sultan's sixtieth birthday featured the railway project on the cover of the magazine, thus making it part of a metaphorical treasure trove as well as a newsworthy item. The cover itself was designed as an *iwan* (readers will recall the description of Tahtawi's travelogue) through which the reader as well as grateful local tribesmen regard the train in a one-point perspective projection (figure 27). Whereas in Europe the popular scientific press was a driving force for secularism, in the Ottoman Empire it further cemented religion to techno-science.

The railway project gained the support of key "Pan-Islamists" like Rashid Rida, who in 1903 promoted it in *al-Manar*, another famous vehicle of transnational Islam whose connection to coalonial infrastructures we discussed in chapter 4.[83] Yet in order to situate this endeavor in its proper Ottoman context, we need to construct a pan-sphere that is not reduced to the homogenizing realm of the two-dimensional map. The notion of buried treasure proves helpful. In 1900, Muḥammad ʿArif al-Dimashqi, another Syrian-born Arab Pan-Islamist, published a short book (or long pamphlet) listing arguments in favor of the project,

FIGURE 27. The Hijaz railway, *Servet-i Funun*, 1902.

in which he claimed that "the railway would facilitate the search for minerals in these deserts and wastelands, which have remained unexplored until now, their treasures [*kunuz*] hidden . . . consequently, hidden coal and minerals would be discovered, and the secrets entrusted there would be unveiled."[84]

The railway's deployment was financed by Muslims worldwide. Much of this international funding came from India, where a central committee for the Hijaz Railway was founded in Hyderabad.[85] As we have seen in chapter 4, this was a major junction where Ottoman and British extraterritoriality intersected in the subcontinent. Indian contributions were solicited by stressing the personal example and personal contributions of sultan Abdulhamit,[86] thus linking the project to the

privy treasure. A complementary source of funding was official conces-
sions that stipulated that coal found on or near the planned route would
be put at the disposal of the Hijaz Railway Authority. In 1903, the sultan gave the coal and iron mines discovered in Seferihisar in the district
of Aydın to the Hijaz Railway Company, to operate directly or lease to
another company, and exempted equipment used there from all duties.[87]
A year later, he gave coal mines discovered at the Kalaycı village of Kas-
tamonu and the coal and iron mines in Payas and Adana to the Hedjaz
Railway Company under similar conditions.[88]

After its inauguration, the railway indeed combined the two aspects
of stimulating new underground excavations and being stimulated and
fueled by those which had already been undertaken. In both senses, the
railway's materiality was charged with spiritual and political meanings
connecting it to hidden realms, and helping it animate visible ones. For
example, in 1914, the Hijaz Railway Authority used resources derived
from coal mines to provide reduced-price tickets for trains and steamers
ahead of ceremonies commemorating the martyrdom of Hamza, the
Prophet's uncle, in Medina during the month of Rajab and commemo-
ration of the Prophet's night journey at the end of the month. Special
train and steamer schedules were created for these occasions and circulated in Istanbul.[89] Or, consider a letter from the Minister of Vaqfs to
the Grandvizirate stating:

> It is necessary to maintain transportation activities in the Hijaz Railway.
> Since railways depend on coal, it is advised that required fuel would be pro-
> vided from the coal mines around the line (if such coal exists) which have not
> been granted to private individuals. This would only be fair as the Hijaz Rail-
> way has been the main driving force behind the development of the region.
> The Sultan's order states that coal mines in several locations of Cebel-i Lub-
> nan, in two locations close to the Trablusşam station, and in Jabal as-Shaikh
> of Syria, and mines that will be discovered in the future in the 50 kilometer
> corridor of the line (25 km each side) will be granted to the Hijaz Railway.[90]

This letter was written during World War I. By the end of the war, Ottoman soldiers were withdrawing from the various destinations of the
railway, often retreating along its track. In some of these places rumors
and stories still circulate about gold buried under the railroad, and people have been known to experience dream visitations instructing them
where to dig to retrieve the buried treasure.

Another lasting set of legacies of such buried treasure facilitates a
deeper understanding of the current age of oil and the contemporary
Middle East. In recent years, several historians have been documenting

the Ottoman prehistories of Saudi policies.[91] The twentieth-century Saudi "turn to geology," as Toby Craig Jones calls it[92]—and especially the sedentarization and control of Bedouin tribes—seems like another Ottoman legacy from this period, traveling with the Hijaz Railway and by other means (recall the visual depiction of those appreciative local tribesmen on the cover of *Servet-i Funun*). That geology was included in the toolkit for centralization and authoritarianism, supporting a civilizing mission that helped translate the subterranean sphere into a source of wealth for the sovereign, is clearly a coalonial legacy.

Such legacies also seem to animate existing Saudi legal frameworks for resource ownership. Article 14 of the Kingdom's Basic Law states, "All natural resources that God has deposited underground . . . shall be the property of the State." As Bernard Haykel has shown, the Saudi code of mineral resources (*nizam al-taʿdin al-Saʿudi*), and more generally the rules stipulating the ownership and handling of minerals in the kingdom, diverge from the otherwise hegemonic Hanbali school, according to which mineral resources are the property of the owner of the land where they are found.[93]

Coal animates the prehistory of oil well beyond the post-Ottoman world. The famous 1904 representation of big oil and the monopolizing Standard Oil corporation as an octopus in the US satirical magazine *Puck* (figure 28) draws on conventions established by Minard, just as petroleum assumed a commodity form which mineral coal tailored. However, if we are indeed still trapped in coal's amber, existing forces are refracted differently through this supposedly transparent medium. As we have seen in chapter 2, Saudi hydro politics and especially oil-fired water desalination policies are anchored in the age of coal and the British-Ottoman imperial struggle. The same is true also for significant attitudes toward the underground and the liquid fuels buried there. Even when these legacies are Islamic or religious, they are not the dictates of a fossilized tradition, but a result of a system in which the fossil was a dynamic medium between science and religion, nature and scripture. Especially as far as the Middle East is concerned, focusing on oil through a supposedly generic and universalistic understanding of this substance runs the risk of ignoring the legacies of coal and the various nonenergy and nonfuel domains—religious, political, and dietary, among others—that this older hydrocarbon created.

Beyond tracing how our own world of oil stands on the shoulders, or rather rides on the tentacles, of that of coal and the nineteenth century,

FIGURE 28. 1904 cartoon suggesting another dimension where coal was an inspiration to understanding the oil industry. *Puck* 56, no. 1436, September 7, 1904.

there might be still broader lessons to be drawn from the stories recounted above, and from the Ottoman engagement with geology and fossil fuels. The textual genealogy excavated here, tracing the work of Boubée, Fa'id, and Fathi, was after all only the first word on Egyptian and Ottoman geology. Reprints, other translations, popular science articles, and gradually also original geology books by Arab and Turkish authors kept readers in the Ottoman and post-Ottoman worlds up to speed with scientific developments, including the emergence of uniformitarianism into a hegemonic role.[94] Opposed to notions of development by catastrophic rupture, uniformitarianism, the cult of continuity that stated "nature does not make leaps," reigned supreme during most of the twentieth century, animated by the statistics of "business as usual." Yet as Amitav Ghosh has recently put it, "nature does certainly jump," a fact supported by the geological record of rapid mass extinctions, now accepted by the scientific community. Indeed, today a synthetic approach of uniformitarianism and catastrophism is all the rage.[95] The notion of the Anthropocene is one of the key symptoms of catastrophism's return to vogue. While at present there are obviously hardly any fossils to mark the supposed watershed between the Holocene and Anthropocene, we nevertheless resort to the language of geology and deep time to signify our current predicament. There are surely advantages in dramatizing climate change by mobilizing this science. However, given geology's history as recounted above, and especially the encroachment of thermodynamics on its fact-setting protocols, the long reign of uniformitarianism in its explanatory schemes, its use of fossils as diagnostic for fuel, and the ways subterranean resources were translated into capital and a-moralized in this tradition, by drawing on geology we might inadvertently push the current debate to the very netherworlds we are trying to escape.

Elaborating on Dipesh Chakrabarty's discussion of the limitations of human history in coming to terms with deep time,[96] Ghosh links geological uniformitarianism, the parallel rise of statistics, and the development of protocols of exclusion from the domain of "serious literature" to explain how our imagination has become fixated on a "deranged" track that prevents a serious grappling with the challenge of climate change. In this sense, the belated and modulated arrival into Ottoman lands of certainties that dominated the twentieth century might be insightful today, when they have been subjected to new forms of questioning. These old genres of narration and coalition-building between science and religion, and between people of different generations, classes, and faiths enrich our impoverished contemporary literary protocols. If global warming is tied to a particular secularized Presbyterian trajectory that separated economic growth (via a thermodynamic work ethic) from its environmental consequences, past points of resistance to, and rearticulations of, the sleights of hand that made this split possible are worth excavating. Seen in this light, strata of men are inseparable from layers of the earth, and fossil fuels, treasure, and ethics are similarly intertwined. There is nothing here that provides a quick fix to contemporary problems. But like "the Anthropocene," other idioms that connect past, present, and future generations to one another and to the world, and especially idioms that resonate with people outside western Europe and North America, are worth amplifying nonetheless. Notably, the particular Ottoman idiom this chapter explored was the making of humanists who reluctantly and serendipitously stumbled upon translation work and on the natural sciences. Continuing this legacy might help infuse the categories of these sciences with the ethics and politics they ignore.

Conclusion

The transnational and transmaritime spaces and circulations explored in *Powering Empire* undermine divides between different empires, nation-states, and ostensibly discrete driving forces. Moreover, the book opens up a narrative form and an analytical language that similarly resist conceptual straitjackets by connecting anew "old" (Marxian) and "new" (Latourian) materialisms, and by uniting Western and non-Western humanities to reforge the link between materialism and humanities. It is now time to expose more fully the roots of these connective thrusts, their origins in the insights of my source material, and their possible implications for the present. As this book reentangles energy and empire, let me begin with the implications of entanglement itself. As the book demonstrates, the corridor leading to the present is wider, more multifaceted, and much more convoluted than we suspected. Submerged layers, unexpected twists and turns, and repressed potentials offer renewed resources for thinking about the present day, especially for those seeking a global politics and notions of justice that should resonate widely in a rapidly warming interconnected world. Indeed, what can this road map of global carbonization offer the project of decarbonization?

EMBRACING ENTANGLEMENT

Circumscribed epistemological and physical terrains are congruent with the cult of energy regimes and energy transitions, the handmaiden of a modernist, technophilic discourse of what are assumed to be clean breaks. In fact, in the name of decarbonization, this perspective leads us in the opposite direction. Such solutions take the form of deus ex machina and increasingly even of deus *as* machina, and entail a belief in expedient fixes rather than painstaking political work. Rather than consulting, educating, agitating, connecting, and mobilizing global publics, we put our hopes in technology, and increasingly also in geoengineering, to mitigate greenhouse gas emissions and capture carbon. In so doing, we are following the lessons the standard history of energy-as-progress teaches us, neglecting other maps of the historical trajectory that led us to where we are now. As long as there are borders, there is also the hope of externality, of a way out. Yet the planetary system is vastly complex and interdependent, constituting an object that requires dense types of description and analysis. An interconnected earth system calls for a historiography that would locate entanglements in the past and a humanities that would condition us to handle and navigate complexities in the present and future. Also, as I demonstrate, solid borders were not easy to detect either in historical time or in space. As far as chronology is concerned, embracing temporal entanglement means acknowledging the importance of historical continuity. First, even if we choose the troubled optics of "energy transition," this book demonstrates that we must qualify assumptions about incremental, unidirectional progress. Moreover, alongside the obvious ruptures introduced by coal-as-energy, from an environmental perspective continuities were no less important and no less injurious: coal invigorated biomass rather than replacing it. Moreover, it was not replaced by oil; rather, oil extended coal, slipping comfortably into its imperial boots on the ground. Spatially speaking, this book shows that imperial projects blended into and fueled one another. Similarly, existing driving forces were neither abandoned nor even simply repurposed. More often than not, reliance on these ostensibly obsolete powers actually increased. In total, there were more people and animal muscle, and more hydropower, after the arrival of coal than before. Remarkably, coal extraction itself grew tremendously in "the age of oil." Our blindness to these facts and to their environmental consequences is tied to the epistemological habit of considering these powers to be part of the same energy pie, under the commensurability created by thermodynamics, and to adopting a relative and statistical ordering of their impacts.

Powering Empire therefore offers to bolster the position of decarbonization via reverse-growth, deceleration, and even by turning our gaze to the past to modes of being and engagement, to passivity rather than knee-jerk activity and activism, and to staying with the trouble. All these need to be explored, defended, articulated, developed, historicized, and democratized. As Walter Benjamin put it, "Marx says that revolutions are the locomotive of world history. But perhaps it is quite otherwise. Perhaps revolutions are an attempt by the passengers on the train— namely, the human race—to activate the emergency brake."[1]

Like Marx several decades earlier, Benjamin was already trapped in the carbonized discursive amber of locomotives and species. And so are we today. The process of breaking away therefore must start from within, from taking stock and using the tools at our disposal. Deceleration will obviously be neither easy nor enough, and the dialectics of acceleration-deceleration is indeed quite limited. However, embarking on this process might buy us the time and clarity necessary to chart a new course and discover hidden intersections and turns.

Focusing on entanglement should not erase difference or remove power from the picture. There are various ways of marking variance while acknowledging interconnectedness, even within the traditions of thermodynamics and interfacing evolutionary science that animate and structure our story. As early as in *On the Origin of Species,* Darwin defined homology as the existence of a common ancestor shared between a pair of structures, in different populations of organisms. The wings of birds and the arms of primates, for example, derive from the same ancestral structure. Unlike analogy, which describes different structures with a similar function, homology combines the comparative and relational differently, highlighting related organs with various functions. To retrace our carbon footprints in this light throughout the book, the shared origins and partly divergent politics of financial capitalism and *rikaz;* of risk and *rizq;* of carbon democracy and carbon authoritarianism; of anticolonial vegetarianism and imperial carnivority; or even the emergence of steam engines from their entanglement with water and muscle, all these are homological structures. Each pair is akin to the human arm bone and the wings of a bird. However, common ancestry is just the beginning. What we have witnessed is more like the collaboration of a falconer and the bird perched on his gauntlet. Human limbs and birds' wings share a common origin and then sometimes intersect at cross-purposes, divide and reconnect. To move from zoological to botanical terminology, we could say that old and new forms of energy repeatedly inosculate, extending rather than replacing one another. Instead of root-and-branch transitions, we found trunk and branch extensions, and tried not to miss the trees for the proverbial forest underground.[2]

These organic principles apply to machines too, and to fuels that were once organic. From the outset, I highlighted the ways in which animals and human workers left their imprints inside engines. The reverse was also the case: engines penetrated animals, and even our own human bodies, giving rise to mechanisms of differentiation and community formation.

From this perspective, our own bodily eating and drinking habits emerge as sites for engagement and resistance to fossil fuels. Rather than thinking about vegetarianism as an individual and privatized lifestyle choice, we have explored Gandhian and Islamist attempts to develop a communal and intercommunal solidarity around these practices and connect them to anti-coalonial politics. Similarly, if the reliance on human muscles and labor has been ignored but not removed, and if steam engines and their progeny did not and will not save work,[3] perhaps rather than setting decarbonization on the horizon of post-work and its politics,[4] or framing it in the false contrast between jobs and the environment, it should be addressed by passing through the existing field of labor politics and organization. For instance, solidifying transnational and intersectoral workers' solidarity might help curb and monitor the thrusts of automation, offshoring, and outsourcing, which, as we have seen, have a historical affinity with evading regulation in carbon-intensive sectors. Be that as it may, by branching out in such seemingly idiosyncratic directions, and by binding together different scales of action and experience, this book offers clear insights into the vast and wild spread of the carbon economy and the fact that we are fully immersed in it, and that it is fully within us all. Some of these aspects, and hence some possibilities for resisting hydrocarbons' further metastasis—like adopting vegetarianism and problematizing water desalination—are more familiar (albeit usually for different reasons than those recounted here) while others, like Islamic dimensions of the coal economy, far less so. It is therefore to these later human and humanist scenarios that we must pay more attention.

The (geo)political map this book charts is equally rhizomatic. Each chapter features an (eager or reluctant) local collaborator in the carbon economy, and usually more than one: Mehmet Ali, Aden's Parsis, Sulayman Qabudan Halawa (the SS *Samannud*'s commanding officer), Sayyid Fadl bin ʿAlawi, the coal heavers of Port Said, the coal miners of Zonguldak and manumitted Yemeni engine stokers, the al-Saqqafs, the various translators of European scientific literature into Arabic and Turkish, and of course Sultan Abdulhamit II. These are only some of the local human collaborators; I have made no mention here of the more numerous nonhuman local forces and actors that coal encountered as it spread. These actors, along with processes like the hajj to Mecca or the need to desalinate water in the arid Arabian environment, significantly promoted the global reach of the hydrocarbon economy.

To insist that local actors, forces, and mediators played a role in the global spread of the hydrocarbon economy/empire/science, and that Egypt, the Ottoman Empire, or the Red Sea world (among others) were not merely passive settings but active participants in their own history, means to apply what we already know about the history of empire and of science (and their coproduction) to the history of fossil fuels. Applying these existing historical insights to yet another field—itself a rather modest achievement—might have significant implications for those seeking to engage with fossil-fueled systems today. Reading energy and empire together makes clear that the global spread of hydrocarbons reverberated in many languages and depended on myriad legal forums and systems, on the manifold spirits of capitalism (Zoroastrian, Islamic, and Presbyterian, among others), and simultaneously on interimperial rivalry and collaboration. It resulted from imperial weakness, not only prowess. Just like empire, "Energy" was thus also a result of feebleness and frailty, of the need for collaboration and external support systems (in both the British and the Ottoman imperial cases). If path dependence is indeed a material reality, and a variegated one at that, current energy policy must acknowledge its slippery and motley historical footings. Global carbonization is the result of the entanglement of energy and empire, and entanglement is thus a necessary departure point for any attempt at decarbonization.

As with the temptation of energy transitions or geoengineering, this complexity presents us with a Gordian knot to sever swiftly with the all-purpose sword of what is called "climate justice." But this is another urge we must resist, if only for a while, because the sword is actually part of the knot. As *Powering Empire* demonstrates, we cannot automatically assume that striving for climate justice would always necessarily converge with anti-imperialism, anticapitalism, or antiracism. The spread of coal simultaneously promoted solidarity in British mining communities and widened rifts among the working class of the global energy sector; it flamed communal and interreligious tensions in some settings along the Aden–Port Said corridor, while creating rapprochement and coalition in others. The history of coal's role in modern food systems and in multispecies environments similarly suggests that neither a human-supremacy approach nor even the common interest and shared purpose of humans and other species can be automatically assumed—as demonstrated by coal's career in the Middle East. If, in a simplified version of the Capitalocene argument, the poorer, exploited, underdeveloped members of the human species cannot be blamed for global warming—they might even offer pathways to emancipation from it—linking coal and colonialism reveals how much the entire world has been both annexed to and grafted onto the fossil fuels complex. While entanglement is by no means culpability, the grave consequences of this affinity are as important as the differences. Further, like carbonization, advancing towards decarbonization will have winners and losers, for example in the exposure of weaker members of the global society to possible rising fuel or food costs. The fact that multiple global entanglements have been fossilized means that fully retracing carbon footprints will often be not only very difficult, but also extremely harmful and unfair.

Stressing historical entanglements in our current predicament, however, does not mean apportioning equal blame for creating the problem, or succumbing to its consequences; one would avoid going gentle into that good night. In fact, the imbroglios of energy and empire suggest that in many arenas, pursuing social justice could significantly advance the crucial goal of decarbonization. Thus, directing attention to the different standards of living and consumption habits of the average North American and sub-Saharan African might curb the carbon footprint of the former. But if the climate crisis cannot be mitigated without environmental justice, then unpacking and democratizing what we mean by "justice" is no less urgent. This is because measured in degrees of temperature or emissions, in market terms or as the right to catch up in the race to industrialize and modernize, climate justice might actually mean more rather than less carbonization, and growing rather than decreasing global warming.

Non-Western involvement in the march of hydrocarbons complicates but neither weakens nor removes questions of justice from the picture. On the contrary, it has the potential of enriching Eurocentric ethics with notions entertained by thus-far ignored actors. This broader history helps expand and inflect the existing Western repertoire of rights-based, market-based, or contract-based notions of justice, and pushes us to consider justice based on duty, and even on sacrifice; thus a Gandhian or one of his fellow-traveling Islamists might put it, in harmony, too, with the Ottoman Hamidian stress on obligation over freedom, and with our *rizq*-seeking protagonists. Whereas rights have been internationalized and even extended to nonhumans (from animals to corporations), a parallel globalization of obligation and responsibility of people, corporations, and nations still awaits articulation and enforcement. A critical history of duty, one that cautions about its conservative and even fascist potential but does not hold back, might begin with Gandhi or his liberal influences, as Samuel Moyn suggests, but can also be grounded in the thought of Gandhi's Muslim interlocutors, as we have seen.[5] Moreover, we must in turn also ask, "whose responsibility to whom?" When factoring the responsibility of previous and future generations into the carbon footprint and future-oriented policies of existing polities, what should be done when potential paths to decarbonization collide with those that lead to justice? How can the historical role of Western powers in planetary catastrophe not license "developing" ones to close the gap? Thus, an Islamic reluctance to gamble on the future can be adopted as an imperative for redistribution of existing nonpolluting resources, rather than the right to enter into carbon trading. We might also recognize the affinities between the diagnoses of the climate scientists and eschatologists, and regard the horizon of human extinction as an impetus to "view the world with dead eyes," as the Islamists put it, that is, to harness the awareness and fear of impending death to take moral collective action in this life. We might begin by writing literature that tells the story of the *Jeddah* from the perspective of the pilgrims onboard. In short, the different forms, meanings, ambassadors, and kinds of politics and ethics that coal animated in different places must be examined in order to begin contemplating a truly global anti-carbon ethics and politics. Among such various potential strands, let us follow one in particular. Far from being a pure, isolated alternative in its own universe, unimplicated by fossil fuels or ideologies of progress, or a perfect mirror image of imperial Western capitalism, our investigation of Islamic engagements with the powers of coal reveals, instead, an integral, if unacknowledged, part of the story.

TOWARDS ISLAMIZING ENERGY HUMANITIES?

Our focus on many things, including climate change, is Atlantic and Eurocentric. *Powering Empire* brings the Indian Ocean and Middle East more forcefully into the conversation, and not only because they are at the receiving end of climate calamity. Energy is a unifying category, one of the key unifying frameworks of the modern age. This book attempts first to expose energy's premodern origins and connections to existing forces, and second to pluralize it—to examine how energy's ostensibly homogeneous unity was and is, in fact, differential. Coal was the first material that energy wrote under erasure. Following it to the marginalized world of the Indian Ocean allows us to appreciate the multiplicity of materials, life-forms, politics, and ethics that it ignited and was animated by. This was especially important because energy and thermodynamics often reduced this multiplicity to sets of mirror images and opposites.

Furthermore, as a unifying framework, energy intersected and competed with other unifying schemes. The literature on the Indian Ocean reveals how Islamic law, kinship, and commercial ties afforded greater integration to this framework during the nineteenth century.[6] The Indian Ocean was an Islamic ecumenical space long before the modern era. But the rapidity, regularity, and intensity of interconnection in the age of the steamship, steam hajj, steam press, and coal-fired submarine telegraphy constituted a real watershed, sometimes eroding, at other times bolstering, and repeatedly modifying Islamic ties. Carbon fibers allowed an intensive cross-fertilization between Arabian, Ottoman, and Indian settings.

Nonetheless, multidirectionality should not lead us to remove power, or Europe, from the picture. To be sure, in some senses, coal's European and transnational power dynamics were repeated on a smaller scale in the colonies, where local elites and modernizers monopolized fossil fuels and the political powers derived from them. But the interconnections of the globalized fossil fuels system created different kinds of political and economic opportunities based on early adoption, proximity to and size of coal mines, means of transportation, and adjacent economic sectors. In the race for defensive developmentalism, wherein modern militaries became the proverbial engines that drove entire economies in the Middle East towards rapid modernization under the shadow of Western encroachment, the original European developers and main suppliers of actual steam engines and their fuel could retain and even expand their competitive edge. In this respect, while fossil fuels were newcomers to an already existing interimperial world, once they had arrived on the scene—initially to serve the goals of empire building— hydrocarbons reshaped empire at least as much if not more than the other way around. The distribution of coal's availability sustained the persistent military advantage of Western over non-Western powers. As colonial and semicolonial armies modernized, European imperial powers could not only use them against one another, but also retain the upper hand by controlling the technologies and fuels of warfare.

Systems of thought and ethics are not beyond or separate from the material transformations to which they supposedly offer independent resources for critique: we have seen how coal interpenetrated Islamic/ Ottoman thought, hybridized with it, and changed it from within. British hydrocarbons thus reshaped the Ottoman Empire through a spectrum of effects that ranged from generic Europeanization to the culturally specific. For example, especially in its early stages, the *tanzimat,* or reform wave that accompanied defensive developmentalism was a project aimed at strengthening the link between shariʿa and state (or rational bureaucracy) across the Ottoman Empire, just when Europeans, and especially the British, were equating modernization with a growing separation of church from state (and of science from both).

This fusion of the physical and cultural also means that material transformations depend on epistemic enablers. If this is indeed the case, then Ottoman or Islamic ethics may offer ways of changing the globalized fossil-fuels economy. The very history that separated the laboratory, church, and state in the West connected the scientific, spiritual, and civic elsewhere. This connection was rooted in a longer history, one scarred by a deep fault line connecting and then dividing *book* and *nature,* as well as Western Europe and some of its Others. Several of the homologies mentioned above are cast from the same two-sided mold. Let me sketch out this *longue durée* in what follows, and situate the Ottoman or Islamic perspective on fossil fuels as Europe's road not taken. This offshoot of the previous chapter's attempt to ground scientific translation in actual earthly strata is just one indication of the potential of the global entanglement of energy and empire to refresh our thinking on fossil fuels.

According to the premodern notion of the Book of Nature, nature is a systematic and structured domain that might be accessible to human reason through *reading,* a domain *authored* by an external entity like God. It originated with Aristotle (whom Muslims call the First Teacher, while rejecting other Greek natural philosophical traditions, such as those of Plato and Pythagoras, for their immoderate reliance on mathematical abstractions rather than experience).[7] This formulation informed Islamic thought from its beginnings. The Qur'an itself uses the same word, *Ayat,* for both textual and natural signs in need of human deciphering.[8] Further, "the parallel (or even the identity) between the revelation of the Qur'an and the revelation of the universe" has been noted by Muslim authors from the Middle Ages onwards.[9] Long before nineteenth-century categories of science and religion came to have a bearing on things Islamic, Muslim thinkers faced the challenge of reconciling what Shahab Ahmad terms "the Pre-Text" and "Text" of revelation,[10] often attempting to do so with the application of rational tools predicated on a coherent hermeneutics and "reading."

The so-called Scientific Revolution in early-modern Europe has been partly attributed to a similar endeavor to reconcile ancient written texts—scripture as well as Greek classical wisdom—with the more directly accessible Book of Nature. In both respects, Islamic thought—represented by formidable early-modern empires, especially the Ottomans—posed significant challenges. In the first instance, considered by Christendom as the youngest Abrahamic religion, the truth-claim of Islamic scripture was predicated on an assertion of falsifications in the two older Testaments. In the second instance, the fact that the corpus of Greek ancient wisdom was available in Europe mainly via Arabic translation was troubling to its European readers.[11] For both these reasons, and in tandem with missionary activities,[12] gentlemen scientists like Robert Boyle—the protagonist (or rather, antagonist) of one of the most important histories of science and its separation from politics[13]—considered Arabic philology an integral part of their scientific work, combining laboratory experimentation with the philological deletion of oriental alterations. But "modest priests of nature" like Boyle and the members of the London Royal Society were nonetheless facing deepening rifts between scripture and the Book of Nature that seemed to offer an unmediated access to divine design.

Exactly such interimperial, interclass, and interfaith tensions—which hastened the gradual secularization of science, and widened the rupture between modern science and old natural history and philosophy in places like the British Isles—bolstered strands of continuity between Greek and later Arabic philosophy, on the one hand, and the modern sciences as translated, taught, and practiced in the Ottoman Empire, on the other; between Islamic scripture and the Book of Nature. During the second part of the nineteenth century, Ottomans of different stripes argued in Arabic and Turkish that modern science was actually Arab or Islamic; that becoming more scientifically oriented meant returning to one's roots or boosting one's piety.[14] Refusing (or "failing") to separate "science" from "religion," or facts from values, is usually taken as a sign of backwardness. But the Ottoman struggle to maintain a coherent ethico-scientific hermeneutics and objective morality during the spread of fossil fuels and their epistemologies is a significant reason why I think they should interest us today. What would environmental or energy humanities look like if they stemmed from such a nexus?

This book draws mainly on the two methodological approaches that have contributed the most to undermining the traditional humanities in recent years: postcolonial studies and science and technology studies.

The modular role of the European human male has come under sustained scrutiny in postcolonial studies. Concomitantly, the history of science and technology has been vigorously promoting post-human actors and perspectives. The so-called "hyperobject" of climate change, with its enormous complexity and magnitude, as we hear repeatedly, similarly calls for a more-than-human, truly *longue durée* perspective, in which human history is just a short chapter. Yet even if in their current form the humanities might prove inadequate to fully explicate the global story of coal, humanistic insights into ethics and storytelling, and the discipline's playful disrespect for conventions, hierarchies, and categories are indispensable, especially when drawing on traditions that have not been conventionally included in the Eurocentric humanities canon. Indeed, what the Ottoman case can offer is a potential nexus through which natural science enters the humanities, creating an energy humanities that can speak to science from both within and without, but always ethically; that speak in more languages than English, and offer a bridge that connects and helps articulate the mind-boggling realities of climate change in more communicable terms.

Of course, Ottoman carbon ethics are themselves ossified, being theological through and through, and mostly inactive for a century. They nevertheless expand the repertoire of narratives, histories, and experiences to distill, reform, secularize, and redeploy. This would not be the first time that Ottoman legacies become subject to dispute in the contemporary Middle East, though so far mainly by the religious right. However, if the façades of this history can be weaponized in the struggles between Erdoğan's Turkey and Saudi Arabia, for example, why not tap its fuller potentials? The Western tradition is informed by an ethos whereby science, politics, and ethics are separate domains.[15] But other traditions, homologic to Western science and connected to it with modern carbon fibers, have not insisted on such a separation of force and power, book and nature. As such, they now offer potential idioms for invigorating and reforming climate ethics and justice, and for injecting existing and new scientific knowledge into politics that might better resonate with publics globally. This book deals with the persistence and relentless efficacy of existing powers, and old and supposedly marginal resources. Coming to terms with the larger-than-human has been one of the humanities' traditional key tasks, and the same applies to the natural sciences. Seeking to combine and broaden these approaches might offer a way forward, even if the intersection lies somewhere behind or on the side of the main road.

Notes

INTRODUCTION

1. Mahan's famous book, *The Influence of Sea Power upon History*, dealt with the period between 1660 and 1783, but its insights were linked to Mahan's own times and to steamer power. Its framing section headings included "Permanence of the teachings of history," "Lessons of history apply especially to strategy," and "Less obvious to tactics, but still applicable." See Alfred Thayer Mahan, *The Influence of Sea Power upon History* (London: Sampson Low Marston, 1890), 2, 7, 9.

2. Alfred Thayer Mahan, *The Interest of America in Sea Power: Present and Future* (1897; repr., Freeport, NY: Books for Libraries Press, 1970); see also Allan Sekula, *Fish Story* (London: Mack, 2018), 108.

3. Alfred Thayer Mahan, *The Persian Gulf and International Relations* (London: Robert Theobald, 1902).

4. These and similar examples are quoted in Crosbie Smith, *Coal, Steam and Ships: Engineering, Enterprise and Empire on the Nineteenth-Century Seas* (Cambridge: Cambridge University Press, 2018), 9.

5. See, for example, "father of thermodynamics" Sadi Carnot in 1824, quoted in Smith, *Coal, Steam and Ships*, 8.

6. Anthony S. Travis, *The Rainbow Makers: The Origins of the Synthetic Dyestuffs Industry in Western Europe* (Bethlehem, PA: Lehigh University Press, 1993). I thank Angela Creager for this point.

7. On Barak, "Outsourcing: Energy and Empire in the Age of Coal, 1820–1911," *International Journal of Middle East Studies* 47, no. 3 (2015): 425–45. The notion of landscapes of intensification was coined by Christopher F. Jones. See Jones, *Routes of Power* (Cambridge, MA: Harvard University Press, 2014).

8. John R. Gillis, *The Human Shore: Seacoasts in History* (Chicago: University of Chicago Press, 2012), 135.

9. Anson Rabinbach, *The Human Motor: Energy, Fatigue, and the Origins of Modernity* (New York: Basic Books, 1990), 45.

10. See Crosbie Smith, "Force, Energy, and Thermodynamics," in *The Modern Physical and Mathematical Sciences*, vol. 5 of *The Cambridge History of Science*, ed. Mary Jo Nye (Cambridge: Cambridge University Press, 2002), 309; Crosbie Smith, *The Science of Energy: A Cultural History of Energy Physics in Victorian Britain* (London: Athlone, 1998); and Allen MacDuffie, *Victorian Literature, Energy, and the Ecological Imagination* (Cambridge: Cambridge University Press, 2014), 17.

11. Cara Daggett, "Energy's Power: Fuel, Work and Waste in the Politics of the Anthropocene" (PhD diss., Johns Hopkins University, 2016).

12. The title of Crosbie Smith and M. Norton Wise's biography of thermodynamic pioneer William Thompson, *Energy and Empire*, promised to address such issues, yet the book seldom leaves the British Isles. Smith and Wise, *Energy and Empire: A Biographical Study of Lord Kelvin* (Cambridge: Cambridge University Press, 1989).

13. Thomas S. Kuhn, "Energy Conservation as an Example of Simultaneous Discovery," in *Critical Problems in the History of Science: Proceedings of the Institute of the History of Science, 1957*, ed. Marshall Clagett (Madison: University of Wisconsin Press, 1959), 321. Social constructivist accounts of science effectively challenged Kuhn's supposition that energy was there to be discovered in nature. See, for example, Smith, "Force, Energy, and Thermodynamics."

14. Kenneth L. Caneva, *Robert Mayer and the Conservation of Energy* (Princeton, NJ: Princeton University Press, 1993).

15. Heinz Otto Sibum, "Reworking the Mechanical Value of Heat: Instruments of Precision and Gestures of Accuracy in Early Victorian England," *Studies in History and Philosophy of Science* 26, no. 1 (1995): 73–106.

16. Dotan Halevy, "Lishtot (bira) meha-Yam shel 'Azah: Tsmihato ve-Shki'ato shel ha-Mishar ha-Yami me-'Azah be-Shalhe ha-Tkufa ha'Othmanit" [Drinking (beer) from the sea at Gaza: The rise and fall of Gaza's maritime trade in the late Ottoman period]. *ha-Mizrah ha-Hadash* 55 (2016): 37–42.

17. Donald Stephen Lowell Cardwell, *James Joule: A Biography* (Manchester: Manchester University Press, 1990), 180–83.

18. Arjun Appadurai called it the "politics of probability." See Appadurai, *The Future as Cultural Fact: Essays on the Global Condition* (London: Verso, 2013).

19. Peter Godfrey-Smith, *Other Minds: The Octopus, the Sea, and the Deep Origins of Consciousness* (New York: Farrar, Straus and Giroux, 2016).

20. Kenneth Pomeranz, *The Great Divergence: China, Europe and the Making of the Modern World Economy* (Princeton, NJ: Princeton University Press, 2000).

21. Over the past decade and a half, scholars of "energopolitics" have been attempting to bring Foucault's work into a dialogue with such attention to the energetic and carbon basis of biological life. See Dominic Boyer, "Energopower: An Introduction," *Anthropological Quarterly* 87 no. 2 (2014): 309–33.

22. Toby Craig Jones, *Desert Kingdom* (Cambridge, MA: Harvard University Press, 2011); Michael Christopher Low, "Ottoman Infrastructures of the Saudi Hydro-State: The Technopolitics of Pilgrimage and Potable Water in the Hijaz," *Comparative Studies in Society and History* 57, no. 4 (2015): 942–74.

For desalination in the UAE, see Gökçe Günel, "The Infinity of Water: Climate Change Adaptation in the Arabian Peninsula," *Public Culture* 28, no. 2 (2016), 291–315.

23. For example, botanist John Hogg catalogued around fifty such new species in 1867. See Hogg, "On the Ballast-Flora of the Coasts of Durham and Northumberland," in *The Annals and Magazine of Natural History, Including Zoology, Botany, and Geology*, vol. 19, 3rd series, conducted by Prideaux John Selby, Charles C. Babington, John Edward Gray, and William Francis (London: Taylor and Francis, 1867), 38–43.

24. Anna Tsing, *The Mushroom at the End of the World: On the Possibility of Life in Capitalist Ruins* (Princeton: Princeton University Press, 2015).

25. Ernesto Schick, *Railway Flora, or Nature's Revenge on Man* (Berlin: Humboldt Books, 2015).

26. Gülnur Ekşi, "The Effects of the Berlin-Baghdad Railway on the Vegetation in Anatolia." Lecture at the Haus der Kulturen der Welt, 2017. www.hkw .de/en/app/mediathek/video/61038.

27. Reviel Netz, *Barbed Wire: An Ecology of Modernity* (Middleton, CT: Wesleyan University Press, 2004).

28. "Prickly Pear and the Plants Sent to Aden," Military Department, 1842–43, vol. 188, comp. 369, 678–711, Maharashtra State Archives, Mumbai.

29. David Laitin and Said Samatar, *Somalia: Nation in Search of a State* (Boulder, CO: Westview Press, 1987), 61.

30. Quoted in Rebecca Woods, "Nature and the Refrigerating Machine: The Politics and Production of Cold in the Nineteenth Century," in *Cryopolitics: Frozen Life in a Melting World*, ed. Joanna Radin and Emma Kowal (Cambridge, MA: MIT Press, 2017), 107.

31. Daniel Stolz, "The Voyage of the Samannud: Pilgrimage, Cholera, and Empire on an Ottoman-Egyptian Steamship Journey in 1865–66," *International Journal of Turkish Studies* 23, nos. 1–2 (2017), 12; Sayyid Ahmed Khan, *A Voyage to Modernism*, trans. and ed. Mushirul Hasan and Nishat Zaidi (Delhi: Primus Books, 2011), 71–72.

32. W.E.B. Du Bois, "Worlds of Color," *Foreign Affairs* 3, no. 3 (April 1925): 423–44.

33. John M. Willis, *Unmaking North and South: Cartographies of the Yemeni Past, 1857–1934* (London: Hurst, 2012).

34. Robert H. MacArthur and Edward O. Wilson, *The Theory of Island Biogeography* (Princeton, NJ: Princeton University Press, 1967).

35. Gary Gutting, *Michel Foucault's Archaeology of Scientific Reason* (Cambridge: Cambridge University Press, 1989), 192.

36. Michel Foucault, *The History of Sexuality, vol. 1: The Will to Knowledge* (New York: Penguin, 2008), 243.

37. This hierarchy was mirrored in views expressed by thinkers like the Egyptian social reformer Qasim Amin, who defined the colonial encounter between "Western civilization speeded by steam and electricity" and "the weak members [of the species]" as a "process of natural selection." Amin, *Tahrir al-Mar'ah* (Cairo, 1899), translated by Samiha Sidhom Peterson as *The Liberation of Women* (Cairo: American University in Cairo Press, 2000), 62–63; see also

Marwa Elshakry, *Reading Darwin in Arabic, 1860–1950* (University of Chicago Press, 2016), 10.

38. This emerges in the subtitle of *The Origin of Species,* "or the Preservation of Favoured Races in the Struggle for Life." The 1871 *The Descent of Man* claims that human "races" belong to the same species.

39. Sven Beckert, *Empire of Cotton: A Global History* (New York: Vintage Books, 2015); Giovanni Arrighi, *The Long Twentieth Century: Money, Power, and the Origins of Our Times* (London: Verso, 2010); Ian Baucom, *Specters of the Atlantic: Finance Capital, Slavery, and the Philosophy of History* (Durham, NC: Duke University Press, 2005); Aaron G. Jakes and Ahmad Shokr, "Finding Value in *Empire of Cotton,*" *Critical Histories Studies* 4, no. 1 (2017): 107–36.

40. Steven Gray, *Steam Power and Sea Power: Coal, the Royal Navy, and the British Empire, c. 1870–1914* (London: Palgrave Macmillan, 2018), chap. 1.

41. Colonial officials like Egypt's Lord Cromer often started their political careers in the Middle East as debt enforcers responsible for securing returns on private speculative capital. Also, as Daniel Stolz has shown, during the 1870s, an association of British lenders developed the wherewithal to persuade their government to send gunboats to assure returns of private loans. Stolz, "'Impossible to Arrive at an Accurate Estimate': Accounting for the Ottoman Debt, 1873–1881" (unpublished manuscript, last modified January 21, 2019).

42. On the Red Sea as a sphere associated with danger, see Alexis Wick, *The Red Sea: In Search of Lost Space* (Berkeley: University of California Press, 2016).

CHAPTER 1. WATER

1. Robert McNeill, *Something New under the Sun: An Environmental History of the Twentieth-Century World* (New York: W. W. Norton & Company, 2000), 297.

2. See Kenneth Pomeranz, *The Great Divergence: China, Europe and the Making of the Modern World Economy* (Princeton, NJ: Princeton University Press, 2000).

3. Karl Marx, *The Poverty of Philosophy* (1847; repr., New York: International Publishers, 1963), 119.

4. See Pomeranz, *Great Divergence,* 43–68.

5. Niche markets have been found to be key in "energy transitions." See Roger Fouquet and Peter Pearson, "Past and Prospective Energy Transitions: Insights from History," *Energy Policy* 50 (November 2012): 1–2.

6. See E. P. Thompson, "Time, Work-Discipline, and Industrial Capitalism," *Past and Present* 38 (December 1967): 56–97.

7. See Andreas Malm, *Fossil Capital: The Rise of Steam-Power and the Roots of Global Warming* (London: Verso, 2016).

8. Wolfgang Schivelbusch, *The Railway Journey: The Industrialization of Time and Space in the Nineteenth Century* (Berkeley: University of California Press, 2014), 3.

9. Smith, "Force, Energy, and Thermodynamics," 291.

10. See Smith and Wise, *Energy and Empire,* 243.

11. Bruno Latour, *Science in Action: How to Follow Scientists and Engineers through Society* (Cambridge, MA: Harvard University Press, 1988), 130.

12. Andreas Malm, "The Origins of Fossil Capital: From Water to Steam in the British Cotton Industry," *Historical Materialism* 21, no. 1 (2013): 33.

13. Bernard Forest de Bélidor, *Architecture Hydraulique* (Paris: C.A. Jombert, 1737–1770), 2:324–25, quoted in Richard Shelton Kirby, Arthur Burr Darling, Sidney Withington, and Frederick Gridley Kilgour, *Engineering in History* (1956; repr., Mineola, NY: Dover, 1990), 165.

14. *Mining: Haulage: The Classic 1907 Mining Engineering Text* (1907; repr., Periscope Film LLC, 2008), 48–55. www.periscopefilm.com/mining-haulage.

15. Roger Fouquet, *Heat, Power and Light: Revolutions in Energy Services* (Cheltenham: Edward Elgar, 2008), 126–27.

16. Edward Baines, *History of the Cotton Manufacture in Great Britain* (London: H. Fisher, R. Fisher, and P. Jackson, 1835), 226.

17. Baines, 202.

18. Baines, 207–8.

19. See On Barak, *On Time: Technology and Temporality in Modern Egypt* (Berkeley: University of California Press, 2013), chap. 1.

20. See Sven Beckert, *Empire of Cotton: A Global History* (New York: Vintage Books, 2015).

21. Alan Mikhail, "Unleashing the Beast: Animals, Energy, and the Economy of Labor in Ottoman Egypt," *American Historical Review* 118, no. 2 (2013): 331–33.

22. The report was written as a letter to John Bowring, who published it in his *Report on Egypt and Candia: Addressed to the Right Hon. Lord Viscount Palmerston, Her Majesty's Principal Secretary of State for Foreign Affairs, &c. &c. &c.* (London: Printed by W. Clowes and Sons for H.M.S.O., 1840), 189.

23. Bowring, 190.

24. Bowring, 190.

25. Barak, *On Time,* 89.

26. Gerald S. Graham, "The Ascendancy of the Sailing Ship, 1850–85," *Economic History Review* 9, no. 1 (1956): 74–88; Robert D. Foulke, "Life in the Dying World of Sail, 1870–1910," *Journal of British Studies* 3, no. 1 (1963): 105–36; and C. Knick Harley, "The Shift from Sailing Ships to Steamships, 1850–1890: A Study in Technological Change and its Diffusion," in *Essays on a Mature Economy: Britain after 1840,* ed. Donald N. McCloskey (Princeton, NJ: Princeton University Press, 1971), 215–34.

27. John Mathew, *Margins of the Market: Trafficking and Capitalism across the Arabian Sea* (Oakland: University of California Press, 2016), 52–82.

28. BEO 3288/246549; DH.MKT 966/64; DH.TMIK.S 22/47, all in Başbakanlık Osmanlı Arşivi (BOA), Ottoman State Archives, Istanbul.

29. Donald Quataert, *The Ottoman Empire, 1700–1922* (New York: Cambridge University Press, 2000), 120; Jonathan Miran, *Red Sea Citizens: Cosmopolitan Society and Cultural Change in Massawa* (Bloomington: Indiana University Press, 2009), 70–71; Halil İnalcık and Donald Quataert, *An Economic and Social History of the Ottoman Empire, 1300–1914,* (Cambridge: Cambridge University Press, 1994), 2:800–802; Salih Özbaran, "İstanbul'da Kayıkçılık ve

Kayık İşletmeciliği," in *Osmanlı İmparatorluğunda Şehircilik ve UlaşımÜzerine Araştırmalar*, ed. Cengiz Orhonlu (İzmir: Ticaret Matbaacılık T.A.Ş., 1984), 83–103.

30. Mikhail, "Unleashing the Beast," 340.

31. Mikhail, 340n111.

32. For the impact of railroad construction on the use of animals for transport and trade purposes in Anatolia, see Donald Quataert, "Limited Revolution: The Impact of the Anatolian Railway on Turkish Transportation and the Provisioning of Istanbul, 1890–1908," *Business History Review* 51, no. 2 (1977): 139–60. For animals in mining see, for example, C.DRB 51/2533 and C.DRB 7/350, BOA.

33. See Barak, *On Time*, 37.

34. On sabotage of animal owners against railways, see Barak, chap. 6.

35. Frederick Ayrton, *Railways in Egypt: Communication with India* (London: Ridgway, 1857), 24.

36. Alan Mikhail, *Nature and Empire in Ottoman Egypt: An Environmental History* (Cambridge: Cambridge University Press, 2011), 260.

37. Mikhail, 286–87.

38. Reuven Aharoni, *The Pasha's Bedouin: Tribes and State in the Egypt of Mehemet Ali, 1805–1848* (London: Routledge, 2007), 10.

39. William Willcocks, *Egyptian Irrigation* (London: E. & F. N. Spon, 1899), 202–4.

40. Willcocks, 370.

41. Willcocks, 476–77.

42. Willcocks, 223.

43. Willcocks, 347.

44. Willcocks, 402, 410.

45. Willcocks, 406–11.

46. See Canay Ozden, "The Pontifex Minimus: William Willcocks and Engineering British Colonialism," *Annals of Science* 71, no. 2 (2014): 183–205.

47. Ozden, 340.

48. Zayn al-ʿAbidin Shams al-Din Najm, *Bur Saʿid: Tarikhuha wa-tatawwuruha mundhu nashʾatiha 1859 ḥata ʿam 1882* (Cairo: al-Hayʾa al-ʿAmma al-Misriyya liʾl-Kitab, 1987), 39–43.

49. Kenneth J. Perkins, *Port Sudan: The Evolution of a Colonial City* (Boulder: Westview Press, 1993), 13.

50. Najm, *Bur Saʿid*, 42–43.

51. See On Barak, "Three Watersheds in the History of Energy," *Comparative Studies of South Asia, Africa and the Middle East* 34, no.3 (2014): 440–53.

52. Bowring, *Report on Egypt and Candia*, 190.

53. Bowring, 191.

54. Foreign Office (FO) 78/343 Campbell-FO 6/8/38, British National Archives (BNA), London. See Thomas E. Marston, *Britain's Imperial Role in the Red Sea Area, 1800–1878* (Hamden, CT: Shoe String Press, 1961), 59–61.

55. Michael Christopher Low, "Ottoman Infrastructures of the Saudi Hydro-State: The Technopolitics of Pilgrimage and Potable Water in the Hijaz," *Comparative Studies in Society and History* 57, no. 4 (2015): 955.

56. See Lauren Benton, *A Search for Sovereignty: Law and Geography in European Empires, 1400–1900* (Cambridge University Press, 2010). For a Middle Eastern example, see the Ottoman map for the 1906 border dispute for the Syria-Egypt border in Yuval Ben Bassat and Yossi Ben-Artzi, "The Collision of Empires as Seen from Istanbul: The Border of British-Controlled Egypt and Ottoman Palestine as Reflected in Ottoman Maps," *Journal of Historical Geography* 50 (2015): 32. See also the map of South Arabia: Anglo-Turkish Boundary from 1914, www.qdl.qa/en/archive/81055/vdc_100023253361.0x000005, last retrieved July 2, 2019. In all these cases, British maps were taken and water sources were added to them in Arabic or Turkish. The existing territorial imaginary was superimposed on the new cartographical one.

57. For a description of these sophisticated waterworks, see Roxani Eleni Margariti, *Aden and the Indian Ocean Trade: 150 Years in the Life of a Medieval Arabian Port* (Chapel Hill: University of North Carolina Press, 2007), 47–53.

58. "Proceedings Relative to the Power of the Steam Engineer Which Is Required from England for Drawing Water from the Wells at Aden," Military Dept., 1845–46, vol. 306, comp. 214, pp. 612–31, esp. 13], National Archives of India, New Delhi.

59. "Water Supply at Aden," Public Works Dept., 1857, vol. 107, comp. 87, pp. 5–148, esp. 32, National Archives of India, New Delhi.

60. Everett D. Howe, *Fundamentals of Water Desalination* (New York: Marcel Dekker, 1974), 8.

61. Joseph P. Pirsson, *Additional and Fresh Evidence of the Practical Working of Pirsson's Steam Condenser* (Washington: Gideon, 1851).

62. Mike McCarthy, *Iron and Steamship Archaeology: Success and Failure on the SS Xantho* (New York: Kluwer Academic, 2002), 19.

63. Howe, *Fundamentals of Water Desalination,* 10–12.

64. "Arrangements for the Water Supply of the Indian Contingent and for the Provision of Condensers at Aden," Military Dept., Pro. A—September 1882, no. 833–59, National Archives of India, New Delhi.

65. J. Birkett and D. Radcliffe, "Normandy's Patent Marine Aerated Fresh Water Company: A Family Business for 60 Years, 1851–1910," *IDA Journal of Desalination and Water Reuse* 6, no. 1 (2014): 24–32.

66. Howe, *Fundamentals.,* 9, 12.

67. "Annual Administration Report of the Bahrain Agency," File 8/8 II, R/15/2/299, India Office Records (IOR), British Library, London.

68. *Military Report on Persia,"* vol. 4, pt. I: "Persian Baluchistan, Kerman and Bandar Abbas," L/PS/20/C201/1, IOR.

69. "Military Report on the Aden Protectorate," L/MIL/17/16/6, IOR.

70. "Muscat: Slave Trade under Cover of French Flag," File 59/15 B (A 14), R/15/1/552, IOR.

71. F. M. Hunter, *An Account of the British Settlement of Aden in Arabia* (1877; repr., London: Frank Cass, 1968), 15.

72. "Field notes: Aden Protectorate," L/MIL/17/16/7, IOR.

73. "Military Report on the Aden Protectorate," L/MIL/17/16/6, IOR.

74. "Kuwait Water Supply," 138/48, IOR.

75. J. Forbes Munro, *Maritime Enterprise and Empire: Sir William Mackinnon and His Business Network, 1823–93* (Woodbridge, Suffolk, UK: Boydell Press, 2003), 66.

76. P/674, IOR.

77. Christaan Snouck Hurgronje, *Mekka in the Latter Part of the Nineteenth Century, 1885–1889*, trans. James Henry Monahan (Leiden: Brill, 1931).

78. Low, "Ottoman Infrastructures," 942–74.

79. For the intraimperial dimensions, see M. Alper Yalçınkaya, *Learned Patriots: Debating Science, State and Society in the Nineteenth-Century Ottoman Empire* (Chicago: University of Chicago Press, 2015). For the interimperial level, see Eileen M. Kane, *Russian Hajj: Empire and the Pilgrimage to Mecca* (Ithaca: Cornell University Press, 2015); Seema Alavi, *Muslim Cosmopolitanism in the Age of Empire* (Cambridge, MA: Harvard University Press, 2015).

80. Low, "Ottoman Infrastructures," 952.

81. For Egypt, see Barak, *On Time*, and Aaron Jakes, "Boom, Bugs, Bust: Egypt's Ecology of Interest, 1882–1914," *Antipode* 49, no. 4 (2017): 1035–59. For Africa, see Mustafa Minawi, *The Ottoman Scramble for Africa: Empire and Diplomacy in the Sahara and the Hijaz* (Stanford: Stanford University Press, 2016).

82. Low, "Ottoman Infrastructures," 961; Birsen Bulmus, *Plague, Quarantines and Geopolitics in the Ottoman Empire* (Edinburgh: Edinburgh University Press, 2012), 166.

83. Low, "Ottoman Infrastructures," 962–63.

84. Bulmus, *Plague, Quarantines and Geopolitics,* 166; Low, "Ottoman Infrastructures," 963.

85. *Gazetteer of Arabia*, vol. 1, L/MIL/17/16/2/1, IOR.

86. *Gazetteer of Arabia*, vol. 1.

87. See Toby Craig Jones, *Desert Kingdom* (Cambridge, MA: Harvard University Press, 2011), 229.

88. F. G. Clemow, "Some Turkish Lazarets and Other Sanitary Institutions in the Near East," *The Lancet*, April 27, 1907, quoted in Bulmus, *Plague, Quarantines and Geopolitics,* 165.

89. Low, "Ottoman Infrastructures," 962.

90. F. G. Clemow, quoted in Bulmus, *Plague, Quarantines and Geopolitics,* 166.

91. Low, "Ottoman Infrastructures," 963; Bulmus, 166.

92. In 1864, for example, 38 steamers visited this port. See Miran, *Red Sea Citizens,* 68.

93. In 1875, 205 steamers visited Jeddah. See Miran, *Red Sea Citizens,* 68.

94. *The Nautical Magazine and Naval Chronicle for 1860: A Journal of Papers on Subjects Connected with Maritime Affairs* (London: Simpkin, Marshall and Co., 1860), 179.

95. Martin Thomas, "Managing the HAJJ: 1918–1930," in *Railways and International Politics: Paths of Empire, 1848–1945*, ed. Keith Neilson and Thomas K. Otte (London: Routledge, 2006), 175, 183.

96. Neilson and Otte, *Railways and International Politics;* "Hedjaz Railway," File 3142/1903, L/PS/10/12, IOR.

97. "Hedjaz Railway."

98. "Hedjaz Railway."

99. Perkins, *Port Sudan*, 47–51, 94.

100. "Steel Pipes Conquer Desert Waste," *Popular Mechanics*, May 1926, 732–35.

101. "Steel Pipes." See photo, www.qdl.qa/en/archive/81055/vdc_100023803156.0x000026 last retrieved June 10, 2019.

102. This shift is examined in chapter 3.

103. A. S. Alsharhan and A. E. M. Nairn, *Sedimentary Basins and Petroleum Geology of the Middle East* (Amsterdam: Elsevier, 1997), 470.

104. l/PS/11, IOR.

105. "Administration Report of the Persian Gulf Political Residency for the Years 1911–1914," R/15/1/711; "Kuwait Water Supply," File 53/47 (D 43), R/15/1/511, IOR.

106. "Abadan Town Planning Report," November 17, 1924, 22–30, ARC 68723, BP (British Petroleum) Archive, University of Warwick.

107. "Steel Pipes," 732–35.

108. William Viscount Weir, *The Weir Group: The History of a Scottish Engineering Legend* (London: Profile Books, 2013).

109. Robert W. Tolf, *The Russian Rockefellers: The Saga of the Nobel Family and the Russian Oil Industry* (Stanford, CA: Hoover Institution Press, 1976), 51.

110. For the Pennsylvania teamsters, see Thomas O. Miesner and William L. Leffler, *Oil & Gas Pipelines in Nontechnical Language* (Tulsa, OK: PennWell Corp. 2006), 18.

111. In addition to Low, *Mechanics of Mecca*, and Jones, *Desert Kingdom*, see Günel, "Infinity of Water."

112. Koray Çalişkan, *Market Threads: How Cotton Farmers and Traders Create a Global Commodity* (Princeton, NJ: Princeton University Press, 2010), 162–66.

CHAPTER 2. ANIMALS

1. David Edgerton, *The Shock of the Old: Technology and Global History since 1900* (London: Profile Books, 2006), 26–32.

2. Charles Montagu Doughty, *Travels in Arabia Deserta* (1888; repr., London: William Clowes & Sons, 1921), 1:9.

3. P. Marcel Kurpershoek, *Oral Poetry and Narratives from Central Arabia*, vol. 5: *Voices from the Desert* (London: Brill, 2005), 326.

4. Charles Montagu Doughty, *Travels in Arabia Deserta* (1888; repr., Charleston, SC: Nabu Press, 2012), 2:322; Kurpershoek, *Oral Poetry*, 28.

5. Kurpershoek, *Oral Poetry*, 14.

6. Kurpershoek, 14.

7. Kurpershoek, 126.

8. Kurpershoek, 262.

9. Doughty, *Arabia Deserta*, 1:307.

10. They remain, however, important in various agricultural peripheries, where they compete cost-effectively with petroleum-guzzling tractors, as is the case in several settings in Egypt and Turkey. See Çalişkan, *Market Threads*.

11. Vaclav Smil, "Eating Meat: Evolution, Patterns, and Consequences," *Population and Development Review* 28, no. 4 (2002): 618.

12. Cattle underwent an enforced bovine nutrition transition, involving high-protein feeds and a mixture of proteins, carbohydrates and fats that shortened its lifespan, reconfigured its bodily form and changed its fat content. See Christopher Otter, "Planet of Meat: A Biological History," in *Challenging (the) Humanities*, ed. Tony Bennett (Canberra: The Australian Academy of the Humanities, 2013), unpaginated.

13. See Susanne Freidberg, *Fresh: A Perishable History* (Cambridge, MA: Harvard University Press, 2010).

14. See Rebecca Woods, "Breed, Culture, and Economy: The New Zealand Frozen Meat Trade, 1880–1914," *Agricultural History Review* 60, no. 2 (2012): 288–302.

15. George Colpitts, *Pemmican Empire: Food, Trade, and the Last Bison Hunts in the North American Plains, 1780–1882* (Cambridge: Cambridge University Press, 2015).

16. See Rebecca J.H. Woods, "Nature and the Refrigerating Machine: The Politics and Production of Cold in the Nineteenth Century," in *Cryopolitics: Frozen Life in a Melting World*, ed. Joanna Radin and Emma Kowal (Cambridge, MA: MIT Press, 2017), 89–116.

17. Herbert Sussman, *Victorian Technology: Invention, Innovation, and the Rise of the Machine* (Santa Barbara, CA: ABC-CLIO, 2009), 51.

18. Christopher Otter, *The Vital State: Food Systems, Nutrition Transitions, and the Making of Industrial Britain* (unpublished manuscript, last modified September 20, 2018), 31.

19. James Troubridge Critchell and Joseph Raymond, *A History of the Frozen Meat Trade: An Account of the Development and Present Day Methods of Preparation, Transport, and Marketing of Frozen and Chilled Meats* (London: Constable, 1912), 32.

20. *Todmorden & District News*, February 6, 1880.

21. *Edinburgh Evening News*, February 7, 1880.

22. "Butchers and Newspaper Wrapping," *Dundee Evening Telegraph*, January 19, 1889.

23. Otter, *Vital State*, 31.

24. *Moreton Bay Courier*, September 22, 1858.

25. Critchell and Raymond, *A History of the Frozen Meat Trade*, 25–27.

26. Critchell and Raymond, 28–31.

27. Critchell and Raymond, 39–42.

28. Critchell and Raymond, 33–34.

29. "The Dock Labourer's Strike in London," *Cumberland Pacquet, and Ware's Whitehaven Advertiser*, August 29, 1889; "Trade Announcements," *Hartlepool Northern Daily Mail*, September 21, 1889.

30. Like those expressed by Campbell in the previous chapter.

31. Edward Alpers, "The Western Indian Ocean as a Regional Food Network in the Nineteenth Century," in *East Africa and the Indian Ocean* (Princeton: Markus Wiener Publishers, 2009), 35.

32. Laitin and Samatar, *Somalia*, 61.

33. F. M. Hunter, *An Account of the British Settlement of Aden in Arabia* (1877; repr., London: Frank Cass, 1968), 47–48.

34. Hunter, 72.

35. Hunter, 34.

36. Hunter, 72.

37. "Levy of Duty on Livestock Exported from the Somali Coast Protectorate, and the Consequent Increase in the Contract Rate for Meat Supplied to the Troops at Aden," Foreign Dept., Proceedings nos. 250–51, August 1899, National Archives of India, New Delhi.

38. S. A. M. Adshead, *Salt and Civilization* (New York: Palgrave, 1992), 302; "Aden Battle for Fresh Meat: Proceedings Regarding the Arrangements That Have Been Made to Supply Fresh Meat to the Troops at Aden," Military Dept., 1846–47, vol. 304: 145–254, Maharashtra State Archives, Mumbai.

39. Mahalal Ha'adani, *Beyn Aden ve-Teyman* [Between Aden and Yemen] (Tel-Aviv: 'Am 'oved, 1947), pt. 1, p. 5, quoted in Reuben Ahroni, *The Jews of the British Crown Colony of Aden: History, Culture, and Ethnic Relations* (Leiden: Brill, 1994), 47.

40. "A Trial to Be Made of the Contract System for Obtaining at Aden and Supplies of Hay and Grain," Military Dept., 1842–43, vol. 184, comp. 279: 354, Maharashtra State Archives, Mumbai.

41. "Aden Battle for Fresh Meat."

42. "Commissariat Contract to Supply Water at Aden," Military Dept., 1847–48, vol. 354, comp. 39: 37–78, Maharashtra State Archives, Mumbai.

43. "Rates of Pay of the Arab Levy at Aden," Foreign Dept., Proceedings nos. 241–245, January 1869, National Archives of India, New Delhi.

44. "[If] any accident [should happen] to the Engines of the steamer and the supplies [be] detained by the Chiefs." Military Dept., no. 399 of 1846, September 29, 1846, Maharashtra State Archives, Mumbai; "Aden Battle for Fresh Meat."

45. "Grass for Aden: Proceedings Relative to the Expediency at Forwarding Grass Forage and Salt Provisions from the Presidency to Aden for the Use of the Troops," Military Dept., 1846–47, vol. 313: 397–492, Maharashtra State Archives, Mumbai.

46. Edward Barrington De Fonblanque, *Treatise on the Administration and Organization of the British Army* (London: Longman, Brown, Green, Longmans, and Roberts, 1858), 364–65.

47. Nutrition and health posed a major concern for the British from the onset of the occupation. Just several months after establishing the outpost, it was observed that "nearly one half of the whole Corps have spongy gums, pain and weakness in the extremities, and other indications of a Scorbutic habit." An increase of vegetables from the Arabian hinterland to the "native" Sepoys proved effective in improving their health and constitution. This, however, was short lived: several years later, a report indicated that many of the Sepoys suffered from "torpidity of the digestive organs . . . swelling of the legs and general lassitude, a well-known effect of indigestion." This was attributed to their unvaried diet of rice and dahl. The colony surgeon suggested that the Native officers "use their influence with the men in general" to convince them to eat meat, rather than food which, "though it satisfies the appetite, is void of the

nutrition property required by the Sepoys in their current state." "The Health of the 24th Native Infantry Lately Returned from Aden and Stationed at Poona," Military Dept., 1839–41, vol. 136: 55–90; "Regarding the Health of the Troops at Aden," Military Dept., 1840–42, vol. 149: 117–44; "Sickness at Aden," Military Dept., 1844–45, vol. 241: 445–53, Maharashtra State Archives, Mumbai.

48. Laitin and Samatar, *Somalia*, 100.

49. Lord George Curzon, *Frontiers* (Oxford: Clarendon Press, 1907), 41.

50. Christian Wolmar, *Railways and the Raj: How the Age of Steam Transformed India* (London: Atlantic Books, 2017).

51. Charles L. Geshekter, "Anti-Colonialism and Class Formation: The Eastern Horn of Africa before 1950," *International Journal of African Historical Studies* 18, no. 1 (1985): 19–20, 23.

52. R. J. Gavin, *Aden under British Rule, 1839–1967* (London: C. Hurst, 1975), 52–53.

53. Otter, *Vital State*, 1.

54. Ahroni, *Jews of the British Crown Colony of Aden*, 41.

55. Gavin, *Aden Under British Rule*, 106.

56. Hunter, *Settlement of Aden*, 86.

57. "Ice-culture," *Chambers's Journal of Popular Literature, Science and Arts*, February 13, 1864, 99–101, quoted in Woods, *Nature and the Refrigerating Machine*, 107.

58. This protocol will be discussed in chapter 3.

59. Stolz, "Voyage of the Samannud," 12.

60. See Stolz, 10.

61. Ramachandra Guha, *Gandhi before India* (London: Penguin, 2013), unpaginated; B. R. Nanda, *Mahatma Gandhi: A Biography* (Oxford: Oxford University Press, 1989), 26.

62. "The Dock Labourer's Strike in London," *Cumberland Pacquet, and Ware's Whitehaven Advertiser*, August 29, 1889; "Trade Announcements," *Cumberland Pacquet*, September 21, 1889.

63. "Indian Vegetarians," *The Vegetarian (London)*, February 7, 1891.

64. *The Vegetarian (London)*, March 14, 1891.

65. *The Vegetarian (London)*, June 13, 1891.

66. Najm, *Bur Sa'id*, 136–37, 165.

67. *Tamworth Herald*, October 1, 1892.

68. *Tamworth Herald*, 158–59.

69. Adshead, *Salt and Civilization*, 160–61.

70. Mary L. Whately, *Letters from Egypt to Plain Folks at Home* (London: Seeley, Jackson & Halliday, 1879), 25.

71. *Dundee Courier*, June 24, 1886.

72. Burcu Kurt, "Buz Temininde Sanayileşme ve Osmanlı İmparatorluğu'nda Kurulan Buz Fabrikaları" [Industrialization in procurement of ice and ice factories established in the Ottoman Empire], *Ankara Üniversitesi Osmanli Tarihi Arastirma ve Uygulama Merkezi Dergisi* (Ankara University Journal of Ottoman History Research and Application Center) 30 (Autumn 2011): 73–98.

73. Kurt, 79.

74. Sam White, *The Climate of Rebellion in the Early Modern Ottoman Empire* (Cambridge: Cambridge University Press, 2011), 34–39.

75. Alan Mikhail, *The Animal in Ottoman Egypt* (Oxford: Oxford University Press, 2013), 46–48.

76. Mikhail, 46–48.

77. Zeinab Abul-Magd, *Imagined Empires: A History of Revolt in Egypt* (Berkeley: University of California Press, 2013), 86.

78. Aharoni, *Pasha's Bedouin*, 102.

79. Aharoni, 106–7.

80. Khaled Fahmy, *All the Pasha's Men: Mehmed Ali, His Army and the Making of Modern Egypt* (Cambridge: Cambridge University Press, 1997), 185.

81. Fahmy, 219.

82. G. N. Sanderson, "The Nile Basin and the Eastern Horn, 1870–1908," in *The Cambridge History of Africa, Vol. 6: From 1870 to 1905*, ed. Roland Oliver and G. N. Sanderson (Cambridge: Cambridge University Press, 1985), 669.

83. Roger Owen, *Cotton and the Egyptian Economy, 1820–1914: A Study in Trade and Development* (Oxford: Clarendon Press, 1969), 309.

84. Robert Tignor, *Modernization and British Colonial Rule in Egypt, 1882–1914* (Princeton, NJ: Princeton University Press, 1966), 230–31.

85. The nineteenth century saw the removal of slaughterhouses from cities for hygienic reasons. See Khaled Fahmy, "An Olfactory Tale of Two Cities: Cairo in the Nineteenth Century," in *Historians in Cairo: Essays in Honor of George Scanlon*, ed. Jill Edwards (Cairo: American University in Cairo Press, 2002), 167, 176; LaVerne Kuhnke, *Lives at Risk: Public Health in Nineteenth-Century Egypt* (Berkeley: University of California Press, 1990), 61, 67. For the definitive analysis of abattoir visuality, see Timothy Pachirat, *Every Twelve Seconds: Industrialized Slaughter and the Politics of Sight* (New Haven, CT: Yale University Press, 2013).

86. Mikhail, *Animal in Ottoman Egypt*; John T. Chalcraft, *The Striking Cabbies of Cairo and Other Stories: Crafts and Guilds in Egypt, 1863–1914* (Albany: State University of New York Press, 2004); Suraiya Faroqhi, *Animals and People in the Ottoman Empire* (Istanbul: Eren, 2010); Roza El-Eini, *Mandated Landscape: British Imperial Rule in Palestine, 1929–1948* (London: Routledge, 2006), 146.

87. Otter, *Planet of Meat*, 2–3.

88. William Cronon, *Nature's Metropolis: Chicago and the Great West* (New York: W. W. Norton, 1991), 209, 211, 226, 231, 236.

89. Otter, *Vital State*, 3.

90. "al-Lahm al-Nafi' lil-Insan," *al-Muqtataf* 3 (1878): 135.

91. "al-Lahm al-Mubarrad," *al-Muqtataf* 41 (1912): 411; "al-Lahm al-Majlud wa al-Sawf," *al-Muqtataf* 15 (1891): 760.

92. "al-Lahm wa ma fihi min al-Ghidha'," *al-Muqtataf* 47 (1915): 5.

93. "al-Ghidha' fi al-Bayd," *al-Muqtataf* 23 (1908): 1055; "al-Ghidha' fi al-Asmak," *al-Muqtataf* 9 (1885): 479; "al-Ghidha' fi al-Ful al-Sudani," *al-Muqtataf* 19 (1895): 204.

94. "Ghidha' al-Tifl," *al-Muqtataf* 26 (1901): 561; "al-Ta'am wa al-Ghidha'," *al-Muqtataf* 27 (1902): 27.

95. "al- Ghidha' fi al-At'imah al-Mukhtalifa," *al-Muqtataf* 54 (1919): 75.

96. Woods, *Nature and the Refrigerating Machine*, 93–96.

97. "Frozen Food," *Chambers's Journal of Popular Literature, Science and Arts*, July 14, 1882, 439, quoted in Woods, *Nature and the Refrigerating Machine*, 96.

98. Seyma Afacan, "Of the Soul and Emotions: Conceptualizing 'the Ottoman Individual' through Psychology" (PhD diss., Oxford University, 2016).

99. "al-Lahm kayfa Yahfazu wa kayfa Yafsidu," *al-Muqtataf* 56 (1920): 64–67.

100. Scott S. Reese, *Imperial Muslims: Islam, Community and Authority in the Indian Ocean, 1839–1937* (Edinburgh: Edinburgh University Press, 2018), 58.

101. Muslims would often hire a Jewish kosher butcher before welcoming a Jewish guest, or they would cook vegetarian meals for members of other communities. Yemeni Jews, for their part sought (and received) rabbinical permits to eat Muslim butter and cheese despite the possibility that milk and meat were not separated when making these foods. See Mark S. Wagner, *Jews and Islamic Law in Early 20th-Century Yemen* (Bloomington, IN: Indiana University Press, 2015), 71–72.

102. Wagner, 72.

103. Sayyid Ahmed Khan, *A Voyage to Modernism*, trans. and ed. Mushirul Hasan and Nishat Zaidi (New Delhi: Primus Books, 2011); Shiblī al-Nu'manī, *Rihlat Shiblī al-Nu'manī ilá al-Qusṭanṭīniyya wa-Bayrut wal-Quds wal-Qahira fi Mustahall al-Qarn al-Rabi' 'Ashar al-Hijrī* (Damascus: Dar al-Qalam, 1967).

104. Khan, *Voyage to Modernism*, 151. Like other steam pilgrims, 'Abdallah al-Sa'udi devotes attention to water infrastructures on board steamers in his own rihlah. See Farid Kioumgi and Robert Graham, *A Photographer on the Hajj: The Travels of Muhammad 'Ali Effendi Sa'udi (1904/1908)* (Cairo: American University in Cairo Press, 2009), 12–13.

105. Bat-Zion Eraqi Klorman, *Yehudey Teyman: Historia, Ḥevra, Tarbut* [Jews of Yemen: History, society, culture], vol. 2 (Ra'anana: The Open University, 2004), 216–17.

106. Roy Bar Sadeh, "Debating Gandhi in al-Manar During the 1920s and 1930s," *Comparative Studies of South Asia, Africa and the Middle East* 38, no. 3 (2018): 491–507.

107. See Noor-Aiman Khan, "The Enemy of My Enemy: Indian Influences on Egyptian Nationalism, 1907–1930" (PhD diss., University of Chicago, 2006).

108. Faisal Devji, *The Terrorist in Search of Humanity: Militant Islam and Global Politics* (New York: Columbia University Press, 2008).

109. Woods, *Nature and the Refrigerating Machine*, 104.

110. Matthew S. Hopper, "The Globalization of Dried Fruit: Transformations in the Eastern Arabian Economy, 1860s–1920s," in *Global Muslims in the Age of Steam and Print*, ed. James Gelvin and Nile Green (Berkeley: University of California Press, 2014), 158–84.

111. European travelers often recounted how water "stained with old filth of camels" or water which is "thick and ill-smelling . . . putrefying with rotten fibres of plants and urea of the nomads' cattle, which have been watered here from the beginning. Of such the Arab (they prefer the thick desert water to pure

water) now boiled their daily coffee, which is not then ill-tasting." See Doughty, *Travels in Arabia Deserta*, 1:306, 242–3.

112. Catherine M. Tucker, *Coffee Culture: Local Experiences, Global Connections* (London: Routledge, 2011), 62.

113. Z.H. Kour, *The History of Aden, 1839–72* (London: Frank Cass, 2005), 9; Miran, *Red Sea Citizens*, 63, 5–75; Thomas E. Marston, *Britain's Imperial Role in the Red Sea Area, 1800–1878* (Hamden, CT: Shoe String Press, 1961), 34–35; Reese, *Imperial Muslims*, chap. 2.

114. *Report from the Select Committee on Steam Communication with India: Together with the Minutes of Evidence, Appendix and Index* (London: Ordered by the House of Commons, Parliament, 1837), 64.

115. Michael Greenberg, *British Trade and the Opening of China, 1800–42* (Cambridge: The University Press, 1951), 81, 106.

116. Greenberg, 653–54.

117. Liquor Contract to Aden Town: Military Dept., 1847–48, vol. 354, comp. 39, 503–22, esp. 22, Maharashtra State Archives, Mumbai.

118. "Farms for Miscellaneous Items: Liquor, Ganja, Opium and Khaat," Revenue Dept., 1847, vol. 90, comp. 450, 95–146, esp. 18. Maharashtra State Archives, Mumbai.

119. Freda Harcourt, *Flagships of Imperialism: The P&O Company and the Politics of Empire from Its Origins to 1867* (Manchester: Manchester University Press, 2013), 89–90.

120. J. Forbes Munro, Maritime Enterprise and Empire: *Sir William Mackinnon and His Business Network, 1823–93* (Woodbridge, Suffolk, UK: Boydell Press, 2003), 82.

121. Munro, 82.

122. Munro, 166.

123. Matthew T. Huber, *Lifeblood: Oil, Freedom, and the Forces of Capital* (Minneapolis: University of Minnesota Press, 2013).

124. Simon Fairlie has argued that given animals' role in manure production, small-scale husbandry is preferable to none at all. See Fairlie, *Meat: A Benign Extravagance* (White River Junction, VT: Chelsea Green, 2010).

CHAPTER 3. HUMANS

1. As William Stanley Jevons put it in *The Coal Question: An Inquiry Concerning the Progress of the Nation, and the Probable Exhaustion of Our Coal-Mines* (London: Macmillan, 1865), viii.

2. Cara Daggett, *Energy's Power: Fuel, Work, and Waste in the Politics of the Anthropocene* (PhD diss., Johns Hopkins University, 2016).

3. Benjamin Disraeli, *Coningsby; Or, the New Generation* (Oxford: Oxford University Press, 1844), 2:7.

4. Anson Rabinbach, *The Human Motor: Energy, Fatigue, and the Origins of Modernity* (New York: Basic Books, 1990).

5. The term was coined by Richard Buckminster Fuller. On fossil fuels and slavery, see Jean-François Mouhot, "Past Connections and Present Similarities in Slave Ownership and Fossil Fuel Usage," *Climatic Change* 105, nos. 1–2

(2011): 329–55; Andrew Nikiforuk, *The Energy of Slaves: Oil and the New Servitude* (Vancouver, BC: Greystone Books, 2012).

6. P.E. Hair, "Slavery and Liberty: The Case of the Scottish Colliers," *Slavery and Abolition* 23, no. 3 (2000): 136–51.

7. House of Commons Parliamentary Papers, *Children's Employment Commission, First Report of the Commissioners, Mines* (1842; repr., Shannon Irish University Press, 1968), 42, 103.

8. See Ivan Hannaford, *Race: The History of an Idea in the West* (Baltimore: John Hopkins University Press, 1996).

9. J. Edward Chamberlin and Sander Gilman, *Degeneration: The Dark Side of Progress* (New York, 1985), xiii. See also Karl Polanyi's argument about how increased foreign trade aggravated pauperization in England: Polanyi, *The Great Transformation* (Boston: Beacon Press, 1957), chap. 8.

10. See, for example, *Morning Chronicle*, January 4, 1833.

11. Adam Smith had argued as much in *The Wealth of Nations* (1776; repr., London: Penguin Books, 1979).

12. M. Savage, "Discipline, Surveillance and the 'Career': Employment on the Great Western Railway 1833–1914," in *Foucault, Management and Organization Theory*, ed. Alan McKinlay and Ken Starkey (London: Sage Publications, 1998), 65–93.

13. On the absence of "careers" in the British-controlled Egyptian railways, see Barak, *On Time*, chap. 2; on the difference between "industrial capitalism" and "war capitalism," see Beckert, *Empire of Cotton*.

14. Thomas C. Holt, *The Problem of Freedom: Race, Labor, and Politics in Jamaica and Britain, 1832–1938* (Baltimore: John Hopkins University Press, 1992); Uday Singh Mehta, *Liberalism and Empire: A Study in Nineteenth-Century British Liberal Thought* (Chicago: University of Chicago Press, 1999); Christopher Alan Bayly, *Recovering Liberties: Indian Thought in the Age of Liberalism and Empire* (Cambridge: Cambridge University Press, 2012); Andrew Stephen Sartori, *Liberalism in Empire: An Alternative Story* (Oakland: University of California Press, 2014).

15. Malm, *Fossil Capital*.

16. Throughout the nineteenth century the most significant component of the price of coal was miners' wages; Malm, 253.

17. On his stance regarding the former question, see Gray, *Steam Power*. On his approach to the latter, see Miloš Ković, *Disraeli and the Eastern Question* (New York: Oxford University Press, 2011). Disraeli's Orientalism is manifest in earlier novels from the 1830s, like *The Rise of Iskander* and *The Wondrous Tale of Alroy*, which anticipated political novels like *Coningsby* and are nowadays read as an allegory for Victorian England's imperial idea. Orientalism also animated these subsequent works, for example the 1847 novel that completed the trilogy begun by *Coningsby, Tancred; or, The New Crusade*.

18. Malm, *Fossil Capital*, 209.

19. A term indexing Bruno Latour's affront at the naïve belief that fetishes—objects invested with mythical powers—are fabricated while scientific facts are not, see Latour, *On the Modern Cult of the Factish Gods* (Durham, NC: Duke University Press, 2010).

20. Dipesh Chakrabarty, *Provincializing Europe: Postcolonial Thought and Historical Difference* (Princeton, NJ: Princeton University Press, 2007).

21. In relation to the coal heavers of Port Said, Zachary Lockman argues that positing a fixed common subject ("workers") diverts attention from the actual subjectivities of the heavers themselves and from the meanings of their actions. Lockman, "'Worker' and 'Working Class' in pre-1914 Egypt," in *Workers and Working Classes in the Middle East: Struggles, Histories, Historiographies*, ed. Zachary Lockman (Albany: State University of New York Press, 1994), 85.

22. Chakrabarty, *Provincializing Europe*, 77–78.

23. And these have been discussed repeatedly since the publication of Jacques Derrida's *Specters of Marx, the State of the Debt, the Work of Mourning, & the New International*, trans. Peggy Kamuf (New York: Routledge, 1994).

24. *Capital*, quoted in Mark Neocleus, "The Political Economy of the Dead: Marx's Vampires," *History of Political Thought* 24, no. 4 (2003): 683.

25. The vampire and Marx's ideas on circulation were informed by the writings of Justus von Liebig, father of organic chemistry, who in 1840 compared Britain, the great importer of mineral fertilizer, to a vampire: "Great Britain seizes from other countries their conditions of their own fertility . . . Vampirelike, it clings to the throat of Europe, one could even say of the whole world, sucking its best blood." *Organic Chemistry*, quoted in Christophe Bonneuil and Jean-Baptiste Fressoz, *The Shock of the Anthropocene: The Earth, History and Us* (London: Verso, 2016), 187.

26. "What remains is [abstract labor's] quality of being an expenditure of human labour-power"; or, "all labour is an expenditure of human labour power, in the physiological sense, and it is in this quality of being equal, or abstract, human labour that it forms the value of commodities." Quotes are in Chakrabarty, *Provincializing Europe*, 53, 57.

27. Philip Mirowski, *More Heat than Light: Economics as Social Physics, Physics as Nature's Economics* (Cambridge: Cambridge University Press, 1991), 19, 174–85.

28. Foucault then leaps to claim, "that is, it is unable to breathe anywhere else." Michel Foucault, *Order of Things: An Archaeology of the Human Sciences* (New York: Psychology Press, 2002), 262. However, at least with respect to current Marxist thinking about abstract labor, this is far from being the case. Michael Heinrich, for example, convincingly anchors Marx's "naturalized or calorific theories of labour" in ambivalences in the critique of political economy which he himself and other thinkers later transcended. See Heinrich, "Ambivalences of Marx's Critique of Political Economy as Obstacles for the Analysis of Contemporary Capitalism," *Historical Materialism Conference*, October 10, 2004, London, revised paper; and Heinrich, *An Introduction to the Three Volumes of Karl Marx's Capital*, trans. Alexander Lucascio (Monthly Review Press, 2004), chap. 3. Nevertheless, even these correctives, stressing that abstract labor is predicated on commodity exchange and not on the physical homogenization of labor power, are arguably still tied to thermodynamic commensurability.

29. Barak, *On Time*, 244–45.

30. Sam Gindin, "Labor," in *Keywords for Radicals: The Contested Vocabulary of Late-Capitalist Struggle*, ed. Kelly Fritsch, Claire O'Connor, and A.K.

Thompson (Oakland, CA: AK Press, 2016), 223; David Spencer, *The Political Economy of Work* (London: Routledge, 2009), chap. 2.

31. Talal Asad, *Formations of the Secular: Christianity, Islam, Modernity* (California: Stanford University Press, 2003), chap. 2; Jacques Donzelot, "Pleasure in Work" in Graham Burchel, Colin Gordon, Peter Miller, eds. *The Foucault Effect: Studies in Governmentality* (Chicago: University of Chicago Press, 1991), 251–81; Michel Foucault, *On the Punitive Society: Lectures at the Collège de France, 1972–1973* (London: Palgrave Macmillan, 2015); Mirowski, *More Heat than Light*, 254–62; Spencer, *A Political Economy of Work*, chap. 5.

32. Quoted in Denver Brunsman, *The Evil Necessity: British Naval Impressment in the Eighteenth-Century Atlantic World* (Charlottesville: University of Virginia Press, 2013), 251.

33. Richard I. Lawless, "Recruitment and Regulation: Migration for Employment of 'Adenese' Seamen in the Late Nineteenth and Early Twentieth Centuries," in *New Arabian Studies*, vol. 2, ed. R.L.Tidwell, G. Rex Smith, and J.R. Smart (Exeter: University of Exeter Press, 1994), 77.

34. Janet J. Ewald, "Bondsmen, Freedmen, and Maritime Industrial Transportation, c. 1840–1900," *Slavery and Abolition* 31, no. 3 (2010): 454.

35. Ewald, 457.

36. This was also the case in the Caribbean, where state subsidy via mail-delivery contracts and emancipation sustained a nascent steamer line. See Anyaa Anim-Addo, "'A Wretched and Slave-like Mode of Labor': Slavery, Emancipation, and the Royal Mail Steam Packet Company's Coaling Stations," *Historical Geography* 39 (2011): 65–84.

37. Johan Mathew, *Margins of the Market: Trafficking and Capitalism across the Arabian Sea* (Oakland: University of California Press, 2016), 34; Ewald, "Bondsmen," 458.

38. Mathew, *Margins of the Market*, 64.

39. Mathew, 54, 56.

40. Michael Ferguson, "Clientship, Social Indebtedness and State-Controlled Emancipation of Africans in the Late Ottoman Empire," in *Debt and Slavery in the Mediterranean and Atlantic Worlds*, ed. Gwyn Campbell & Alessandro Stanziani (London: Pickering and Chatto, 2013), 49–62.

41. *Portsmouth Evening News*, October 23, 1894.

42. *Edinburgh Evening News*, August 29, 1888.

43. *The Dewsbury Reporter*, July 23, 1881.

44. Edwin Arnold, "India Revisited," *Daily Telegraph*, December 4, 1885, discussed in Ewald, "Bondsmen."

45. Lawless, "Recruitment and Regulation," 76.

46. See Marika Sherwood, "Race, Nationality and Employment among Lascar Seamen, 1660 to 1945," *New Community* 17 (1991): 229–44; G. Balachandran, *Globalizing Labour?: Indian Seafarers and World Shipping, c.1870–1945* (New Delhi: Oxford University Press, 2012).

47. Mohammad Siddique Seddon, *The Last of the Lascars: Yemeni Muslims in Britain, 1836–2012* (Leicestershire: Kube, 2014).

48. By 1856, over one thousand escaped slaves lived there, many of whom engaged in maritime work. Between 1865 and 1870, more than double that

number reached Aden en route to other ports, and more than half of them were reported to make a living in maritime work. Ewald, "Bondsmen," 457, 459.

49. R. J. Gavin, *Aden under British Rule, 1839–1966* (London: C Hurst, 1975), 59, 191.

50. Lawless, "Recruitment and Regulation," 80–81.

51. As the reports of the *Hugh Lindsey* reveal; Captain J. H. Wilson, *On Steam Communication between Bombay and Suez, with an Account of the Hugh Lindsay's Four Voyages* (Bombay: American Press, 1833), 341–42, quoted in Ewald, "Bondsmen," 460; see also House of Commons, *Report from the Select Committee on Steam Communication with India* (London: Ordered by the House of Commons, Parliament, 1834), Appendix 17.

52. Ewald, 461.

53. *Hampshire Telegraph,* November 14, 1891.

54. Delhi, "Decision That Arab Seamen, Who Are Native of Aden, Should Sign on Indian Articles of Agreement and Not on European Articles as Hitherto," Finance and Commerce Dept., proceedings nos. 166–68, April 1904, National Archives of India, New Delhi.

55. Ravi Ahuja, "Mobility and Containment: The Voyages of South Asian Seamen, c.1900–1960," *International Review of Social History* 51 (2006): 116–19.

56. Ahuja, 113.

57. See, for example, the summary of findings from the RN Blue Book of 1908, "Heat in Stokeholds: Interesting Experiments in the Red Sea," *London Daily News,* September 22, 1909.

58. "Heat in Stokeholds."

59. Timothy Mitchell, *Colonising Egypt* (Berkeley: University of California Press, 1991); Barak, *On Time,* chap. 2.

60. See Malm, *Fossil Capital,* 245, 247.

61. Steven Gray, "Black Diamonds: Coal, the Royal Navy, and British Imperial Coaling Stations, circa 1870–1914" (PhD diss., University of Warwick, 2014) 224, 230.

62. John Chalcraft, "Popular Protest, the Market and the State in Nineteenth and Early Twentieth Century Egypt," in *Subalterns and Social Protest: History from Below in the Middle East and North Africa,* ed. Stephanie Cronin (London: Routledge 2008), 71.

63. Mrs. Postans, "Characteristics of Aden, with the Passage of the Red Sea," *The Illuminated Magazine* 1 (London, 1843): 78; see also Ewald, "Bondsmen," 460–61.

64. See also Valeska Huber, *Channelling Mobilities: Migration and Globalisation in the Suez Canal Region and Beyond, 1869–1914* (Cambridge: Cambridge University Press, 2013), 119–21; Gray, *Black Diamonds,* 230–31.

65. E. G. Anning, F. J. Bentley, and Lionel Yexley, *The Log of H.M.S. Argonaut, 1900–1904* (London: Westminster Press, 1904), 9, quoted in Gray, *Black Diamonds,* 231.

66. H. M. Fowler, *The Log of H.M.S. Encounter, Australian Station, 1908–1910* (London: Westminster Press, 1910), 3, quoted in Gray, *Black Diamonds,* 232.

67. House of Commons, *Report from the Select Committee on Steam Navigation to India* (1834), 104.

68. Robert J. Blyth, "Aden, British India and the Development of Steam Power," in *Maritime Empires: British Imperial Maritime Trade in the Nineteenth Century*, ed. David Killingray, Margarette Lincoln, and Nigel Rigby (Suffolk: Boydell Press, 2004) 73–74; Huber, *Channelling Mobilities*, 119.

69. Blyth, "Aden, British India and the Development of Steam Power," 73–74.

70. A. O. Lamplough, *Egypt and How to See It* (London, 1908), 14.

71. Najm, *Bur Sa'id*, 78.

72. Najm, 78–79n2; *Nelson Evening Mail* (New Zealand), December 21, 1907.

73. Quoted in Joel Beinin and Zachary Lockman, *Workers on the Nile: Nationalism, Communism, Islam, and the Egyptian Working Class, 1882–1954* (Princeton, NJ: Princeton University Press, 1987), 27.

74. Anna Green, "The Work Process," in *Dock Workers: International Explorations in Comparative Labour History, 1790–1970*, ed. Sam Davies, Colin J. Davis, David de Vries, Lex Heerma van Voss, Lidewijb Hesselink, and Klaus Weinhauer (Burlington, VT: Ashgate, 2000), 2:575–76.

75. Mariam Dossal Panjwani, "*Godis, Tolis* and *Mathadis:* Dock Workers of Bombay," in *Dock Workers: International Explorations in Comparative Labour History, 1790–1970*, ed. Sam Davies, Colin J. Davis, David de Vries, Lex Heerma van Voss, Lidewijb Hesselink, and Klaus Weinhauer (Burlington, VT: Ashgate, 2000), 1:428; Green, "Work Process," 561–62.

76. For a reading that does highlight these aspects, see Huber, *Channelling Mobilities*, 121–22.

77. Najm, *Bur Sa'id*, 79.

78. Najm, 79n1.

79. Najm, 79.

80. See Beinin and Lockman, *Workers on the Nile*, 29.

81. Huber, *Channelling Mobilities*, 121–22.

82. *London Evening Standard*, April 10, 1882.

83. *Sunderland Daily Echo and Shipping Gazette*, April 3, 1882.

84. Beinin and Lockman, *Workers on the Nile*, 29–30.

85. Gavin, *Aden under British Rule, 1839–1966*, 191.

86. *Aberdeen Journal*, September 29, 1882.

87. See references in Bonneuil and Fressoz, *Shock of the Anthropocene*, 239.

88. Beinin and Lockman, *Workers on the Nile*, 30.

89. *London Evening Standard*, May 23, 1894.

90. *Daily Advertiser* (Wagga Wagga, Australia), June 23, 1894.

91. *London Evening Standard*, May 23, 1894; a similar line was taken in June. See *Daily Advertiser,* June 23, 1894.

92. *Daily Advertiser,* June 23, 1894.

93. *Morning Post* (London), October 1, 1894. The beginning of the strike was reported in the *Dundee Courier,* August 22, 1894.

94. The historiography on Greek labor militancy in Egypt, and particularly on organizing efforts in 1894, tended to view these workers as a "labor elite," a term reminiscent of disapproving characterizations of certain segments of the

English proletariat as "labor aristocracy." See Anthony Gorman, "Foreign Workers in Egypt 1882–1914: Subaltern or Labour Elite?," in *Subalterns and Social Protest: History from Below in the Middle East and North Africa,* ed. Stephanie Cronin (London: Routledge 2008), 237–61.

95. Quoted in John Chalcraft, "The Coal Heavers of Port Sa'id: State-Making and Worker Protest, 1869–1914," *International Labor and Working-Class History,* no. 60 (2001): 120.

96. FO 141/322 (May 1896) and FO 633/8 (Cromer Papers, May 1896), British National Archives (BNA), London.

97. Juan Cole, *Colonialism and Revolution in the Middle East: Social and Cultural Origins of Egypt's 'Urabi Movement* (Princeton, NJ: Princeton University Press, 1993), 164–89; Chalcraft, "Popular Protest," 77–86.

98. G.R. Parker, *The Commission of H.M.S. Impeccable, Mediterranean Station, 1901–1904* (London: Westminster Press, 1904), quoted in Gray, *Black Diamonds,* 222.

99. Gray, *Black Diamonds,* 222–23.

100. *Commission of H.M.S. Impeccable,* 97, quoted in Gray, *Black Diamonds,* 223.

101. *Coaling, Docking, and Repairing Facilities of the Ports of the World* (Washington, DC: Government Printing Office, 1909), 52, 53.

102. D.G. Lance, "Interview with Arthur Lilly" (London: Imperial War Museum, 1976), 750, quoted in Gray, *Black Diamonds* 245.

103. Quoted in Gray, *Black Diamonds,* 246.

104. Étienne Balibar and Immanuel Maurice Wallerstein, *Race, Nation, Class: Ambiguous identities* (London: Verso, 2011).

105. Chalcraft, "Popular Protest," 71.

106. *Huddersfield Chronicle* (Yorkshire), April 19, 1882.

107. Roy Church, *Strikes and Solidarity: Coalfield Conflict in Britain, 1889–1966* (Cambridge: Cambridge University Press, 1998), 38, 264.

108. *St James's Gazette* (London), Monday April 17, 1882.

109. Y.A. HUS 484/60, BOA.

110. Huber, *Channelling Mobilities,* 121–22.

111. *Morning Post,* August 23, 1893.

112. *South Wales Daily News* (Cardiff), September 21, 1893.

113. During that tumultuous year, strikes by the Istanbul porters and lighter boatmen guild against the French-owned Istanbul Quay Company (the capital's quays were transformed from timber to mainly coal repositories) were seen in this transnational context. Yıldız Sadaret Hususi Maruzat Evrakı (Y.A.HUS), 298/83 (1311 [1893]), BOA. On the strikes see Beinin and Lockman, *Workers on the Nile,* 79; Cem Behar, *A Neighborhood in Ottoman Istanbul: Fruit Vendors and Civil Servants in the Kasap Ilyas Mahalle* (Albany: State University of New York Press, 2003), 155–58.

114. A.MKT.MVL 74/26 (1271 [1854]), 28, BOA.

115. Gordon Wilson, *Alexander McDonald: Leader of the Miners* (Aberdeen: Aberdeen University Press, 1982), 32.

116. By the mid-1860s, when much of the colonial and semicolonial world had already become a captured market for English coal, Jevons estimated that

"the best coal is *put on board* at Newcastle for 9s. per ton. Before it reaches France, it is about trebled; in the Mediterranean ports, Genoa, or Leghorn, it is quadrupled, while in many remote parts of the world coal cannot be purchased for less than 3*l*. or 3*l*. 10s. per ton." Jevons, *Coal Question*, 251. While towards century's end a global trend towards price convergence was observable, this accelerated only after World War I. Alexis Wegerich, "Digging Deeper: Global Coal Prices and Industrial Growth, 1840–1960" (PhD diss., University of Oxford, 2016). Alongside attempts to regulate a quoted price, there were constant efforts to sell both English and local coal at a higher price, by hoarding it and manipulating its regulation. Sadaret Mektubi Kalemi Nezaret ve Deva'ir Evrakı (A.MKT.NZD), 427/91 (1278 [1861]); Sadaret Mektubi Kalemi Umum Vilayat Evrakı (A.MKT.UM), 214/84 (1272 [1855–1856]), BOA.

117. Such as racketeering and contraband starting in the 1840s. Sadaret Mektubi Kalemi Evrakı (A.MKT.), 111/94 (1260 [1844]), BOA.

118. Babıali Evrak Odası Evrakı (BEO), 1315/98616 (1317 [1899]); BEO 1380/103485 (1317 [1899]); BEO 2676/200646 (1323 [1905]), BOA.

119. Christopher Alan Bayly, *The Birth of the Modern World, 1780–1914: Global Connections and Comparisons* (Malden, MA: Blackwell, 2004), 191–93.

120. Can Nacar, "Labor Activism and the State in the Ottoman Tobacco Industry," *International Journal of Middle East Studies* 46, no. 3 (2014): 533.

121. *Sheffield Daily Telegraph*, September 10, 1896.

122. Taner Akan, "Does Political Culture Matter for Europeanization? Evidence from the Ottoman Turkish Modernization in State-Labor Relations," *Employee Relations* 33, no. 3 (2011): 221–48.

123. A. Baran Dural and Ertem Uner, "Development of the Worker Class in the Ottoman Empire," *European Scientific Journal* 8, no. 11 (2012): 215–35.

124. "Dahiliye Emniyet-i Umumiye Tahrirat Kalemi Evrakı" (DH.EUM. THR) 94/6 (1327 [1909]), BOA.

125. Zachary Lockman, "'Worker' and 'Working Class' in pre-1914 Egypt: A Rereading," in *Workers and Working Classes in the Middle East: Struggles, Histories, Historiographies,* ed. Zachary Lockman (Albany: State University of New York Press, 1994), 84.

126. Dahiliye Nezareti Emniyet-i Umumiye Kısm-ı Adli Kalemi (DH.EUM. KADL), 3/44 (1328 [5 January 1911]), BOA.

127. *Aberdeen Journal*, Aberdeen, SCT, December 15, 1914.

128. Joel Beinin, *Workers and Peasants in the Modern Middle East* (Cambridge: Cambridge University Press, 2010), 78.

129. Kadir Yıldırım, *Osmanlılar'da İşçiler (1870–1922)* (Istanbul: İletişim, 2013), 233, 235; Zabtiye Nezareti Evrakı (ZB) 312/16, November 23, 1908, BOA.

130. Akan, "Does Political Culture Matter?"

131. M. Şükrü Hanioğlu, *Preparation for a Revolution: The Young Turks, 1902–1908* (London: Oxford University Press, 2001), 311.

132. DH.EUM.KADL, 3/44 5 (1328 [January 1911]), BOA.

133. Neomi Levy-Aksu, "Institutional Cooperation and Substitution: The Ottoman Police and Justice System at the Turn of the 19th and 20th Centuries," in *Order and Compromise: Government Practices in Turkey from the Late*

Ottoman Empire to the Early 21st Century, ed. Marc Aymes, Benjamin Gourisse, and Elise Massicard (Leiden: Brill, 2014), 146–68.

134. ZB 491/75, (1324 [1906]); DH.MKT 2616/37 (1326 [1908]); Dahiliye Nezareti İdare Evrakı (DH.İD), 112–1/21 (1332 [1914]); DH.EUM.KADL, 3/44 (1329 [1911]), BOA.

135. Clive Emsley, *The English Police: A Political and Social History* (London: Longman, 1996), chap. 4.

136. See Beckert, *Empire of Cotton.*

137. Herbert Stanley Jevons, *Foreign Trade in Coal* (London: P.S. King & Son, 1909).

138. See copy of Jevons's *Foreign Trade in Coal* in CHAR 11/21 (February 1904–December 24, 1909), and correspondence by Jevons in CHAR 2/597A-B (Dec. 1936), both in Churchill Archives, University of Cambridge, Cambridge.

139. Timothy Mitchell, *Carbon Democracy: Political Power in the Age of Oil* (London: Verso, 2011), 36.

140. Donald Quataert, *Miners and the State in the Ottoman Empire: The Zonguldak Coalfield, 1822–1920* (New York: Berghahn Books, 2006), 53.

141. Quataert, chap. 4.

142. Shadid Amin and Marcel van den Linden, eds., *Peripheral Labour: Studies in Partial Proletarianization* (Cambridge: Cambridge University Press, 1997). A critique in this vein has also been developed by historians of Ottoman labor; see Touraj Atabaki and Gavin Brockett, "Ottoman and Republican Turkish Labour History: An Introduction," in *Ottoman and Republican Turkish Labour History*, ed. Touraj Atabaki and Gavin Brockett (Cambridge: Cambridge University Press, 2009), 1–19. For Egypt, see Lockman, "'Worker' and 'Working Class.'"

143. Marx, *Capital*, vol. 1, chap. 26.

144. BEO 4019/301377 (1330 [1912]), R 5; and DH.İD 112–1/15 (1330 [1912]), § 6, BOA.

145. DH.İD 112–1/15 (1330 [1912]), § 6, BOA.

146. Consider a November 1910 strike of about one thousand coal heavers in the Port of Istanbul, who explicitly took advantage of the leverage afforded them by the situation in Ereğli and in England. While an official inquiry found little evidence of foreign interference, the port officer explained that coal merchants incited the porters to strike so that the merchants might gain from the rise in coal prices. The strike had to be general, otherwise merchants would be obliged to pay indemnities to delayed steamer companies. DH.EUM.KADL, 3/44 (1328 [5 January 1911]), BOA.

147. DH.EUM.KADL, 3/44 (1328 [5 January 1911]), BOA. See BEO 4092/306870 for a decree exempting Ereğli miners from conscription to prevent delays in coal supply to Ottoman steamers.

148. Quataert, *Miners and the State*, 43.

149. Yıldırım, *Osmanlılar'da İşçiler*, 282.

150. Quataert, *Miners and the State*, 43.

151. Nurşen Gürboğa, "Mine Workers, the State and War: The Ereğli-Zonguldak Coal Basin as the Site of Contest, 1920–1947" (PhD diss., Boğaziçi University, Istanbul, 2005).

152. Georg Borgstrom, *Hungry Planet: The Modern World at the Edge of Famine* (New York: Collier Books, 1967), 70–86; William Catton, *Overshoot: The Ecological Basis of Revolutionary Change* (Urbana: University of Illinois Press, 1980), 44. See also Kenneth Pomeranz, *The Great Divergence: China, Europe, and the Making of the Modern World Economy* (Princeton, NJ: Princeton University Press, 2011).

153. Ellis Goldberg, "Mobilizing Coal for War: The Rise and Decline of English Socialism," unpublished paper online at SSRN (2015). https://ssrn.com/abstract = 2566110 or http://dx.doi.org/10.2139/ssrn.2566110.

154. Isador Lubin and Helen Everett, *The British Coal Dilemma* (New York: Institute of Economics, 1927). 24–25.

155. Wilson, *Alexander McDonald*, 9–10, 25; Church, *Strikes and Solidarity*, 29–30; Alan Campbell, "18th Century Legacies and 19th Century Traditions: The Labour Process, Work Culture and Miners' Unions in the Scottish Coalfield before 1914," in *Vom Bergbau zum Industrierevier*, ed. Ekkehard Westermann (Stuttgart: Franz Steiner Verlag, 1995), 233–34.

CHAPTER 4. ENVIRONMENT

1. In 1895, half of the British coal delivered to Istanbul was used in the railways and some factories in the city, and the other half was consumed by the foreign steamers. See Alaaddin Tok, "From Wood to Coal: Energy Economy in Ottoman Anatolia and the Balkans (1750–1914)" (PhD diss., Boğaziçi University, Istanbul, 2017), 198.

2. Jevons, *British Coal Trade*, 31.

3. Sarah Searight, "The Charting of the Red Sea," *History Today* 53, no. 3 (March 2003): 40–46.

4. Wilson, *On Steam Communication*, 40.

5. *Reports from Committees: Eighteen Volumes; East India Company's Affairs* (Parliamentary Papers), vol. 10, pt. 2 (London: H.M. Stationery Office, 1832), 742.

6. Caesar E. Farah, *The Sultan's Yemen: 19th-Century Challenges to Ottoman Rule* (London: I.B. Tauris, 2002), 1–14. On Yemen under Ottoman rule and British infringement, see also Thomas Kuehn, *Empire, Islam, and Politics of Difference: Ottoman Rule in Yemen, 1849–1919* (Leiden: Brill, 2011).

7. *Report from the Select Committee on Steam Communication with India: Together with the Minutes of Evidence, Appendix and Index* (London: Ordered by the House of Commons, 1837), 43.

8. *Reports from Committees*, 10:2:743.

9. *Reports from Committees*, 10:2:752.

10. See Nile Green, *Bombay Islam: The Religious Economy of the West Indian Ocean, 1840–1915* (Cambridge: Cambridge University Press, 2011), 31–34; Khan, *Voyage to Modernism*, 68–69; see also Robert E. Kennedy, "The Protestant Ethic and the Parsis," *American Journal of Sociology* 68., no. 1 (1962): 11–20.

11. William Fancy, "The Red Sea: The Wind Regime and Location of Ports," in *Trade and Travel in the Red Sea Region: Proceedings of Red Sea Project I,*

Held in the British Museum, October 2002, ed. Paul Lunde and Alexandra Porter (Oxford: Tempus Reparatum, 2004), 7–17.

12. Satyindra Singh, *Blueprint to Bluewater, the Indian Navy, 1951–1965* (New Delhi: Lancer International, 1992), 252.

13. Adam Kirkaldy, *British Shipping: Its History, Organisation and Importance* (London: Kegan, Paul, Trench, Trübner, 1914), 582.

14. Kirkaldy, 582.

15. F. E. Chadwick, U.S. Navy, "The Development of the Steamship, and the Liverpool Exhibition of 1886." *Scribner's Magazine* 1, no. 5 (May 1887): 515–43.

16. Michael Pearson, *The Indian Ocean* (London: Taylor & Francis, 2003), 193.

17. "Proceedings and Consultations of the Government of India and of its Presidencies and Provinces: Sanitary," IOR.

18. G. D. Urquhart, *Dues and Charges on Shipping in Foreign Ports: A Manual of Reference for the Use of Shipowners, Shipbrokers, and Shipmasters* (London: George Philip, 1872).

19. For example, the Perim Coal Company arranged that as soon as a steamer was reported, day or night, the settlement was aroused by electric bells. Coal coolies would rush to the loaded lighters and tow the coal in quantities signaled by the steamer's whistle, viz: for 50 tons, one short whistle; for 75 tons, two short whistles; for 100 tons, one long whistle; for 150 tons, one long and one short whistle; for 200 tons, two long whistles; for 250 tons, two long and one short whistle; for 300 tons, three long whistles; etc. See United States Office of Naval Intelligence, *Coaling, Docking, and Repairing Facilities of the Ports of the World, with Analyses of Different Kinds of Coal,* 4th ed. (Washington: Government Printing Office, 1900), 102.

20. Jevons, *Coal Question,* 260–61.

21. Galina Harlaftis, *A History of Greek-Owned Shipping: The Making of an International Tramp Fleet, 1830 to the Present Day* (London: Routledge, 1995), 22.

22. Jevons, *Coal Question,* 265, 278.

23. Jevons, 253.

24. David Alfred Thomas, *The Growth and Direction of Our Foreign Trade in Coal during the Last Half Century* (London: Harrison and Sons, 1903): 16.

25. Kirkaldy, *British Shipping,* 461–62.

26. See, for example, the comparison between different ballasting arrangements in E. E. Allen, "On the Comparative Cost of Transit by Steam and Sailing Colliers, and on the Different Modes of Ballasting," *Minutes of Proceedings of the Institution of Civil Engineers,* 14:328–36. www.icevirtuallibrary.com/doi/abs/10.1680/imotp.1855.23910.

27. *The Land We Live In: A Pictorial and Literary Sketch-Book of the British Empire* (London 1856), 24–26; B. Martel, "On Water Ballast," *Scientific American: Supplement* 4, no. 97 (November 10, 1877): 1536–38.

28. Erich W. Zimmerman, "Why the Export Coal Business of America Should Be Built Up—III," *Coal Age* 19 (1921): 18.

29. Michael Pearson, *The Indian Ocean* (London: Taylor & Francis, 2003), 192–93, 200; Frank Broeze, "The Ports and Port System of the Asian Seas: An

Overview with Historical Perspective from c. 1750," *The Great Circle* 18, no. 2 (1996): 78–79.

30. Quoted in Pearson, *Indian Ocean,* 212.

31. Najm, *Bur Saʿid,* 246–47, 303.

32. Gavin, *Aden under British Rule,* 186.

33. Sarah Palmer, "Current Port Trends in an Historical Perspective," *Journal of Maritime Research* 1, no. 1 (1999): 102; Gavin, *Aden under British Rule,* 181–84.

34. "Question of the Manufacture of Salt on Account of Govt. at Aden, Perim and Zaila; Development of Salt Trade from Karachi to Calcutta," Finance & Commerce Dept., Separate Revenue, proceedings nos. 171–200, April 1891, enclosed in Foreign Dept., External, proceeding no. 177, June 1891, National Archives of India, New Delhi.

35. S. A. M. Adshead, *Salt and Civilization* (London: Palgrave Macmillan, 2016), 160.

36. "Report on the Working of the Salt Department at Aden for 1884–85," Finance & Commerce Dept., Separate Revenue, proceedings nos. 608–611, June 1885; "Question of the Manufacture of Salt"; "Tenders for the Lease of Certain Unoccupied Land for the Construction of Salt Works at Aden," Finance & Commerce Dept., Separate Revenue—A, proceedings nos. 502–3, December 1907; "Salt: Lease of Certain Lands at Aden for the Manufacture of Salt," Revenue Dept., 1908; National Archives of India, New Delhi.

37. See John Smeaton, *A Narrative of the Building and a Description of the Construction of the Edystone Lighthouse with Stone: To Which is Subjoined, an Appendix, Giving Some Account of the Lighthouse on the Spurn Point, Built upon a Sand* (London: by the author, 1791).

38. Naguib Amin, Raymond Collet, Marie-Laure Crosnier Leconte, Arnaud Du Boistesselin, and Ǧamāl al- Ġīṭānī, *Port-Saïd: Architectures XIXe - XXe siècles = Būr Saʿīd: 'imārat al-qarn at-tāsiʿ 'ašar wa-'l-qarn al-'išrīn* (Le Caire: Institut Français d'Archéologie Orientale, 2006).

39. "Further Correspondence Respecting the Lighthouses in the Red Sea and Persian Gulf, 1886–1888," Foreign Office (FO) 881/5825, BNA.

40. "Affidavit on an Attempted Piratical Attack upon the British Steamship Rocket by Arab Dhows on Her Voyage from Aden to Cochin," Political Dept., proceedings nos. 221–22, July 1870, National Archives of India, New Delhi; for the British Isles, see Bella Bathurst, *The Lighthouse Stevensons: The Extraordinary Story of the Building of the Scottish Lighthouses by the Ancestors of Robert Louis Stevenson* (London: HarperCollins, 1999).

41. *British Documents on Foreign Affairs, 1856–1914* (Frederick, MD: University Publications of America, 1984), 19:280–82.

42. For the Ottoman and French perspectives see, for example, "Written Proceedings of the Council of Ministers," BEO 1090/81697, March 7, 189?, BOA. For the British perspective, see "Correspondence respecting the Lighthouses in the Red Sea and Persian Gulf, 1881–1886," FO 407/81, and "Further Correspondence respecting the Lighthouses in the Red Sea and Persian Gulf, 1886–1888," FO 881/5825, BNA.

43. James Taylor, "Private Property, Public Interest, and the Role of the State in Nineteenth-Century Britain: The Case of the Lighthouses," *Historical Journal* 44, no. 3 (2001): 749–71.

44. Avner Wishnitzer, "Into the Dark: Power, Light and Nocturnal Life in 18th-Century Istanbul," *International Journal of Middle East Studies* 46 (2014) 513–31; Avner Wishnitzer, "Shedding New Light: Outdoor Illumination in Late Ottoman Istanbul," in *Urban Lighting, Light Pollution and Society,* ed. Josiane Meier, Ute Hasenöhrl, Katharina Krause, and Merle Pottharst (New York: Routledge, 2015), 66–88.

45. See for example, ŞD 1905/25, March 27, 1880, BOA.

46. DH.MKT 1633/63, June 30, 1889, BOA.

47. "Memorandum about Red Sea," February 4, 1898, BEO 1076/80656, BOA.

48. For a history of the Ottoman lighthouse administration, see Jacques Thobie, *L'administration générale des phares de l'Empire ottoman et la société Collas et Michel, 1860–1960: Un siècle de coopération économique et financière entre la France, l'Empire ottoman et les états successeurs* (Paris: L'Harmattan, 2004).

49. See, for example, correspondence between Istanbul and the governor of Crete about a British steamer captain engaged in mapping the Mediterranean calling to build two lighthouses there: A.MKT.MHM 189/68, July 28, 1860, BOA.

50. Ha-'Adani, *Beyn 'Aden ve-Teman,* 4–5, quoted in Ahroni, *Jews of the British Crown Colony* (Leiden: E. J. Brill,1994), 43–44.

51. Such an account, of Shaykh Uways Muhammad, is discussed in Scott Steven Reese, "Patricians of the Benaadir: Islamic Learning, Commerce and Somali Urban Identity in the Nineteeenth Century" (PhD diss., University of Pennsylvania, 1996), 281.

52. Khan, *Voyage to Modernism,* 94.

53. See *al-Manar* 3 (1900): 1, and *al-Manar* 10 (1907): 628; Rida, *Ta'rikh al-'ustadh al-Imam,* 1000; and Stephane Dudoignon, Hisao Komatsu, and Yasushi Kosugi, eds., *Intellectuals in the Modern Islamic World: Transmission, Transformation and Communication* (London: Routledge, 2006), 10.

54. 'Abd al-Rahman Kawakibi, *Yesodot ha-Islam ha-Salafi: Protokol ye'idat ha-tehiyah ha-Islamit bam he-'arim Makah,* trans. Shosh Ben-Ari (1899; Hebrew trans., Tel Aviv: Resling, 2015), 174.

55. Quoted in W. C. Jacob, "Of Angels and Men: Sayyid Fadl bin Alawi and Two Moments of Sovereignty," *Arab Studies Journal* 20, no. 1 (Spring 2012): 51. See Engseng Ho, *The Graves of Tarim: Genealogy and Mobility across the Indian Ocean* (Berkeley: University of California Press, 2010), for how Fadl and his ancestors conceived of their genealogy from their forefather the 'Adani in terms of transmission of light.

56. Barak, *On Time,* chap. 1; *Report from the Select Committee on Steam Navigation to India* (1834), 42, 72, 224.

57. S. Smith, ed., *The Red Sea Region: Sovereignty Boundaries and Conflict, 1839–1967* (Cambridge: Cambridge Archive Editions, 2008), 34–35.

58. Smith, 34–35.

59. As was the case in Jeddah in 1858. See Ulrike Freitag, "Symbolic Politics and Urban Violence in Late Ottoman Jeddah," in *Urban Violence in the Middle East,* ed. Ulrike Freitag, Nelida Fuccaro, Claudia Ghrawi, and Nora Lafi (New York: Berghahn Books, 2015), 111–38.

60. C. W. Adye, "Eastern Telegraph Company Limited: Financial Memoranda, Sep. 1907." I thank Pauline Lewis for sharing these materials with me.

61. Daniel Defoe, *A Tour Thro' the Whole Island of Great Britain Divided into Circuits or Journeys* (London: S. Birt, T. Osborne, D. Browne, J. Hodges, J. Osborn, A. Miller and J. Robinson, 1748), 220.

62. Barbara Freese, *Coal: A Human History* (New York: Penguin Books, 2004), 41–42.

63. Ralph Waldo Emerson, *The Conduct of Life,* 1860, quoted in Freese, *Coal,* 10.

64. The book was first translated anonymously in 1835, then again in 1861, by Butrus al-Bustanī as *Kitab al-tuhfah al-Bustanīyah fī al-asfar al-Kuruzīyah, aw Rihlat Rubinsun Kruzī,* and again in 1923, by Ahmad Abbas. These Arabic translations spurred fifteen translations into Ottoman Turkish between 1864 to 1927, first from Arabic, and then from English.

65. Linda Colley, *Captives: Britain, Empire and the World 1600–1850* (New York: Anchor Books, 2004), introduction.

66. By the turn of the twentieth century, with bidirectional flows to and from Europe impossible to ignore, this anxiety would be compounded by its mirror image—the fear of brown matter, disease, and people from the East washed up on European shores, as exemplified by Thomas Mann's *Death in Venice* and Joseph Conrad's *Amy Foster.* I thank Virginia Richter for drawing my attention to this literary reversal.

67. Reese, *Imperial Muslims,* 47–55.

68. "Establishment of a French Coal Depot at Aden," Foreign Department, Proceedings 746/62, nos. 176–87, 1861–62, National Archives of India, New Delhi.

69. Y.EE.KP.86–38/3790, April 24, 1879, BOA.

70. Nelida Fuccaro, *Histories of City and State in the Persian Gulf: Manama since 1800* (New York: Cambridge University Press, 2009), 72.

71. *Asiatic Journal and Monthly Miscellany* 11 (1821): 161–63.

72. Reese, *Imperial Muslims,* 54.

73. Reese, 56, 63.

74. Freitag, "Symbolic Politics and Urban Violence," 125.

75. Freitag, "Symbolic Politics and Urban Violence"; and Ulrike Freitag, "The Changing Faces of the 'Bride of the Red Sea,' Jeddah c. 1840–1947" (unpublished manuscript, last modified November 24, 2018), chap. 6.

76. W. W. Hunter, *Orissa,* 2 vols. (London: Smith, Elder, 1872), 1:145–67.

77. On sanitation in the Hijaz, see Michael Christopher Low, "The Mechanics of Mecca: The Technopolitics of the Late Ottoman Hijaz and the Colonial Hajj" (PhD diss., Columbia University, 2015), chaps. 2 and 5; William Roff, "Sanitation and Security: The Imperial Powers and the Nineteenth Century Hajj," Arabian Studies 6 (1982): 143; Huber, *Channelling Mobilities,* 204–38;

Eric Tagliacozzo, "The Longest Journey: Southeast Asians and the Pilgrimage to Mecca" (New York: Oxford University Press, 2013), 109–200.

78. *The Spectator,* July 17, 1858.

79. Betty Joseph, *Reading the East India Company 1720–1840: Colonial Currencies of Gender* (Chicago: University of Chicago Press, 2004), chap. 1.

80. As demonstrated in the correspondence between Safvet Pacha and Lord Derby about whether or not Aden is located on the shores of the Indian Ocean; HR.SYS. 90/11, 15.05.1877, BOA.

81. "Relief of the 10th Native Infantry from Aden and a Vegetable Garden at That Place," Military Dept., 1839–41, vol. 140, comp. 511, 579–636, Maharashtra State Archives, Mumbai.

82. "Regarding the Health of the Troops at Aden," Military Dept., 1840–42, vol. 149, 117–44, Maharashtra State Archives, Mumbai.

83. "Sickness at Aden," Military Dept., 1844–45, vol. 241, 445–534, Maharashtra State Archives, Mumbai.

84. See, for example, John Ovington, *A Voyage to Surat in the Year 1689* (New Delhi: Asian Educational Services, 1994), 216; for an extended account of the first decades of Parsi presence in India, see Andre Wink, *Al-Hind: The Making of the Indo-Islamic World* (Leiden: Brill, 2002), 104–8.

85. Nile Green, *Bombay Islam: The Religious Economy of the West Indian Ocean, 1840–1915* (Cambridge: Cambridge University Press, 2011), 32–33.

86. Reese, *Imperial Muslims,* 89–91.

87. David L. White, "Parsis in the Commercial World of Western India, 1700–1750," *The Indian Economic and Social History Review* 24, no. 2 (1987): 188.

88. James Onley, *The Arabian Frontier of the British Raj: Merchants, Rulers, and the British in the Nineteenth Century Gulf* (Oxford: Oxford University Press, 2007), 74–75.

89. White, "Parsis in the Commercial World," 193–99.

90. See his *Voyage to Modernism,* 68–70.

91. See Athnasiyus Aghnatiyus Nuri, *Rihlat al-Hind: 1899–1900* (Abu Dhabi: Dar al-Suwaydī lil-Nashr wa-al-Tawzīʻ, 2003), 54. I thank Roy Bar Sadeh for this reference.

92. *Al-Manar* 15 (1912): 450. For Rida's India writing, see Roy Bar Sadeh, "Constructing a Pan-Islamic Sphere: Muhammad Rashid Rida's Journey to British India in 1912" (unpublished paper, Tel Aviv University, 2015).

93. *Al-Manar* 15 (1912): 457.

94. See, for example, Political Dept., 1859, vol. 35, comp. 1681, 96–107, Maharashtra State Archives, Mumbai.

95. Alavi, *Muslim Cosmopolitanism,* 93–98.

96. Winston S. Churchill, *Churchill by Himself* (Rosetta Books, 2013).

97. *Al-Manar* 15 (1912): 457–58.

98. FO 881/848; FO 195/579, June 15, 1858, BNA, quoted in Low, *Mechanics,* 109.

99. Lauren Benton, *A Search for Sovereignty: Law and Geography in European Empires, 1400–1900* (New York: Cambridge University Press, 2010).

100. Alavi, *Muslim Cosmopolitanism*, 151–52.

101. See Low, *Mechanics*, chap. 5.

102. Alavi, *Muslim Cosmopolitanism*, 101, 109; and see Freitag "Symbolic Politics and Urban Violence" about his involvement in the 1858 Jeddah anti-European violence.

103. Y.EE 35/10, 18 Ramadan 1296 (September 5, 1879); Y/A Resmi 4/59, BOA, quoted in S. Tufan Buzpinar, "Abdulhamid II and Sayyid Fadl Pasha of Hadramawt: An Arab Dignitary's Ambitions (1876–1900)," *Journal of Ottoman Studies* 13 (1993): 231.

104. Y.EE 35/10, 18 Ramadan 1296 (September 5, 1879), BOA.

105. Anne K. Bang, *Sufis and Scholars of the Sea: Family Networks in East Africa, 1860–1925* (New York: Psychology Press, 2003), 85. See also Muhammad 'Arif ibn Ahmad Munayyir, *The Hejaz Railway and the Muslim Pilgrimage: A Case of Ottoman Political Propaganda*, trans. Jacob M. Landau (Detroit: Wayne State University Press, 1971), 12.

106. See Alavi, *Muslim Cosmopolitanism*, 217–18, for a description of other members of this milieu.

107. Where he writes that "whoever despises the Sultan is despised by God. Whoever betrays the Sultan is betrayed by God." Quoted in Alavy, *Muslim Cosmopolitanism*, 115.

108. Quoted in Michael Christopher Low, "Unfurling the Flag of Extraterritoriality: Autonomy, Foreign Muslims, and the Capitulations in the Ottoman Hijaz," *Journal of the Ottoman and Turkish Studies Association* 3, no. 2 (2016): 304.

109. Cemil Aydin, *The Idea of the Muslim World: A Global Intellectual History* (Cambridge, MA: Harvard University Press, 2017), 5–6, 72.

110. Low, *Mechanics*, chap. 4. For the infrastructural turn, see also Mustafa Minawi, *The Ottoman Scramble for Africa: Empire and Diplomacy in the Sahara and the Hijaz* (Stanford: Stanford University Press, 2016).

111. İ.DH 1237/96875, 16 Zulhijjah 1308 (July 23, 1891), BOA.

112. Low, *Mechanics*, chap. 2

113. Fadl, *Idah al-Asrar*, 80, quoted in Jacob, "Of Angels and Men," 56, with minor translation variations.

114. 'Abd al-Rahman Kawakibi, *Umm al-Kura*, in *Al-A'mal al-Kamilah lil-Kawakibi* (Beirut: Markaz Dirasat al-Wahdah al-Arabiyya, 1995).

115. Samir Ben-Layashi, "Discourse and Practice of Medicine, Hygiene and Body in Morocco (1880–1912)" (PhD diss., Tel Aviv University, 2013), 89.

116. Contra Low's understanding of the process, as one "rooted in the secular logics of governmentality, sovereignty, and the materialization of territory" covered by a veneer of pan-Islamic rhetoric. Low, *Mechanics*, 24–25.

117. Andreas Malm, "Who Lit This Fire? Approaching the History of the Fossil Economy," *Critical Historical Studies* 2, no. 2 (Fall 2016): 215–28.

118. "Coal Mining in India: Magnificent Results of Enterprise," *Dundee Courier*, August 22, 1917.

119. See, for example, reports about Chinese efforts in this direction and proclamations about its "intention of using no foreign coal in the future . . . but to depend entirely on native supply"; "Coal Mining in China," *Dover Express*, April 3, 1891. On coal mining in China, see Shellen Xiao Wu, *Empires of Coal:*

Fueling China's Entry into the Modern World Order, 1860–1920 (Stanford, CA: Stanford University Press, 2015).

120. Blair B. Kling, *Partner in Empire: Dwarkanath Tagore and the Age of Enterprise in Eastern India* (Berkeley: University of California Press, 1976), 91.

121. Alaaddin Tok, *The Ottoman Mining Sector in the Age of Capitalism: An Analysis of State-Capital Relations (1850–1908)* (Saarbrücken: Lambert Academic Publishing, 2011), 109–11.

122. Ali İhsan Gencer, *Türk Denizcilik Tarihi Araştırmaları* (Istanbul: Türkiye Denizciler Sendikası, 1986), 13–32; İrade Hariciye (İ.HR.) 335/21582 (1298 [March 27, 1881]), BOA.

123. E.P. Thompson, *The Making of the English Working Class* (Vintage Books, 1963), 245.

124. Nile Green, *The Love of Strangers: What Six Muslim Students Learned in Jane Austen's London* (Princeton, NJ: Princeton University Press), 2016.

125. Douglas Dakin, *The Greek Struggle for Independence, 1821–1833* (Berkeley: University of California Press, 1973), 169–70.

126. Charles MacFarlane, *Constantinople in 1828: A Residence of Sixteen Months in the Turkish Capital and Provinces* (London: Saunders and Otley, 1829), 353. For the history of the *Swift,* see Nurcan Bal, "İlk Buharlı Gemimiz Buğ Gemisi," *Yedikita* 31 (2011): 18–24. On the battle of Navarino, hailed as the last great battle of the age of sail, see C.M. Woodhouse, *The Battle of Navarino* (London: Hodder and Stoughton, 1965).

127. Donald Quataert, *Miners and the State in the Ottoman Empire: The Zonguldak Coalfield, 1822–1920* (New York: Berghahn Books, 2006), 9–18.

128. On Ottoman mining before the nineteenth century, see Tok, *Ottoman Mining Sector,* 11–17.

129. See Emir Yener, *From the Sail to the Steam: Naval Modernization in the Ottoman, Russian, Chinese and Japanese Empires, 1830–1905* (Saarbrücken: Lambert Academic Publishing, 2010), 65–69; and Nurcen Bal, "XIX. Yüzyılda Osmanlı Bahriyesi'nde Gemi İnşa Teknolojisinde Değişim: Buharlı Gemiler Dönemi" (Master's thesis, Mimar Sinan Fine Arts University, 2010), 16–18.

130. Richard Cowling Taylor, *Statistics of Coal: The Geographical and Geological Distribution of Fossil Fuel* (Philadelphia: J.W. Moore, 1848), 631–32.

131. Bruce Podobnik, *Global Energy Shifts: Fostering Sustainability in a Turbulent Age* (Philadelphia: Temple University Press, 2008), 22.

132. Jevons, *Coal Question,* 248.

133. See Jevons, 74–79. Yener's description does not go into detail regarding the move to acquisitions, but from the details of the ships making up the Ottoman fleet, as well as from archival documents, it is possible to discern that circa 1847, such a move took place. See, for example, Sadaret Mektubi Kalemi Meclis-i Vala Evrakı (A.MKT.MVL) 2/2 (1262 [1846]), BOA.

134. Yener, *From the Sail to the Steam,* 88–93.

135. A.MKT.MVL 74/26 (1271 [1854]), 28, BOA.

136. Selçuk Dursun, "Forest and the State: History of Forestry and Forest Administration in the Ottoman Empire" (PhD diss., Sabanci University, 2007), 7.

137. Muhammad Mahmud Ibrahim, *Iktishaf al-Fahm fi Misr* (Cairo : Matba'at Costa Thomas wa-Shuraka'ihi, 1957); Muhammad Fathi 'Awad

Allah, *Qissat al-Fahm fi Misr* (Cairo: Dar al-Kitab al-'Arabi li-l-Tiba'a wa-l-Nashr, 1968).

138. Alan Mikhail, *Nature and Empire in Ottoman Egypt: An Environmental History* (New York: Cambridge University Press, 2011).

139. *Proceedings of the American Philosophical Society* (Philadelphia: M'Calla and Stavely, 1876), 14:243; Taylor, *Statistics of Coal*, 623; J.G. Long, "Foreign Markets for American Coal," in *United States Bureau of Foreign Commerce, Special Consular Reports* (Washington, DC: Government Printing Office, 1900), 21:272; John Timbs, *The Year-Book of Facts in Science and Art* (London: David Bogue, 1850), 265.

140. Sadaret Mektubi Mühimme Kalemi Evrakı (A.MKT.MHM) 305/7 (1281 [1864]), S1, BOA.

141. Mikhail, *Nature and Empire*, 165–67.

142. Robert Burford, *Description of a View of the Bombardment of St. Jean d'Acre* (London: Geo. Nichols, 1841), 2–4.

143. Gencer, *Türk Denizcilik Tarihi Araştırmaları*, 23–24.

144. W. H. Hall and W. D. Bernard, *The Nemesis in China*, 3rd ed. (London: Henry Colburn, 1846), 1.

145. Henry Wise, *An Analysis of One Hundred Voyages to and from India, China &c.* (London: J. W. Norie, 1839), xiv–xv.

146. Tim Wright, *Coal Mining in China's Economy and Society, 1895–1937* (New York: Cambridge University Press, 1984).

CHAPTER 5. RISK

1. See, for example, Kapil Raj, "Mapping Knowledge Go-Betweens in Calcutta, 1770–1820," in *The Brokered World: Go-betweens and Global Intelligence, 1770–1820*, ed. Simon Schaffer, Lissa Roberts, Kapil Raj, and James Delbourgo (Sagamore Beach, MA: Science History Publications, 2009), 105–50.

2. As Perry Miller and later David Nye described the common feeling of awe inspired by technological prowess. See Nye, *American Technological Sublime* (Cambridge, MA: MIT Press, 1994).

3. See, for example, *Times of India*, November 9, 1885, quoted in Low, *Mechanics*, 303.

4. The connection is creatively explored in Frances Wilson, *How to Survive the Titanic: The Sinking of J. Bruce Ismay* (New York: Bloomsbury, 2011).

5. The novel was originally published in serialized form in *Blackwood's Magazine* between October 1899 and November 1900. It was published in book form in 1900.

6. On September 17, 1882, Joseph Conrad, a second mate aboard the SS *Palestine*, sailed from Falmouth to Bangkok. On March 11, 1883, smoke began to rise from the coals, the vessel burst into flames, and the crew abandoned ship. Zdzisław Najder, *Joseph Conrad: A Life* (Rochester NY: Camden House, 2007), 90–96.

7. Francis Ewald, "Insurance and Risk," in *The Foucault Effect: Studies in Governmentality*, ed. Graham Burchell, Colin Gordon, and Peter Miller (Chicago: University of Chicago Press, 1991), 199.

8. Bruce Masters, "Hajj," in *The Encyclopedia of the Ottoman Empire,* ed. Gábor Ágoston and Bruce Alan Masters (New York: Facts on File, 2009), 246–48.

9. See Michael Francis Laffan, *Islamic Nationhood and Colonial Indonesia: The Umma below the Winds* (London: Routledge, 2003). For more on the division of lands below and above the winds, see P.J. Rivers, "Negeri below and above the Wind: Malacca and Cathay," *Journal of the Malaysian Branch of the Royal Branch of the Royal Asiatic Society* 78, no. 2 (2005): 1–32; Anthony Reid, "Understanding Melayu (Malay) as a Source of Diverse Modern Identities," *Journal of Southeast Asian Studies* 32, no. 3 (2001), 295–313; Anthony Reid, "An 'Age of Commerce' in Southeast Asia History," *Modern Asian Studies* 24, no. 1 (Feb., 1990), 1–30; Ibn Muhammad Ibrahim, *The Ship of Sulaiman,* trans. John O'Kane (London: Routledge, 1972).

10. Slight, "The Hajj and the Raj: From Thomas Cook to Bombay's Protector of Pilgrims," in *The Hajj: Collected Essays,* ed. Venetia Porter and Liana Saif (London: British Museum Press, 2013), 115.

11. Barak, *On Time,* 4.

12. On the latter, see Nile Green, "The Rail *Hajjis:* The Trans-Siberian Railway and the Long Way to Mecca," in *The Hajj: Collected Essays,* ed. Venetia Porter and Liana Saif (London: British Museum Press, 2013), 102–3.

13. Freitag, "Changing Faces," chap. 6.

14. See Freitag, "Changing Faces," chap. 6, and Christiaan Snouck Hurgronje, *Mekka: Aus dem heutigen Leben,* 2 vols. (Den Hague: Martinus Nijhoff, 1888–1889). On brokers' intimidation and information networks see Low, *Mechanics,* 296, 301.

15. Green, "Rail *Hajjis,*" 100.

16. Michael B. Miller, "Pilgrims' Progress: The Business of the Hajj," *Past and Present,* no. 191 (May 2006): 192.

17. Slight, "The Hajj and the Raj," 118–19.

18. See, for example, Ziauddin Sardar, "The Destruction of Mecca," *New York Times,* September 30, 2014.

19. On the conception of tourism as pilgrimage, see Michael Stausberg, *Religion and Tourism: Crossroads, Destinations and Encounters* (London: Routledge, 2011). Religion and tourism are indeed closely linked: if popular tourism traces its history back to Thomas Cook's early rail and steamer excursions, then this former Baptist preacher saw himself first and foremost as a social and religious reformer when organizing such coal-fueled temperance rallies in the early 1840s.

20. Huber, *Channeling Mobilities,* 22.

21. William Ochsenwald, *Religion, Society and the State in Arabia: The Hijaz under Ottoman Control, 1840–1908* (Columbus: Ohio State University Press, 1984), 99.

22. See, for example, Huber, *Channeling Mobilities,* 249, on the scheme of "passage in quarantine" tested on pilgrim ships in the Suez Canal in 1875.

23. John Armstrong and David M. Williams, "The Steamboat and Popular Tourism," in *The Impact of Technological Change: The Early Steamship in Britain* (St. John's, Newfoundland: International Maritime Economic History Association, 2011), 78.

24. As evident, for example, in the different chapters of Torsten Feys, Lewis R. Fischer, Stephane Hoste, and Stephen Vanfraechem, eds., *Maritime Transport and Migration: The Connections between Maritime and Migration Networks* (Oxford: Oxford University Press, 2017).

25. Drew Keeling, "The Transportation Revolution and Transatlantic Migration, 1850–1914," *Research in Economic History* 19 (1999), 39–74.

26. Abderrahmane El Moudden, "The Ambivalence of *Rihla*: Community Integration, and Self-Definition in Moroccan Travel Accounts," in *Muslim Travellers: Pilgrimage, Migration, and the Religious Imagination*, ed. Dale F. Eickelman and James Piscatori (Berkeley: University of California Press, 1990), 75.

27. Jan Just Witkam, "The Islamic Pilgrimage in the Manuscript Literatures of Southeast Asia," in *The Hajj: Collected Essays*, ed. Venetia Porter and Liana Saif (London: British Museum Press, 2013), 216.

28. For more information on the subject, see Barak, *On Time.*

29. Ochsenwald, *Religion*, 100.

30. Freitag, "Changing Faces," chap. 6.

31. FO 78/3334, BNA, cited in Huber, 206–7.

32. Based on figures provided by Ochsenwald, *Religion*, 61.

33. John H. White, Jr., *Wet Britches and Muddy Boots: A History of Travel in Victorian America* (Bloomington: Indiana University Press, 2012), 351.

34. Slight, "The Hajj and the Raj," 118.

35. Board of Trade, "Wreck Report for 'Jeddah' 1881," Southampton City Council Library. https://southampton.spydus.co.uk/cgi-bin/spydus.exe/TRN/OPAC/BIBENQ/34737664/?SEL=26596146.

36. P/525, IOR. British regulation stipulated a minimum space of nine superficial feet, raised in a 1984 convention to twenty-one. See Low, *Mechanics*, 295. To put the Jeddah's passenger list in perspective, we can compare it to the two most famous steamers of the period, the far larger *Kaiser Wilhelm II* and *Titanic*, each transporting about seven hundred third-class passengers.

37. See, for example, a letter from the Ottoman lieutenant governor of Mecca to the Ministry of Interior, requesting that they circumvent the maximum passenger numbers decreed by the Ministry of Health. "Since long accommodation periods are unhealthy [local officers] ask for the boarding of these people to the steamers even if the ships exceed their limits." DH.MUİ 44–2/5, January 24, 1910, BOA.

38. See, for example, an 1885 account of two steamers crashing into one another in the port of Jeddah in Michael Wolfe, *One Thousand Roads to Mecca: Ten Centuries of Travelers Writing about the Muslim Pilgrimage* (New York: Grove Press, 1998), 289–90.

39. See, for example, Christiaan Snouck Hurgronje, *Advice on Governmental Matters*, Weltevreden, July 17, 1893, in Christiaan Snouck Hurgronje and E. Gobée, *Ambtelijke adviezen van C. Snouck Hurgronje: 1889–1936, Deel 3* ('s-Gravenhage: Nijhoff, 1965), 1381–82.

40. Snouck Hurgronje, *Advice*, August 26, 1896, 1445.

41. See Snouck Hurgronje's 1895 "Letter to the Head of the Department of Education, Worship and Industry," *Advice*, 1413–16. The al-Saqqafs held land

at Cocob (or Kukub) granted to them by Sultan Abu Baker of Johor, where they could even print their own money.

42. Low, *Mechanics*, 298, 317–21.

43. Hurgronje, *Advice*, 1413–16. A letter to the editor of the *Times*, by a captain formerly involved in the pilgrim trade, expressing his opinion on the *Jeddah* affair, stated that "there are horrors on board such a ship . . . worse, by far, than was ever found on board a slaver." Quoted in Tagliacozzo, *Longest Journey*, 120. See also *Straits Times Overland Journal*, July 28, 1881, and Eric Tagliacozzo, "Crossing the Great Water: The Hajj and Commerce from Pre-Modern Southeast Asia," in *Religion and Trade: Cross-Cultural Exchanges in World History, 1000–1900*, ed. Francesca Trivellato, Leor Halevi, and Cátia Antunes (New York: Oxford University Press, 2014), 229.

44. Ulrike Freitag, *Indian Ocean Migrants and State Formation in Hadhramaut: Reforming the Homeland* (Leiden: Brill, 2003), 53; Janet Ewald and W. G. Clarence-Smith, "The Economic Role of the Hadrami Diaspora in the Red Sea and Gulf of Aden, 1820s to 1930s," in *Hadhrami Traders, Scholars and Statesmen in the Indian Ocean, 1750s to 1960s,* ed. Ulrike Freitag and William G. Clarence-Smith (Leiden: Brill, 1997), 289–291.

45. Mathew, *Margins of the Market*, 69.

46. Steamers previously used as cattle ships were sometimes refitted to take pilgrims, who were themselves seen as cattle. As Isabel Burton put it: "The *Calypso* was one of the first to leave, having stowed away eight hundred pilgrims. In short, the old cattle-ship seemed to have returned to her original trade,—not that I mean to compare men and women with divine souls to cattle, but because they are stowed away like, and have the habits of, cattle." See Burton, *Arabia, Egypt, India: A Narrative of Travel* (1879; repr., Cambridge: Cambridge University Press, 2012), 93. On the other hand, as Stolz suggests, pilgrim steamships were sometimes converted to carry cattle. See Stolz, "Voyage of the *Samannud*"; on the pilgrim as animal, see also Sanjay Krishnan's critique of Conrad in "Seeing the Animal: Colonial Space and Movement in Joseph Conrad's *Lord Jim*," *Novel: A Forum on Fiction* 37, no. 3 (Summer 2004): 326–31.

47. Similarly, names from the category of *ta'abid,* including the prefix *abd* (slave) before one of the ninety-nine names of God, are among the most common Islamic names.

48. W. G. Clarence-Smith, *Islam and the Abolition of Slavery* (New York: Oxford University Press, 2006), 51.

49. Greg Grandin, *The Empire of Necessity: Slavery, Freedom, and Deception in the New World* (New York: Henry Holt, 2014), 214–15.

50. Carl A. Trocki, "Singapore as a Nineteenth Century Migration Node," in *Connecting Seas and Connected Ocean Rims: Indian, Atlantic, and Pacific Oceans and China Seas Migrations from the 1830s to the 1930s,* ed. Donna R. Gabaccia and Dirk Hoerder (Leiden: Brill, 2011), 204, 209.

51. Trocki, 209.

52. Radhika Singha, "Passport, Ticket, and India-Rubber Stamp: 'The Problem of the Pauper Pilgrim' in Colonial India c. 1882–1925," in *The Limits of British Colonial Control in South Asia: Spaces of Disorder in the Indian Ocean*

Region, ed. Ashwini Tambe and Harald Fischer-Tiné (New York: Routledge, 2008), 79n149.

53. Armstrong and Williams, "The Steamboat, Safety and the State: Government Reaction to New Technology in a Period of Laissez-Faire," in *Impact of Technological Change,* 54–55.

54. Including *Proceedings of the Legislative Council of Singapore,* September 14, 1880 (recorded in the *Straits Times Overland Journal,* September 20, 1880); "Action for Salvage Brought against the *Jeddah* in the Vice Admiralty Court of the Straits Settlements," *Straits Times Overland Journal,* October 22, 1881; and the Aden court case.

55. See Bruno Latour, *Pandora's Hope: Essays on the Reality of Science Studies* (Cambridge, MA: Harvard University Press, 1999), 183, in which Latour describes blackboxing as "a process that makes the joint production of actors and artifacts entirely opaque."

56. Low, *Mechanics,* 295–96.

57. Farid Kioumgi and Robert Graham, *A Photographer on the Hajj: The Travels of Muhammad 'Ali Effendi Sa'udi (1904/1908)* (Cairo: American University in Cairo Press, 2009), 7, 9, 12–13.

58. See Freitag, *Changing Faces,* chap. 6; see also Saurabh Mishra, *Pilgrimage, Politics, and Pestilence: The Haj from the Indian Subcontinent, 1860–1920* (Oxford: Oxford University Press, 2011), 57f.

59. Burton, *Arabia, Egypt, India,* 84.

60. *Sunderland Daily Echo,* August 7, 1880.

61. *Tamworth Herald,* November 12, 1881.

62. See, for example, *Morpeth Herald,* August 5, 1882; *Morning Post,* August 25, 1884; *Edinburgh Evening News,* August 11, 1885; *Lancashire Evening Post,* February 26, 1887; *Standard,* September 10, 1900.

63. See, for example, "The Terrible Accident at Sea," *Evening News,* August 11, 1880.

64. Beyond the *Jeddah* case, for analysis of maritime accidents for assessing water ballasting systems, see, for example, the minutes of the Royal Shipping Committee in the *Shipping and Mercantile Gazette,* November 18, 1874.

65. F.E. Chadwick, *Ocean Steamships* (1891; repr., Paderborn: Salzwasser Verlag, 2011), 243–45.

66. Khan, *Voyage to Modernism,* 101.

67. *Daily Times,* September 10, 1880, in Gene M. Moore, comp., "Newspaper Accounts of the *Jeddah* Affair," in *Lord Jim: Centennial Essays,* ed. Allan Simmons and J.H. Stape (Amsterdam: Rodopi, 2003), 116.

68. *The Nautical Magazine and Naval Chronicle for 1873: A Journal of Papers on Subjects Connected with Maritime Affairs* (London: Simpkin, Marshall and Co., 1873), 876. See also 1035 regarding the SS *London* stranded off Cape Guardafui.

69. See Necla Geyikdagi, *Foreign Investment in the Ottoman Empire: International Trade and Relations, 1854–1914* (London: Tauris Academic Studies, 2011), 62; and Thobie, *L'administration générale.*

70. *British Documents on Foreign Affairs 1856–1914,* 19:280–82.

71. Alexander George Findlay, *A Description and List of Lighthouses of the World* (London: R.H. Laurie, 1861), 99, 103.

72. *British Documents on Foreign Affairs 1856–1914*, 19:280–82.

73. *Daily Times*, September 10, 1880, in Moore, "Newspaper Accounts," 116–17.

74. See Alexis Wick, *The Red Sea: In Search of Lost Space* (Oakland: University of California Press, 2016), 134–41.

75. See chapter 4.

76. R\20\E\143, IOR.

77. J.D. Gordan, *Joseph Conrad: The Making of a Novelist* (1940; repr., New York: Russell & Russell, 1963), 61.

78. On the reforms enforcing the marking of a load line, which helped prevent the hazardous overloading of ships, to Plimsoll's fight against overinsured and unseaworthy ships, see Nicolette Jones, *The Plimsoll Sensation: The Great Campaign to Save Lives at Sea* (London: Abacus, 2007). On overloading and insurance in the black Atlantic, see Ian Baucom, *Specters of the Atlantic: Finance Capital, Slavery, and the Philosophy of History* (Durham, NC: Duke University Press, 2005), and Marcus Rediker, *The Slave Ship: A Human History* (New York: Penguin Books, 2014).

79. Baucom, *Specters of the Atlantic*.

80. Jonathan Levy, *Freaks of Fortune: The Emerging World of Capitalism and Risk in America* (Cambridge, MA: Harvard University Press, 2014), 28–29.

81. Kris Manjapra, "Plantation Dispossessions: The Global Travel of Agricultural Racial Capitalism," in *American Capitalism: New Histories*, ed. Sven Beckert and Christine Desan (Columbia University Press, 2018), 370.

82. Association of Average Adjusters, *Report of the General Meeting Held at Lloyd's* (London, 1872), 5.

83. Frederick Martin, *The History of Lloyd's and of Marine Insurance in Great Britain* (Whitefish, MT: Kessinger, 2010), 383.

84. H.A.L. Cockerell and Edwin Green, *The British Insurance Business: A Guide to Its History and Records* (London: Sheffield Academic Press, 1976), 20.

85. Guy Chet, *The Ocean Is a Wilderness: Atlantic Piracy and the Limits of State Authority, 1688–1856* (Amherst: University of Massachusetts Press, 2014).

86. Ewald, "Insurance and Risk," 200.

87. Albert Chaufton, *Les assurances, leur passé, leur présent, leur avenir* (Paris: Librairie A. Marescq Ainé, 1884), quoted in Ewald, 208.

88. "Insurance, through the category of risk, objectifies every event as an accident." Ewald, "Insurance and Risk," 199.

89. Armstrong and Williams, "Steamboat, Safety, and the State," 55–56.

90. House of Commons Parliamentary Papers, *Report on Steam Vessel Accidents to the Committee of the Privy Council for Trade* (London, 1839), quoted in Armstrong and Williams, "Steamboat, Safety, and the State," 56.

91. Theodor M. Porter, *Trust in Numbers: The Pursuit of Objectivity in Science and in Life* (Princeton, NJ: Princeton University Press, 1996).

92. On the importance of Labuan, see Malm, "Who Lit This Fire?"

93. Cockerell and Green, *British Insurance Business*, 86.

94. Benjamin Martel, "On Water Ballast," *Scientific American: Supplement* 4, no. 97 (November 10, 1877): 1536–38.

95. Thomas Curson Hansard, *Hansard's Parliamentary Debates, 3rd series: Commencing with the Accession of William . . . Iv, 43 and 44 Victoriae, 1880* (n.p.: Forgotten Books, 2017), 309.

96. "Lighting of the Red Sea and Gulf of Aden," Finance & Commerce Dept., proceeding no. 1659, dated August 10, 1880; proceedings nos. 1659–66, October 1881; National Archives of India, New Delhi.

97. See, for example, FO 195/1827, BNA, on pressure the company applied for reforms in the port of Istanbul in 1894.

98. Insurance policies transformed harbors into "risk-traps" by allocating high premiums to loading cargo from wharf to ship, the maneuvers of leaving port, coastal navigation, etc. See Frank C. Spooner, *Risks at Sea: Amsterdam Insurance and Maritime Europe, 1766–1780* (Cambridge: Cambridge University Press, 1983), 249–50.

99. And thus gave rise to new complications and accidents: a steamer hitting a dredger in 1885 caused it to collapse into the canal, interrupting passage for ten days and creating a jam of 123 ships waiting for the dredger to be blown up, and conflicts with quarantine arrangements caused by the fact that pilots breached the prohibition against contact between ship and shore. See Huber, *Channeling Mobilities*, 111, 250.

100. Janet Abu-Lughod, *Before European Hegemony* (Oxford: Oxford University Press, 1991), 115.

101. Abu-Lughod, 119.

102. Bruce Masters, *Christians and Jews in the Ottoman Arab World: The Roots of Sectarianism* (Cambridge: Cambridge University Press, 2001), 69.

103. Cockerell and Green, *British Insurance Business*, 23.

104. *Straits Times Overland Journal,* October 22, 1881, 3–4, and *Daily Times,* October 20, in Moore, "Newspaper Accounts," 26–36.

105. Tan Twan Eng, "The Law of Salvage" (PhD diss., University of Cape Town, 2014), 15.

106. Kevin Doran, "Adrift on the High Sea: The Application of Maritime Salvage Law to Historic Shipwrecks in International Waters," *Southwestern Journal of International Law* 18 (2013): 102–3.

107. R. J. Gavin, *Aden under British Rule, 1839–1967* (London: C. Hurst, 1975), 181.

108. Reginald G. Marsden, *A Digest of Cases Relating to Shipping and Marine Insurance to the End of 1897, with Supplement of Cases to the End of 1910* (London: Sweet & Maxwell, 1911), 653.

109. See Levy, *Freaks of Fortune*, 5–6.

110. Ewald, "Insurance and Risk," 208.

111. Ewald, 201.

112. Pierre-Joseph Proudhon, quoted in Ewald, "Insurance and Risk," 206.

113. Arjun Appadurai, *The Future as Cultural Fact: Essays on the Global Condition* (London: Verso, 2013).

114. Ulrich Beck, *Risk Society: Towards a New Modernity* (1986; repr., London: Sage, 2010).

115. Fatih Kahya, "Osmanlı Devleti'inde Sigortacılığın Ortaya Çıkışı ve Gelişimi" [The emergence and the development of insurance in the Ottoman Empire] (Master's thesis, Marmara Üniversitesi, Sosyal Bilimler Enstitüsü, Istanbul, 2007), 27, 38.

116. Kahya, 39–40, 54.

117. David M. Kohen, "Insurance in the Ottoman Empire," *Middle East Insurance Review*, May 2009; Kahya, 10.

118. Kahya, 10.

119. The *fatwa* is presented in Syed Junaid A. Quadri, "Transformations of Tradition: Modernity in the Thought of Muḥammad Bakhit al-Muti'i" (PhD diss., McGill University, 2013), 30–41.

120. See Ayesha Jalal, *Partisans of Allah: Jihad in South Asia* (Cambridge, MA: Harvard University Press, 2008), 66–68, 114–16, 136–39.

121. Quadri, "Transformations of Tradition," 41–57.

122. Valentino Cattelan, "In the Name of God: Managing Risk in Islamic Finance," EABH Papers no. 14–07 (July 2014). www.econstor.eu/bitstream/10419/100004/1/791365034.pdf.

123. Murat Çizakça, *A Comparative Evolution of Business Partnerships: The Islamic World and Europe, with Specific Reference to the Ottoman Archives* (Leiden: E.J. Brill, 1996), chap. 2.

124. Bruce Masters, *The Origins of Western Economic Dominance in the Middle East: Mercantilism and the Islamic Economy in Aleppo, 1600–1750* (New York: NYU Press, 1988), 52. As Fahad Ahmad Bishara has recently demonstrated, Muslim merchants trading in the western Indian Ocean during the nineteenth century developed schemes that ran counter to this rule, and in the face of opposition by coreligionists. The al-Saqqafs described in this chapter can be seen as part of this trend. See Bishara, *A Sea of Debt: Law and Economic Life in the Western Indian Ocean, 1780–1950* (Cambridge: Cambridge University Press, 2017).

125. See, for example, Mona Atia, *Building a House in Heaven: Pious Neoliberalism and Islamic Charity in Egypt* (Minneapolis: University of Minnesota Press, 2013).

126. Freitag, *Changing Faces,* chap. 6.

127. See Krishnan's critique of Conrad along these lines in "Seeing the Animal."

128. Joseph Conrad, *"Typhoon" and Other Stories,* ed. J.H. Stape (London: Penguin, 2007), 27; see also Andreas Malm, "'This is the Hell I Have Heard Of': Some Dialectical Images in Fossil Fuel Fiction," *Forum for Modern Language Studies* 53, no. 2 (2017): 121–41.

129. Holt, *Fictitious Capital,* 6.

130. Holt, 123–24.

131. Yaʿqub Sarruf, *Fatat Misr, Riwayah* (Cairo, Matbaʿat al-muqtataf, 1922), 162–64.

132. Thomas Arnold was a prominent British Orientalist who taught at several of Sayyid Ahmad Khan's institutions.

133. Shiblī al-Nuʿmanī, *Riḥlat Shiblī al-Nuʿmanī ilá al-Qusṭanṭīniyya wa-Bayrut wal-Quds wal-Qahira fi Mustaḥall al-Qarn al-Rabiʿ ʿAshar al-Hijrī* (Damascus: Dar al-Qalam, 1967), 32.

134. Samir Ben-Layashi, "Discourse and Practice of Medicine, Hygiene and Body in Morocco (1880–1912)" (PhD diss., Tel Aviv University, 2013), 85–93.

135. Fadl, *Idah al-Asrar,* 107; and see Jacob, "Of Angels and Men," 58.

136. Green, *Bombay Islam,* 112.

137. See Charles Hirschkind's analysis of this principle in *The Ethical Soundscape: Cassette Sermons and Islamic Counterpublics* (New York: Columbia University Press, 2009).

138. Peter Linebaugh and Marcus Rediker's *The Many-Headed Hydra: Sailors, Slaves, Commoners, and the Hidden History of the Revolutionary Atlantic* (London: Verso, 2000), to take a familiar example, opens with such an account of a motley crew coming together to pump water from a sinking sail ship near Bermuda.

139. Steve Mentz, *Shipwreck Modernity: Ecologies of Globalization, 1550–1719* (Minneapolis: University of Minnesota Press, 2015); see also Fredric Jameson, *The Political Unconscious: Narrative as a Socially Symbolic Act* (London: Routledge, 1989), 219–20, on Conrad's "allegory of the ship as the civilized world on its way to doom."

140. Elizabeth A. Povinelli, *Economies of Abandonment: Social Belonging and Endurance in Late Liberalism* (Durham, NC: Duke University Press, 2011); Wael Abu-'Uksa, *Freedom in the Arab World: Concepts and Ideologies in Arabic Thought in the Nineteenth Century* (Cambridge: Cambridge University Press, 2016).

CHAPTER 6. FOSSIL

1. John Holland, *The History and Description of Fossil Fuel, the Collieries and Coal Trade of Great Britain* (London: Cass, 1841).

2. Brian C. Black, *Crude Reality: Petroleum in World History* (Lanham, MD: Rowman and Littlefield, 2012), 6–25.

3. Donna Haraway, *Staying with the Trouble: Making Kin in the Chthulucene* (Durham, NC: Duke University Press, 2016).

4. Clive Hamilton, *Defiant Earth: The Fate of Humans in the Anthropocene* (Cambridge: Polity Press, 2017); Andreas Malm, *The Progress of the Storm: Nature and Society in a Warming World* (London: Verso, 2018).

5. See, for example, Jason W. Moore, *Capitalism in the Web of Life: Ecology and the Accumulation of Capital* (Verso, 2016).

6. Martin J.S. Rudwick, "Charles Lyell's Dream of a Statistical Paleontology," in *Lyell and Darwin, Geologists: Studies in the Earth Sciences in the Age of Reform* (Farnham: Ashgate, 2005), 229, 237–38.

7. On the derivative nature of geology, considered an offshoot of nineteenth-century physics, along with geologists' acceptance of Lord Kelvin's calculations of the age of the world, see Jeff Dodick and Nir Orion, "Geology as an Historical Science: Its Perception within Science and the Education System," *Sci-*

ence & Education 12 (2003): 197–211; on thermodynamics' imperialism and encroachment on biology and ecology and the centrality of a reconfigured "organism" for this process, see Cara Daggett, "Energy's Power: Fuel, Work and Waste in the Politics of the Anthropocene" (PhD diss., Johns Hopkins University, 2016), chaps. 3 and 4.

8. Anson Rabinbach, *The Human Motor: Energy, Fatigue, and the Origins of Modernity* (Berkeley: University of California Press, 1992); Philip Mirowski, *More Heat than Light: Economics as Social Physics, Physics as Nature's Economics* (Cambridge: Cambridge University Press, 1989).

9. Hugh Torrens, *The Practice of British Geology, 1750–1850* (Farmham: Ashgate, 2002); see also Martin J. S. Rudwick, *Bursting the Limits of Time: The Reconstruction of Geohistory in the Age of Revolution* (Chicago: University of Chicago Press, 2007).

10. Rudwick, *Bursting the Limits of Time,* 194.

11. Dipesh Chakrabarty, "The Climate of History: Four Theses," *Critical Inquiry* 35, no. 2 (Winter 2009): 197–222.

12. Sir Thomas Wyse, *Education Reform; or, The Necessity of a National System of Education* (London: Longman, Rees, Orme, Brown, Green & Longman, 1836), 161.

13. M. Alper Yalçınkaya, *Learned Patriots: Debating Science, State, and Society in the Nineteenth-Century Ottoman Empire* (Chicago: University of Chicago Press, 2015); Miri Shefer-Mossensohn, *Science among the Ottomans: The Cultural Creation and Exchange of Knowledge* (Austin: University of Texas Press, 2015).

14. See Latour, *Pandora's Hope,* chap. 1.

15. Rifā'ah Rāfi' Ṭahṭāwī. *Ta'rīb kitāb al-mu'allim Firād fī al-ma'ādin al-nāfi'ah li-tadbīr ma'āyish al-khalāyiq* (Būlāq: Maṭba'at Būlāq, 1833), 1. http://nrs.harvard.edu/urn-3:FHCL:24426959.

16. Ṭahṭāwī, 2.

17. Such as *Al-Tuhfah al-fakhirah fī hay'at al-a'ḍa' al-zahirah* (1835); *Kanz al-bara'ah fī mabadi fann al-zira'ah* (1838); *Tuhfat al-riyaḍ fī kullīyat al-amraḍ* (1839); *Tuhfat al-qalam fī amraḍ al-qadam* (1842); *al-Shudhur al-Dhahabiyya fī'l Muṣṭalaḥat al-Tibbiyya* (1851).

18. Especially through the notion of *Hazine-yi Evrak.* "Ottoman Archives," Turkish Cultural Foundation. www.turkishculture.org/general/museums/ottoman-archives/ottoman-archives-190.htm?type = 1.

19. İ.MVL 62/1180, 25 Z 1260 (05.01.1845), BOA.

20. The following description is based on Celal Şengör, "Osmanlı'nın İlk Jeoloji Kitabı ve Osmanlı'da Jeolojinin Durumu Hakkında Öğrettikleri," *Osmanlı Bilimi Araştırmaları* 11, nos. 1–2 (2009–10), 119–58.

21. Salih Kış, "First Mining School in the Ottoman Empire and Mining Engineer Training," *Journal of History Studies* 7, no. 3 (September 2015): 113n9.

22. Mehmet Ali Fethi, *İlm-i tabakat-i arz* (Istanbul: Dar üt-Tibaat ül-Amire, 1852), 2–8. See also Yalçınkaya, *Learned Patriots,* 63–64.

23. Fethi, *İlm-i tabakat-i arz,* 2–7.

24. See Shefer-Mossensohn, *Science among the Ottomans,* chap. 4.

25. Ibn Ishaq, quoted in Brannon Wheeler, *Mecca and Eden: Ritual, Relics, and Territory in Islam* (Chicago: University of Chicago Press, 2006), 19.

26. Wheeler, *Mecca and Eden,* 34, 43.

27. "Imperial Treasury," Topaka Palace Museum. http://topkapisarayi.gov .tr/en/content/imperial-treasury.

28. For example, scholarship on the popular prayer litany *Dala'il al-khayrat* points to how this book orientated Muslims to the passage between the Prophet's grave and one of the gardens of paradise. See Jan Just Witkam, "The Battle of the Images: Mekka vs. Medina in the Iconography of the Manuscripts of al-Jazūlī's Dala'il al-Khayrat," *Beiruter Texte und Studien* 111 (2007), 67–82.

29. Narrated by Abu Huraira, *Sahih al-Bukhari,* vol. 9, bk. 83, no. 47.

30. Ömer Faruk Şentürk, *Charity in Islam: A Comprehensive Guide to Zakat* (Clifton, NJ: The Light, 2007), 79.

31. *The Ottoman Land Laws: With a Commentary on the Ottoman Land Code, of 7th Ramadan 1274* (April 21, 1858), 101.

32. I thank Selim Karlitekin for opening my eyes to this connection and for providing the references about *arz-arazi-arziyat.*

33. It appears as such in İshak Efendi's multivolume encyclopedia *Mecmua-i Ulum-ı Riyaziye* (Collection of mathematical sciences), published between 1831 and 1834 as a course book for the Mühendihane-i Berri-i Humâyun's (Royal Engineering School), where the author served as headmaster.

34. As it appears in 1856 in the course offerings of Mekteb-i Tıbbiye (medical school), which also had a chair for geology and minerology (*ilm-ül arz ve'l maadin*). Between 1862 and 1874, Abdullah Bey (Karl Eduard Hammerschmidt) taught stratigraphy in the school. His lectures were translated by İbrahim Lütfi Pasha as *İlm-i Tabakatü'l-Arz ve'l Maadin.*

35. This was the case in Iskenderun, under an Egyptian occupation, and in Chukurova, "the Anatolian Egypt"—see Chris Gratien, "The Ottoman Quagmire: Malaria, Swamps, and Settlement in the Late Ottoman Mediterranean," *International Journal of Middle East Studies* 49, no. 4 (2017): 583–604.

36. Donald Quataert, *Miners and the State in the Ottoman Empire: The Zonguldak Coalfield, 1822–1920* (New York: Berghahn Books, 2006), 112–16.

37. See, for example, A.} MKT. NZD 362/39; i.HUS 100 1320 Ş 51; İ.HUS 148/1324 L 039, BOA.

38. Marinos Sariyannis, "Ruler and State, State and Society in Ottoman Political Thought," *Turkish Historical Review* 4 (2013): 92–126.

39. Carter V. Findley, *Bureaucratic Reform in the Ottoman Empire: The Sublime Porte, 1789–1922* (Princeton, NJ: Princeton University Press, 1980), 238; Hakan T. Karateke, *An Ottoman Protocol Register: Containing Ceremonies from 1736 to 1808, BEO Sadaret Defterleri 350 in the Prime Ministry Ottoman State Archives, Istanbul* (Istanbul: Ottoman Bank Archive and Research Centre, 2007), 76.

40. In China during the same period, coal and similar minerals were seen as a "hidden treasure" and geology as the key to tap it. There too, mining was connected to a long tradition of statecraft. See Shellen Xiao Wu, *Empires of*

Coal: Fueling China's Entry into the Modern World Order, 1860–1920 (Stanford, CA: Stanford University Press, 2015), 165.

41. Christophe Bonneuil and Jean-Baptiste Fressoz, *The Shock of the Anthropocene: The Earth, History and Us* (London: Verso, 2016), 204.

42. Quoted in Bonneuil and Fressoz, 204.

43. On the development of this legislation, see George Hill, *Treasure Trove in Law and Practice: From Earliest Time to the Present Day* (Oxford: Clarendon Press, 1936).

44. Peter Brown, *Through the Eye of a Needle: Wealth, the Fall of Rome, and the Making of Christianity in the West, 350–550 AD* (Princeton, NJ: Princeton University Press, 2014), chap. 4.

45. Crosbie Smith, *The Science of Energy: A Cultural History of Energy Physics in Victorian Britain* (London: Athlone, 1998), 308–9.

46. Even before the proper category of "energy" would assume this position of nature's gift away from the messiness of coal, economists like David Ricardo would ask in 1817: "Are the powers of wind and water, which move our machinery, and assist navigation, nothing? The pressure of the atmosphere and the elasticity of steam, which enable us to work the most stupendous engines— are they not the gifts of nature?" Quoted in Margaret Schabas, *The Natural Origins of Economics* (Chicago: University of Chicago Press, 2007), 114.

47. Barak, *On Time*, 11; Avner Wishnitzer, *Reading Clocks Alla Turca: Time and Society in the Late Ottoman Empire* (Chicago: University of Chicago Press, 2016), 160–66.

48. Khan, *Voyage to Modernism*, 68.

49. See Barak, *On Time*, 248.

50. Albert Vandal, *Une ambassade française en Orient sous Louis XV: La mission du Marquis de Villeneuve* (Paris: E. Plon, Nourrit, 1887): 145; Alaaddin Tok, "Fuelling the Empire: Energy Economics in the Ottoman Anatolia and the Balkans (1750–1914)" (PhD diss., Boğaziçi University, Istanbul, 2018), 109.

51. *Ceride-i Havadis*, 1 C 1256 (July 31, 1840).

52. Quataert, *Miners and the State*, 9–15.

53. Macgregor, *Commercial Statistics*, 182–84.

54. Rıfat Önsoy, "Osmanlı İmparatorluğu'nun Katıldığı İlk Uluslararası Sergiler ve Sergi-i Umumi-i Osmânî (1863 İstanbul Sergisi)," *Belleten* 185 (January 1983): 206, discussed in Tok, "Fuelling the Empire," 25.

55. Y.PRK.TNF 2 /68 30 Z 1307, BOA, quoted in Alaaddin Tok, "The Ottoman Mining Sector in the Age of Capitalism: An Analysis of State-Capital Relations, 1850–1908" (MA thesis, Boğaziçi University, Istanbul, 2010), 25–26, with minor translation changes.

56. Tok, "Fuelling the Empire," 56, 100.

57. See Daggett, "Energy's Power."

58. Madeline C. Zilfi, *Women and Slavery in the Late Ottoman Empire: The Design of Difference* (Cambridge: Cambridge University Press, 2010), 157.

59. Bonneuil and Fressoz, *The Shock of the Anthropocene*, 202.

60. Bonneuil and Fressoz, 202.

61. Schabas, *Natural Origins of Economics*, 115–19.

62. E. J. Marey, *La méthode graphique dans les sciences expérimentales et principalement en physiologie et en médecine* (Paris: Librairie de l'Académie de Médecine, 1878), 75.

63. William Gourlie, "Notice of the Fossil Plants in Glasgow Geological Museum," in *Proceedings of the Royal Philosophical Society of Glasgow: 1841–1844* (London: Richard Griffin, 1844), 1:112.

64. Jamal al-Din Afghani, *Kitab Tatimmat al-bayan fi ta'rikh al-afghan* (Cairo, 1899).

65. Charles Joseph Minard, *Des tableaux graphiques et des cartes figuratives* (Paris: E. Thunot, 1862).

66. On the specific transformations discussed here, see Lorraine Daston and Peter Galison, *Objectivity* (Cambridge, MA: Zone Books, 2007), 16.

67. Rudyard Kipling, "The Deep-Sea Cables," in *The Seven Seas* (London: Methuen, 1896).

68. Martin J. S. Rudwick, *Worlds before Adam: The Reconstruction of Geohistory in the Age of Reform* (Chicago: University of Chicago Press, 2010), 25–34.

69. *Organic Chemistry,* quoted in Bonneuil and Fressoz, *Shock of the Anthropocene,* 187.

70. See Andreas Malm, *Fossil Capital: The Rise of Steam-Power and the Roots of Global Warming* (London: Verso, 2016), 209; Daggett, "Energy's Power," chap. 4.

71. And these have been discussed repeatedly since the publication of Jacques Derrida's *Specters of Marx.*

72. Jeffrey Jerome Cohen, "Monster Culture (Seven Theses)," in *Monster Theory,* ed. Jeffrey Jerome Cohen (Minneapolis: University of Minnesota Press, 1996), 3–25; Mark Neocleous, "The Political Economy of the Dead: Marx's Vampires," *History of Political Thought* 24, no. 4 (2003): 684. The process was not confined to the (social and natural) sciences, and beginning with the 1818 publication of Mary Shelly's *Frankenstein* we see the emergence of what Ezra Pound called the Cult of Ugliness in various visual and literary art fields.

73. Quoted in Barak, *On Time,* 105.

74. Surat al-Baqarah, Quran, 2:286; see also Butrus al-Bustanī, *Kitab Muhīṭ al-Muhīṭ* (Beirut, 1870), 1304.

75. Karen C. Pinto, *Medieval Islamic Maps: An Exploration* (Chicago: University of Chicago Press, 2016), 158–59.

76. See Homayra Ziad, "The Return of Gog: Politics and Pan-Islamism in the Hajj Travelogue of 'Abd al-Majid Daryabadi," in *Global Muslims in the Age of Steam and Print,* ed. James L. Gelvin and Nile Green (California: University of California Press, 2014), 227–49.

77. DH.MKT 2396/100, August 27, 1900; DH.MKT 2323/123, March 26, 1900; DH.MKT 540/33, January 7, 1903; DH.UMVM 118/29, August 21, 1920, BOA.

78. The people of İzmit were so accustomed to synchronizing their *iftar* and *suhur* during Ramadan to the sound of the cannon of the navy steamer that in 1908, when steamers were taken for maintenance to Istanbul, religious authorities in town petitioned the Sultan to make sure a ship would be sent before the next month of fast. DH.MKT 1297/47, September 18, 1908, BOA.

79. DH.MKT 1259/6, June 5, 1908, BOA.

80. DH.MKT 2760/91, March 8, 1909, BOA.

81. İ.MLU 4/23, March 13, 1911, BOA.

82. As they still are, under Erdoğan.

83. William Ochsenwald, *The Hijaz Railroad* (Charlottesville: University Press of Virginia, 1980), 75.

84. Ochsenwald, 127–28, 214–15.

85. See Low, *Mechanics,* 341.

86. Low, 341.

87. İ.HUS 112/31, December 26, 1903, BOA.

88. İ.İMT 4/17, January 11, 1904, BOA.

89. ŞD 200/24, April 30, 1914, BOA.

90. İ.MMS 196/29, May 18, 1915, BOA.

91. For example, Michael Christopher Low has argued that the Saudi resort to desalination and the particular ways the regime translated the provision of water into political dependence should be seen as an Ottoman legacy from the turn of the twentieth century. See Low, "Ottoman Infrastructures of the Saudi Hydro-State: The Technopolitics of Pilgrimage and Potable Water in the Hijaz," *Comparative Studies in Society and History* 57, no. 4 (2015): 942–74.

92. Jones, *Desert Kingdom: How Oil and Water Forged Modern Saudi Arabia* (Cambridge, MA: Harvard University Press, 2010).

93. Bernard Haykel, *Saudi Arabia in Transition* (Cambridge: Cambridge University Press, 2015).

94. *İlm-i tabaqatul arz* was reprinted in a new edition in 1899, absorbed as a section inside Esad Feyzi's *İlmü'l arz ve'l maadin,* and again in 1911 and used as a textbook.

95. Amitav Ghosh, *The Great Derangement: Climate Change and the Unthinkable* (Chicago: University of Chicago Press, 2017), 20.

96. In Chakrabarty, "Climate of History."

CONCLUSION

1. Walter Benjamin, *Walter Benjamin: Selected Writings, 1938–1940,* ed. Howard Eiland and Michael W. Jennings (Cambridge, MA: Harvard University Press, 2006), 402.

2. Rolf Peter Sieferle, *The Subterranean Forest: Energy Systems and the Industrial Revolution,* trans. Michael P. Osman (Cambridge: White Horse Press, 2001). And see Malm's critique of the Malthusian underpinnings of this argument in *Fossil Capital.*

3. Mechanization and robotization might very well make various occupations redundant. It is their emancipatory potential that I seek to cast doubt on.

4. As a certain reading of Kathi Weeks's *The Problem with Work: Feminism, Marxism, Antiwork Politics, and Postwork Imaginaries* (Durham, NC: Duke University Press, 2001) might suggest. See Daggert, *Energy's Power.*

5. Samuel Moyn, "Rights vs. Duties: Reclaiming Civic Balance," *Boston Review,* May 16 2016.

6. See, for example, Ho, *Graves of Tarim*; Bishara, *Sea of Debt*; Mathew, *Margins of the Market*.

7. The importance of Neoplatonism notwithstanding, as evident in Matthew Melvin-Koushki's work on Neoplatonism in early modern Persianate writing, for example.

8. See 41:53 for example.

9. See Fazlur Rahman quoted in Shahab Ahmed, *What Is Islam?* (Princeton, NJ: Princeton University Press, 2015), 349.

10. Ahmed, *What Is Islam?*, 431. Key in this pre- or ur-text was the notion of "The Preserved Tablet" (*al-lawh al-mahfuz*)—the original and pure record of divine scripture, and of the history of heaven and earth, made of red rubies, written with light, and preserved in the highest echelons of heaven.

11. Steven Shapin, *The Scientific Revolution* (Chicago: University of Chicago Press, 2018), 90–93.

12. For example, Robert Boyle was deeply involved in financing the first translation of the New Testament into Ottoman Turkish.

13. Steven Shapin and Simon Schaffer, *Leviathan and the Air Pump: Hobbes, Boyle, and the Experimental Life* (Princeton, NJ: Princeton University Press, 2011).

14. M. Alper Yalçınkaya, *Learned Patriots: Debating Science, State, and Society in the Nineteenth-Century Ottoman Empire* (Chicago: University of Chicago Press, 2015); Marwa Elshakry, *Reading Darwin in Arabic, 1860–1950* (Chicago: University of Chicago Press, 2016).

15. Latour, *An Inquiry into Modes of Existence: An Anthropology of the Moderns* (Cambridge, MA: Harvard University Press, 2018).

Bibliography

PERIODICALS

Aberdeen Journal
Al- ʿAsifa (Beirut)
Al-Manar (Cairo)
Al-Muqtataf (Cairo)
Asiatic Journal and Monthly Miscellany
Ceride-i Havadis (Istanbul)
Chambers's Journal of Popular Literature, Science and Arts
Cumberland Pacquet, and Ware's Whitehaven Advertiser
Daily Advertiser (Wagga Wagga, Australia)
Daily Telegraph (London)
Dewsbury Reporter
Dover Express
Dundee Courier
Dundee Evening Telegraph
Edinburgh Evening News
Evening News
Hampshire Telegraph
Hartlepool Northern Daily Mail
Huddersfield Chronicle (Yorkshire)
Lancashire Evening Post
London Daily News
London Evening Standard
Moreton Bay Courier
Morning Chronicle (London)
Morning Post (London)
Morpeth Herald

Nautical Magazine
Nelson Evening Mail (New Zealand)
Portsmouth Evening News
Sheffield Daily Telegraph
Shipping and Mercantile Gazette
South Wales Daily News (Cardiff)
Spectator
St James's Gazette (London)
Standard
Straits Times Overland Journal
Sunderland Daily Echo
Sunderland Daily Echo and Shipping Gazette
Tamworth Herald
Times of India
Todmorden & District News
Vegetarian

ARCHIVES

Başbakanlık Osmanlı Arşivi (BOA), Ottoman State Archives, Istanbul
BP Archive, University of Warwick
British National Archives (BNA), London
Churchill Archives, University of Cambridge
India Office Records (IOR), British Library, London
Maharashtra State Archives, Mumbai
National Archives of India, New Delhi

WORKS CITED

Abu-Lughod, Janet. *Before European Hegemony.* Oxford: Oxford University Press, 1991.
Abu-'Uksa, Wael. *Freedom in the Arab World: Concepts and Ideologies in Arabic Thought in the Nineteenth Century.* Cambridge: Cambridge University Press, 2016.
Abul-Magd, Zeinab. *Imagined Empires: A History of Revolt in Egypt.* Berkeley: University of California Press, 2013.
Adshead, S. A. M. *Salt and Civilization.* London: Palgrave Macmillan, 2016.
Afacan, Seyma. "Of the Soul and Emotions: Conceptualizing 'the Ottoman Individual' through Psychology." PhD diss., Oxford University, 2016.
Afghani, Jamal al-Din. *Kitab Tatimmat al-bayan fi ta'rikh al-afghan.* Cairo, 1899.
Aharoni, Reuven. *The Pasha's Bedouin: Tribes and State in the Egypt of Mehemet Ali, 1805–1848.* London: Routledge, 2007.
Ahmed, Shahab. *What Is Islam?* Princeton, NJ: Princeton University Press, 2015.
Ahroni, Reuben. *The Jews of the British Crown Colony of Aden: History, Culture, and Ethnic Relations.* Leiden: E. J. Brill, 1994.

Ahuja, Ravi. "Mobility and Containment: The Voyages of South Asian Seamen, c.1900–1960." *International Review of Social History* 51 (2006): 111–41.

Akan, Taner. "Does Political Culture Matter for Europeanization? Evidence from the Ottoman Turkish Modernization in State-Labor Relations." *Employee Relations* 33, no. 3 (2011): 221–48.

Al-Bustanī, Butrus. *Kitab al-tuhfah al-Bustanīyah fī al-asfar al-Kuruzīyah, aw Rihlat Rubinsun Kruzī* [translated from Daniel Defoe, *Robinson Crusoe*]. Beirut, 1861.

———. *Kitab Muhiṭ al-Muhiṭ*. Beirut, 1870.

Al-Nuʿmanī, Shiblī. *Rihlat Shiblī al-Nuʿmanī ilá al-Qusṭanṭīniyya wa-Bayrut wal-Quds wal-Qahira fi Mustahall al-Qarn al-Rabiʿ ʿAshar al-Hijrī*. Damascus: Dar al-Qalam, 1967.

Alavi, Seema. *Muslim Cosmopolitanism in the Age of Empire*. Cambridge, MA: Harvard University Press, 2015.

Allah, Muhammad Fathi ʿAwad. *Qissat al-Fahm fi Misr*. Cairo: Dar al-Kitab al-ʿArabi li-l-Tibaʿa wa-l-Nashr, 1968.

Allen, E. E. "On the Comparative Cost of Transit by Steam and Sailing Colliers, and on the Different Modes of Ballasting (Including Plate)." *Minutes of Proceedings of the Institution of Civil Engineers* 14 (1855): 318–48. www.icevirtuallibrary.com/doi/abs/10.1680/imotp.1855.23910.

Alpers, Edward. "The Western Indian Ocean as a Regional Food Network in the Nineteenth Century." In *East Africa and the Indian Ocean*. Princeton: Markus Wiener Publishers, 2009.

Alsharhan, A. S. and A. E. M. Nairn. *Sedimentary Basins and Petroleum Geology of the Middle East*. Amsterdam: Elsevier, 1997.

Amin, Naguib, Raymond Collet, Marie-Laure Crosnier Leconte, Arnaud Du Boistesselin, and Ǧamāl al-Ġīṭānī. *Port-Saïd: Architectures XIXe – XXe siècles = Būr Saʿīd; ʿimārat al-qarn at-tāsi' ʿašar wa-'l-qarn al-ʿišrīn*. Le Caire: Institut Français d'Archéologie Orientale, 2006.

Amin, Qasim. *Tahrir al-Marʾah*. Cairo, 1899. Translated by Samiha Sidhom Peterson as *The Liberation of Women*. Cairo: American University in Cairo Press, 2000.

Amin, Shadid, and Marcel van den Linden, eds. *Peripheral Labour: Studies in Partial Proletarianization*. Cambridge: Cambridge University Press, 1997.

Anim-Addo, Anyaa. "'A Wretched and Slave-like Mode of Labor': Slavery, Emancipation, and the Royal Mail Steam Packet Company's Coaling Stations." *Historical Geography* 39 (2011): 65–84.

Anning, E. G., F. J. Bentley, and Lionel Yexley. *The Log of H.M.S. Argonaut, 1900–1904*. London: Westminster Press, 1904.

Appadurai, Arjun. *The Future as Cultural Fact: Essays on the Global Condition*. London: Verso, 2013.

Armstrong, John, and David M. Williams. *The Impact of Technological Change: The Early Steamship in Britain*. St. John's, Newfoundland: International Maritime Economic History Association, 2011.

Arrighi, Giovanni. *The Long Twentieth Century: Money, Power, and the Origins of Our Times*. London: Verso, 2010.

Asad, Talal. *Formations of the Secular: Christianity, Islam, Modernity.* Stanford, CA: Stanford University Press, 2003.

Association of Average Adjusters. *Report of the General Meeting Held at Lloyd's.* London, 1872.

Atabaki, Touraj, and Gavin Brockett. "Ottoman and Republican Turkish Labour History: An Introduction." In *Ottoman and Republican Turkish Labour History,* edited by Touraj Atabaki and Gavin Brockett, 1–19. Cambridge: Cambridge University Press, 2009.

Atia, Mona. *Building a House in Heaven: Pious Neoliberalism and Islamic Charity in Egypt.* Minneapolis: University of Minnesota Press, 2013.

Aydin, Cemil. *The Idea of the Muslim World: A Global Intellectual History.* Cambridge, MA: Harvard University Press, 2017.

Ayrton, Frederick. *Railways in Egypt: Communication with India.* London: Ridgway, 1857.

Baines, Edward. *History of the Cotton Manufacture in Great Britain.* London: H. Fisher, R. Fisher, and P. Jackson, 1835.

Bal, Nurcan. "İlk Buharlı Gemimiz Buğ Gemisi." *Yedikita* 31 (2011): 18–24.

———. "XIX. Yüzyılda Osmanlı Bahriyesi'nde Gemi İnşa Teknolojisinde Değişim: Buharlı Gemiler Dönemi." Master's thesis, Mimar Sinan Fine Arts University, Istanbul, 2010.

Balachandran, G. *Globalizing Labour?: Indian Seafarers and World Shipping, c.1870–1945.* New Delhi: Oxford University Press, 2012.

Balibar, Étienne, and Immanuel Maurice Wallerstein. *Race, Nation, Class: Ambiguous Identities.* London: Verso, 2011.

Bang, Anne K. *Sufis and Scholars of the Sea: Family Networks in East Africa, 1860–1925.* New York: Psychology Press, 2003.

Barak, On. *On Time: Technology and Temporality in Modern Egypt.* Berkeley: University of California Press, 2013.

———. "Outsourcing: Energy and Empire in the Age of Coal, 1820–1911." *International Journal of Middle East Studies* 47, no. 3 (2015): 425–45.

———. "Three Watersheds in the History of Energy." *Comparative Studies of South Asia, Africa and the Middle East* 34, no. 3 (2014): 440–53.

Bar Sadeh, Roy. "Debating Gandhi in al-Manar During the 1920s and 1930s." *Comparative Studies of South Asia, Africa and the Middle East* 38, no. 3 (2018): 491–507.

Bathurst, Bella. *The Lighthouse Stevensons: The Extraordinary Story of the Building of the Scottish Lighthouses by the Ancestors of Robert Louis Stevenson.* London: HarperCollins, 1999.

Baucom, Ian. *Specters of the Atlantic: Finance Capital, Slavery, and the Philosophy of History.* Durham, NC: Duke University Press, 2005.

Bayly, Christopher Alan. *The Birth of the Modern World, 1780–1914: Global Connections and Comparisons.* Malden, MA: Blackwell Publishing, 2004.

———. *Recovering Liberties: Indian Thought in the Age of Liberalism and Empire.* Cambridge: Cambridge University Press, 2012.

Beck, Ulrich. *Risk Society: Towards a New Modernity.* 1986. Reprint, London: Sage, 2010.

Beckert, Sven. *Empire of Cotton: A Global History*. New York: Vintage Books, 2015.

Behar, Cem. *A Neighborhood in Ottoman Istanbul: Fruit Vendors and Civil Servants in the Kasap Ilyas Mahalle*. Albany: State University of New York Press, 2003.

Beinin, Joel. *Workers and Peasants in the Modern Middle East*. Cambridge: Cambridge University Press, 2010.

Beinin, Joel, and Zachary Lockman. *Workers on the Nile: Nationalism, Communism, Islam, and the Egyptian Working Class, 1882–1954*. Princeton, NJ: Princeton University Press, 1987.

Ben Bassat, Yuval, and Yossi Ben-Artzi. "The Collision of Empires as Seen from Istanbul: The Border of British-Controlled Egypt and Ottoman Palestine as Reflected in Ottoman Maps." *Journal of Historical Geography* 50 (2015): 25–36.

Ben-Layashi, Samir. "Discourse and Practice of Medicine, Hygiene and Body in Morocco (1880–1912)." PhD diss., Tel Aviv University, 2013.

Benjamin, Walter. *Walter Benjamin: Selected Writings, 1938–1940*. Edited by Howard Eiland and Michael W. Jennings. Cambridge, MA: Harvard University Press, 2006.

Benton, Lauren. *A Search for Sovereignty: Law and Geography in European Empires, 1400–1900*. New York: Cambridge University Press. 2010.

Birkett, J., and D. Radcliffe. "Normandy's Patent Marine Aërated Fresh Water Company: A Family Business for 60 Years, 1851–1910." *IDA Journal of Desalination and Water Reuse* 6, no. 1 (2014): 24–32.

Bishara, Fahad Ahmad. *A Sea of Debt: Law and Economic Life in the Western Indian Ocean, 1780–1950*. Cambridge: Cambridge University Press, 2017.

Black, Brian C. *Crude Reality: Petroleum in World History*. Lanham, MD: Rowman and Littlefield, 2012.

Blyth, Robert J. "Aden, British India and the Development of Steam Power." In *Maritime Empires: British Imperial Maritime Trade in the Nineteenth Century*, edited by David Killingray, Margarette Lincoln, and Nigel Rigby. Suffolk: Boydell Press, 2004.

Board of Trade. "Wreck Report for 'Jeddah' 1881." Southampton City Council Library. https://southampton.spydus.co.uk/cgi-bin/spydus.exe/TRN/OPAC /BIBENQ/34737664/?SEL=26596146.

Bonneuil, Christophe, and Jean-Baptiste Fressoz. *The Shock of the Anthropocene: The Earth, History and Us*. London: Verso, 2016.

Borgstrom, Georg. *Hungry Planet: The Modern World at the Edge of Famine*. New York: Collier Books, 1967.

Bowring, John. *Report on Egypt and Candia: Addressed to the Right Hon. Lord Viscount Palmerston, Her Majesty's Principal Secretary of State for Foreign Affairs, &c. &c. &c.* London: Printed by W. Clowes and Sons for H.M.S.O., 1840.

Boyer, Dominic. "Energopower: An Introduction." *Anthropological Quarterly* 87, no. 2 (2014): 309–33.

"Brief van Sajid Jahja Weltevreden aan Christiaan Snouck Hurgronje (1857–1936)." Publisher unknown, 1931.

British Documents on Foreign Affairs 1856–1914. Vol. 19. Frederick, MD: University Publications of America, 1984.

Broeze, Frank. "The Ports and Port System of the Asian Seas: An Overview with Historical Perspective from c. 1750." *The Great Circle* 18, no. 2 (1996): 73–96.

Brown, Peter. *Through the Eye of a Needle: Wealth, the Fall of Rome, and the Making of Christianity in the West, 350–550 AD.* Princeton, NJ: Princeton University Press, 2014.

Brunsman, Denver. *The Evil Necessity: British Naval Impressment in the Eighteenth-Century Atlantic World.* Charlottesville: University of Virginia Press, 2013.

Bulmus, Birsen. *Plague, Quarantines and Geopolitics in the Ottoman Empire.* Edinburgh: Edinburgh University Press, 2012.

Burford, Robert. *Description of a View of the Bombardment of St. Jean d'Acre.* London: Geo. Nichols, 1841.

Burton, Isabel. *Arabia, Egypt, India: A Narrative of Travel.* 1879. Reprint, Cambridge: Cambridge University Press, 2012.

Buzpinar, S. Tufan. "Abdulhamid II and Sayyid Fadl Pasha of Hadramawt: An Arab Dignitary's Ambitions (1876–1900)." *Journal of Ottoman Studies* 13 (1993): 227–39.

Çalişkan, Koray. *Market Threads: How Cotton Farmers and Traders Create a Global Commodity.* Princeton, NJ: Princeton University Press, 2010.

Campbell, Alan. "18th Century Legacies and 19th Century Traditions: The Labour Process, Work Culture and Miners' Unions in the Scottish Coalfield before 1914." In *Vom Bergbau zum Industrierevier,* edited by Ekkehard Westermann, 239–55. Stuttgart: Franz Steiner Verlag, 1995.

Caneva, Kenneth L. *Robert Mayer and the Conservation of Energy.* Princeton, NJ: Princeton University Press, 1993.

Cardwell, Donald Stephen Lowell. *James Joule: A Biography.* Manchester: Manchester University Press, 1990.

Cattelan, Valentino. "In the Name of God: Managing Risk in Islamic Finance." EABH Papers no. 14–07 (July 2014). www.econstor.eu/bitstream/10419/100004/1/791365034.pdf.

Catton, William. *Overshoot: The Ecological Basis of Revolutionary Change.* Urbana: University of Illinois Press, 1980.

Chadwick, F. E. "The Development of the Steamship, and the Liverpool Exhibition of 1886." *Scribner's Magazine* 1, no. 5 (May 1887): 515–43.

———. *Ocean Steamships.* 1891. Reprint, Paderborn: Salzwasser Verlag, 2011.

Chakrabarty, Dipesh. "The Climate of History: Four Theses." *Critical Inquiry* 35, no. 2 (Winter 2009): 197–222.

———. *Provincializing Europe: Postcolonial Thought and Historical Difference.* Princeton, NJ: Princeton University Press, 2007.

Chalcraft, John. "The Coal Heavers of Port Sa'id: State-Making and Worker Protest, 1869–1914." *International Labor and Working-Class History,* no. 60 (2001): 110–24.

———. "Popular Protest, the Market and the State in Nineteenth- and Early Twentieth-Century Egypt." In *Subalterns and Social Protest: History from*

Below in the Middle East and North Africa, edited by Stephanie Cronin, 83–104. London: Routledge, 2008.

———. *The Striking Cabbies of Cairo and Other Stories: Crafts and Guilds in Egypt, 1863–1914.* Albany: State University of New York Press, 2004.

Chamberlin, J. Edward, and Sander Gilman. *Degeneration: The Dark Side of Progress.* New York: Columbia University Press, 1985.

Chaufton, Albert. *Les assurances, leur passé, leur présent, leur avenir.* Paris: Librairie A. Marescq Aîné, 1884.

Chet, Guy. *The Ocean Is a Wilderness: Atlantic Piracy and the Limits of State Authority, 1688–1856.* Amherst: University of Massachusetts Press, 2014.

Church, Roy. *Strikes and Solidarity: Coalfield Conflict in Britain, 1889–1966.* Cambridge: Cambridge University Press, 1998.

Churchill, Winston S. *Churchill by Himself.* New York: Rosetta Books, 2013.

Çizakça, Murat. *A Comparative Evolution of Business Partnerships: The Islamic World and Europe, with Specific Reference to the Ottoman Archives.* Leiden: E. J. Brill, 1996.

Clarence-Smith, W. G. *Islam and the Abolition of Slavery.* New York: Oxford University Press, 2006.

Clemow, F. G. "Some Turkish Lazarets and Other Sanitary Institutions in the Near East." *The Lancet,* April 27, 1907.

Cockerell, H. A. L., and Edwin Green. *The British Insurance Business 1547–1970: An Introduction and Guide to Historical Records in the United Kingdom.* London: Heinemann, 1976.

Cohen, Jeffrey Jerome. "Monster Culture (Seven Theses)." In *Monster Theory,* edited by Jeffrey Jerome Cohen, 3–25. Minneapolis: University of Minnesota Press, 1996.

Cole, Juan. *Colonialism and Revolution in the Middle East: Social and Cultural Origins of Egypt's 'Urabi Movement.* Princeton, NJ: Princeton University Press, 1993.

Colley, Linda. *Captives: Britain, Empire and the World, 1600–1850.* New York: Anchor Books, 2004.

Colpitts, George. *Pemmican Empire: Food, Trade, and the Last Bison Hunts in the North American Plains, 1780–1882.* Cambridge: Cambridge University Press, 2015.

Conrad, Joseph. *"Typhoon" and Other Stories.* Edited by J. H. Stape. London: Penguin, 2007.

Critchell, James Troubridge, and Joseph Raymond. *A History of the Frozen Meat Trade: An Account of the Development and Present Day Methods of Preparation, Transport, and Marketing of Frozen and Chilled Meats.* London: Constable, 1912.

Cronon, William. *Nature's Metropolis: Chicago and the Great West.* New York: W. W. Norton, 1991.

Curzon, Lord George. *Frontiers.* Oxford: Clarendon Press, 1907.

Daggett, Cara. "Energy's Power: Fuel, Work and Waste in the Politics of the Anthropocene." PhD diss., Johns Hopkins University, 2016.

Dakin, Douglas. *The Greek Struggle for Independence, 1821–1833*. Berkeley: University of California Press, 1973.

Daston, Lorraine, and Peter Galison. *Objectivity*. Cambridge MA: Zone Books, 2007.

Davies, Sam, Colin J. Davis, David de Vries, Lex Heerma van Voss, Lidewijb Hesselink, and Klaus Weinhauer, eds. *Dock Workers: International Explorations in Comparative Labour History, 1790–1970*. Burlington, VT: Ashgate, 2000.

De Bélidor, Bernard Forest. *Architecture Hydraulique*. 4 vols. Paris: C. A. Jombert, 1737–1770.

De Fonblanque, Edward Barrington. *Treatise on the Administration and Organization of the British Army*. London: Longman, Brown, Green, Longmans, and Roberts, 1858.

Defoe, Daniel. *A Tour Thro' the Whole Island of Great Britain Divided into Circuits or Journeys*. London: S. Birt, T. Osborne, D. Browne, J. Hodges, J. Osborn, A. Miller and J. Robinson, 1748.

Derrida, Jacques. *Specters of Marx, the State of the Debt, the Work of Mourning, & the New International*. Translated by Peggy Kamuf. New York: Routledge, 1994.

Devji, Faisal. *The Terrorist in Search of Humanity: Militant Islam and Global Politics*. New York: Columbia University Press, 2008.

Disraeli, Benjamin. *Coningsby; Or, the New Generation*. Vol. 2. Oxford: Oxford University Press, 1844.

Dodick, Jeff, and Nir Orion. "Geology as an Historical Science: Its Perception within Science and the Education System." *Science & Education* 12 (2003): 197–211.

Donzelot, Jacques. "Pleasure in Work." In *The Foucault Effect: Studies in Governmentality*, edited by Graham Burchel, Colin Gordon, and Peter Miller. Chicago: University of Chicago Press, 1991.

Doran, Kevin. "Adrift on the High Sea: The Application of Maritime Salvage Law to Historic Shipwrecks in International Waters." *Southwestern Journal of International Law* 18 (2013): 101–20.

Doughty, Charles. *Travels in Arabia Deserta*. Vol. 1. 1888. Reprint, London: William Clowes & Sons, 1921.

———. *Travels in Arabia Deserta*. Vol. 2. 1888. Reprint, Charleston, SC: Nabu Press, 2012.

Du Bois, W. E. B. "Worlds of Color." *Foreign Affairs* 3, no. 3 (April 1925): 423–44.

Dudoignon, Stephane, Hisao Komatsu, and Yasushi Kosugi, eds. *Intellectuals in the Modern Islamic World: Transmission, Transformation and Communication*. London: Routledge, 2006.

Dural, A. Baran, and Ertem Uner. "Development of the Worker Class in the Ottoman Empire." *European Scientific Journal* 8, no. 11 (2012): 215–35.

Dursun, Selçuk. "Forest and the State: History of Forestry and Forest Administration in the Ottoman Empire." PhD diss., Sabanci University, Istanbul, 2007.

Edgerton, David. *The Shock of the Old: Technology and Global History since 1900*. London: Profile Books, 2006.

Ekşi, Gülnur. "The Effects of the Berlin-Baghdad Railway on the Vegetation in Anatolia." Lecture at the Haus der Kulturen der Welt, 2017. www.hkw.de /en/app/mediathek/video/61038.

El-Eini, Roza. *Mandated Landscape: British Imperial Rule in Palestine, 1929– 1948*. London: Routledge, 2006.

El Moudden, Abderrahmane. "The Ambivalence of *Rihla*: Community Integration and Self-Definition in Moroccan Travel Accounts." In *Muslim Travellers*, edited by Dale F. Eickelman and James Piscatori, 69–84. Berkeley: University of California Press, 1990.

Elshakry, Marwa. *Reading Darwin in Arabic, 1860–1950*. Chicago: University of Chicago Press, 2016.

Emsley, Clive. *The English Police: A Political and Social History*. London: Longman, 1996.

Eng, Tan Twan. "The Law of Salvage." PhD diss., University of Cape Town, 2014.

Ewald, Francis. "Insurance and Risk." In *The Foucault Effect: Studies in Governmentality*, edited by Graham Burchell, Colin Gordon, and Peter Miller, 197–210. Chicago: University of Chicago Press, 1991.

Ewald, Janet J. "Bondsmen, Freedmen, and Maritime Industrial Transportation, c. 1840–1900." *Slavery and Abolition* 31, no. 3 (2010): 451–66.

Ewald, Janet, and William G. Clarence-Smith. "The Economic Role of the Hadrami Diaspora in the Red Sea and Gulf of Aden, 1820s to 1930s." In *Hadhrami Traders, Scholars and Statesmen in the Indian Ocean, 1750s to 1960s*, edited by Ulrike Freitag and William G. Clarence-Smith, 281–96. Leiden: Brill, 1997.

Faḍl ibn ʿAlawī. *Kitāb al-Īḍāḥ al-asrār al-ʿAlawīyah wa-minhāj al-sādah al-ʿAlawīyah*. Miṣr: Maṭbaʿat al-Ādāb wa-al-Muʾayyid, 1898.

Fahmy, Khaled. *All the Pasha's Men: Mehmed Ali, His Army and the Making of Modern Egypt*. Cambridge: Cambridge University Press, 1997.

———. "An Olfactory Tale of Two Cities: Cairo in the Nineteenth Century." In *Historians in Cairo: Essays in Honor of George Scanlon*, edited by Jill Edwards, 155–87. Cairo: American University in Cairo Press, 2002.

Fairlie, Simon. *Meat: A Benign Extravagance*. White River Junction, VT: Chelsea Green, 2010.

Fancy, William. "The Red Sea: The Wind Regime and Location of Ports." In *Trade and Travel in the Red Sea Region: Proceedings of Red Sea Project I, Held in the British Museum, October 2002*, edited by Paul Lunde and Alexandra Porter, 7–17. Oxford: Tempus Reparatum, 2004.

Farah, Caesar E. *The Sultan's Yemen: 19th-Century Challenges to Ottoman Rule*. London: I. B. Tauris, 2002.

Faroqhi, Suraiya. *Animals and People in the Ottoman Empire*. Istanbul: Eren, 2010.

Fethi, Mehmet Ali. *İlm-i tabakat-i arz*. Istanbul: Dar üt-Tibaat ül-Amire, 1852.

Ferguson, Michael. "Clientship, Social Indebtedness and State-Controlled Emancipation of Africans in the Late Ottoman Empire." In *Debt and Slavery in the Mediterranean and Atlantic Worlds*, edited by Gwyn Campbell and Alessandro Stanziani, 49–62. London: Pickering and Chatto, 2013.

Feys, Torsten, Lewis R. Fischer, Stephane Hoste, and Stephen Vanfraechem, eds. *Maritime Transport and Migration: The Connections between Maritime and Migration Networks.* Oxford: Oxford University Press, 2017.

Findlay, Alexander George. *A Description and List of Lighthouses of the World.* London: R.H. Laurie, 1861.

Findley, Carter V. *Bureaucratic Reform in the Ottoman Empire: The Sublime Porte, 1789–1922.* Princeton, NJ: Princeton University Press, 1980.

Foucault, Michel. *The History of Sexuality.* Vol. 1, *The Will to Knowledge.* New York: Penguin, 2008.

———. *On the Punitive Society: Lectures at the Collège de France, 1972–1973.* London: Palgrave Macmillan, 2015.

———. *Order of Things: An Archaeology of the Human Sciences.* New York: Psychology Press, 2002.

Foulke, Robert D. "Life in the Dying World of Sail, 1870–1910." *Journal of British Studies* 3, no. 1 (1963): 105–36.

Fouquet, Roger. *Heat, Power and Light: Revolutions in Energy Services.* Cheltenham: Edward Elgar, 2008.

Fouquet, Roger, and Peter Pearson. "Past and Prospective Energy Transitions: Insights from History." *Energy Policy* 50 (November 2012): 1–2.

Fowler, H.M. *The Log of H.M.S. Encounter, Australian Station, 1908–1910.* London: Westminster Press, 1910.

Freese, Barbara. *Coal: A Human History.* New York: Penguin Books, 2004.

Freidberg, Susanne. *Fresh: A Perishable History.* Cambridge, MA: Harvard University Press, 2010.

Freitag, Ulrike. "The Changing Faces of the 'Bride of the Red Sea,' Jeddah c. 1840–1947." Unpublished manuscript, last modified November 24, 2018.

———. *Indian Ocean Migrants and State Formation in Hadhramaut: Reforming the Homeland.* Leiden: Brill, 2003.

———. "Symbolic Politics and Urban Violence in Late Ottoman Jeddah." In *Urban Violence in the Middle East,* edited by Ulrike Freitag, Nelida Fuccaro, Claudia Ghrawi, and Nora Lafi, 111–38. New York: Berghahn Books, 2015.

Fuccaro, Nelida. *Histories of City and State in the Persian Gulf: Manama since 1800.* New York: Cambridge University Press, 2009.

Gavin, R.J. *Aden under British Rule, 1839–1966.* London: C. Hurst, 1975.

Gencer, Ali İhsan. *Türk Denizcilik Tarihi Araştırmaları.* Istanbul: Türkiye Denizciler Sendikası, 1986.

Geshekter, Charles L. "Anti-Colonialism and Class Formation: The Eastern Horn of Africa before 1950." *International Journal of African Historical Studies* 18, no. 1 (1985): 1–32.

Geyikdağı, V. Necla. *Foreign Investment in the Ottoman Empire: International Trade and Relations, 1854–1914.* London: Tauris Academic Studies, 2011.

Ghosh, Amitav. *The Great Derangement: Climate Change and the Unthinkable.* Chicago: University of Chicago Press, 2017.

Gillard, David, ed. *British Documents on Foreign Affairs.* Series B. *The Near and Middle East: 1856–1914.* Vol. 19, *The Ottoman Empire: Nationalism and Revolution, 1885–1908.* London: United Kingdom Foreign Office; Frederick, MD: University Publications of America, 1985.

Gillis, John R. *The Human Shore: Seacoasts in History.* Chicago: University of Chicago Press, 2012.

Gindin, Sam. "Labor." In *Keywords for Radicals: The Contested Vocabulary of Late-Capitalist Struggle,* edited by Kelly Fritsch, Claire O'Connor and A.K. Thompson. Oakland, CA: AK Press, 2016.

Godfrey-Smith, Peter. *Other Minds: The Octopus, the Sea, and the Deep Origins of Consciousness.* New York: Farrar, Straus and Giroux, 2016.

Goldberg, Ellis. "Mobilizing Coal for War: The Rise and Decline of English Socialism." Unpublished paper online at SSRN, 2015. https://ssrn.com /abstract = 2566110; http://dx.doi.org/10.2139/ssrn.2566110.

Gordan, J.D. *Joseph Conrad: The Making of a Novelist.* 1940. Reprint, New York: Russell & Russell, 1963.

Gorman, Anthony. "Foreign Workers in Egypt 1882–1914: Subaltern or Labour Elite?" In *Subalterns and Social Protest: History from Below in the Middle East and North Africa,* edited by Stephanie Cronin, 237–61. London: Routledge, 2008.

Gourlie, William. "Notice of the Fossil Plants in Glasgow Geological Museum." In *Proceedings of the Royal Philosophical Society of Glasgow: 1841–1844.* Vol. I. London: Richard Griffin, 1844.

Graham, Gerald S. "The Ascendancy of the Sailing Ship, 1850–85." *Economic History Review* 9, no. 1 (1956): 74–88.

Grandin, Greg. *The Empire of Necessity: Slavery, Freedom, and Deception in the New World.* New York: Henry Holt, 2014.

Gratien, Chris. "The Ottoman Quagmire: Malaria, Swamps, and Settlement in the Late Ottoman Mediterranean." *International Journal of Middle East Studies* 49, no. 4 (2017): 583–604.

Gray, Steven. "Black Diamonds: Coal, the Royal Navy, and British Imperial Coaling Stations, circa 1870-1914." PhD diss., University of Warwick, 2014.

———. *Steam Power and Sea Power: Coal, the Royal Navy, and the British Empire, c. 1870–1914.* London: Palgrave Macmillan, 2018.

Green, Anna. "The Work Process." In *Dock Workers: International Explorations in Comparative Labour History, 1790–1970,* edited by Sam Davies, Colin J. Davis, David de Vries, Lex Heerma van Voss, Lidewijb Hesselink, and Klaus Weinhauer, 2:560–579. Burlington, VT: Ashgate, 2000.

Green, Nile. *Bombay Islam: The Religious Economy of the West Indian Ocean, 1840–1915.* Cambridge: Cambridge University Press, 2011.

———. *The Love of Strangers: What Six Muslim Students Learned in Jane Austen's London.* Princeton, NJ: Princeton University Press, 2016.

———. "The Rail *Hajjis:* The Trans-Siberian Railway and the Long Way to Mecca." In *The Hajj: Collected Essays,* edited by Venetia Porter and Liana Saif, 100–107. London: British Museum Press, 2013.

Greenberg, Michael. *British Trade and the Opening of China, 1800–42.* Cambridge: The University Press, 1951.

Guha, Ramachandra. *Gandhi before India.* London: Penguin, 2013.

Günel, Gökçe. "The Infinity of Water: Climate Change Adaptation in the Arabian Peninsula." *Public Culture* 28, no. 2 (2016): 291–315.

Gürboğa, Nurşen. "Mine Workers, the State and War: The Ereğli-Zonguldak Coal Basin as the Site of Contest, 1920–1947." PhD diss., Boğaziçi University, Istanbul, 2005.

Gutting, Gary. *Michel Foucault's Archaeology of Scientific Reason*. Cambridge: Cambridge University Press, 1989.

Hair, P.E. "Slavery and Liberty: The Case of the Scottish Colliers." *Slavery and Abolition* 23, no. 3 (2000): 136–51.

Halevy, Dotan. "Lishtot (bira) meha-Yam shel 'Azah: Tsmihato ve-Shki'ato shel ha-Mishar ha-Yami me-'Azah be-Shalhe ha-Tkufa ha'Othmanit" [Drinking (beer) from the sea at Gaza: The rise and fall of Gaza's maritime trade in the late Ottoman period]. *ha-Mizrah ha-Hadash* 55 (2016): 37–42.

Hall, W.H., and W.D. Bernard. *The Nemesis in China*. 3rd ed. London: Henry Colburn, 1846.

Hamilton, Clive. *Defiant Earth: The Fate of Humans in the Anthropocene*. Cambridge: Polity Press, 2017.

Hanioğlu, M. Şükrü. *Preparation for a Revolution: The Young Turks, 1902–1908*. London: Oxford University Press, 2001.

Hannaford, Ivan. *Race: The History of an Idea in the West*. Baltimore: John Hopkins University Press, 1996.

Hansard, Thomas Curson. *Hansard's Parliamentary Debates, 3rd series: Commencing with the Accession of William . . . Iv, 43 and 44 Victoriae, 1880*. N.p.: Forgotten Books, 2017.

Ha'adani, Mahalal. *Beyn 'Aden ve-Teyman* [Between Aden and Yemen]. Tel-Aviv: 'Am 'oved, 1947.

Haraway, Donna. *Staying with the Trouble: Making Kin in the Chthulucene*. Durham, NC: Duke University Press, 2016.

Harcourt, Freda. *Flagships of Imperialism: The P&O Company and the Politics of Empire from Its Origins to 1867*. Manchester: Manchester University Press, 2013.

Harlaftis, Galina. *A History of Greek-Owned Shipping: The Making of an International Tramp Fleet, 1830 to the Present Day*. London: Routledge, 1995.

Harley, C. Knick. "The Shift from Sailing Ships to Steamships, 1850–1890: A Study in Technological Change and its Diffusion." In *Essays on a Mature Economy: Britain after 1840*, edited by Donald N. McCloskey, 215–34. Princeton, NJ: Princeton University Press, 1971.

Haykel, Bernard. *Saudi Arabia in Transition*. Cambridge: Cambridge University Press, 2015.

Heinrich, M. "Ambivalences of Marx's Critique of Political Economy as Obstacles for the Analysis of Contemporary Capitalism." Paper presented at the Historical Materialism Conference, London, October 10, 2004.

———. *An Introduction to the Three Volumes of Karl Marx's Capital*. Translated by Alexander Lucascio. New York: Monthly Review Press, 2004.

Hill, George. *Treasure Trove in Law and Practice: From Earliest Time to the Present Day*. Oxford: Clarendon Press, 1936.

Hirschkind, Charles. *The Ethical Soundscape: Cassette Sermons and Islamic Counterpublics*. New York: Columbia University Press, 2009.

Ho, Engseng. *The Graves of Tarim: Genealogy and Mobility across the Indian Ocean*. Berkeley: University of California Press, 2010.

Hogg, John. "On the Ballast-Flora of the Coasts of Durham and Northumberland." In *The Annals and Magazine of Natural History, Including Zoology, Botany, and Geology*. Vol. 19, 3rd series. Conducted by Prideaux John Selby, Charles C. Babington, John Edward Gray, and William Francis, 38–43. London: Taylor and Francis, 1867.

Holland, John. *The History and Description of Fossil Fuel, the Collieries and Coal Trade of Great Britain*. London: Cass, 1841.

Holt, Elizabeth M. *Fictitious Capital: Silk, Cotton, and the Rise of the Arabic Novel*. New York: Fordham University Press, 2017.

Holt, Thomas C. *The Problem of Freedom: Race, Labor, and Politics in Jamaica and Britain, 1832–1938*. Baltimore: Johns Hopkins University Press, 1992.

Hopper, Matthew S. "The Globalization of Dried Fruit: Transformations in the Eastern Arabian Economy, 1860s-1920s." In *Global Muslims in the Age of Steam and Print*, edited by James Gelvin and Nile Green, 158–84. Berkeley: University of California Press, 2014.

House of Commons Parliamentary Papers. *Children's Employment Commission: First Report of the Commissioners, Mines*. 1842. Reprint, Shannon: Irish University Press, 1968.

———. *Report on Steam Vessel Accidents to the Committee of the Privy Council for Trade*. London, 1839,

Howe, Everett D. *Fundamentals of Water Desalination*. New York: Marcel Dekker, 1974.

Huber, Matthew T. *Lifeblood: Oil, Freedom, and the Forces of Capital*. Minneapolis: University of Minnesota Press, 2013.

Huber, Valeska. *Channelling Mobilities: Migration and Globalisation in the Suez Canal Region and Beyond, 1869–1914*. Cambridge: Cambridge University Press, 2013.

Hunter, F.M. *An Account of the British Settlement of Aden in Arabia*. 1877. Reprint, London: Frank Cass, 1968.

Hunter, W.W. *Orissa*. 2 vols. London: Smith, Elder and Co., 1872.

Ibrahim, Muhammad Mahmud. *Iktishaf al-Fahm fi Misr*. Matbaʿat Costa Thomas wa-Shurakaʾihi, Cairo, 1957.

Ibrahim, Ibn Muhammad. *The Ship of Sulaiman*. Translated by John O'Kane. London: Routledge, 1972.

"Ice-culture." *Chambers's Journal of Popular Literature, Science and Arts*, February 13, 1864, 99–101.

"Imperial Treasury." Topaka Palace Museum. http://topkapisarayi.gov.tr/en/content/imperial-treasury.

İnalcık, Halil, and Donald Quataert. *An Economic and Social History of the Ottoman Empire, 1300–1914*. Vol. 2. Cambridge: Cambridge University Press, 1994.

Jacob, W.C. "Of Angels and Men: Sayyid Fadl bin Alawi and Two Moments of Sovereignty." *Arab Studies Journal* 20, no. 1 (Spring 2012): 40–73.

Jakes, Aaron. "Boom, Bugs, Bust: Egypt's Ecology of Interest, 1882–1914." *Antipode* 49, no. 4 (2017): 1035–59.

Jakes, Aaron G., and Ahmad Shokr. "Finding Value in Empire of Cotton." *Critical Histories Studies* 4, no. 1 (2017): 107–36.

Jalal, Ayesha. *Partisans of Allah: Jihad in South Asia.* Cambridge, MA: Harvard University Press, 2008.

Jameson, Fredric. *The Political Unconscious: Narrative as a Socially Symbolic Act.* London: Routledge, 1989.

Jevons, Herbert Stanley. *Foreign Trade in Coal.* London: P. S. King & Son, 1909.

Jevons, William Stanley. *The Coal Question: An Inquiry Concerning the Progress of the Nation, and the Probable Exhaustion of Our Coal-Mines.* London: Macmillan, 1865.

Jones, Christopher F. *Routes of Power.* Cambridge: Harvard University Press, 2014.

Jones, Nicolette. *The Plimsoll Sensation: The Great Campaign to Save Lives at Sea.* London: Abacus, 2007.

Jones, Toby Craig. *Desert Kingdom.* Cambridge, MA: Harvard University Press, 2011.

Joseph, Betty. *Reading the East India Company 1720–1840: Colonial Currencies of Gender.* Chicago: University of Chicago Press, 2004.

Kahya, Fatih. "Osmanlı Devleti'inde Sigortacılığın Ortaya Çıkışı ve Gelişimi" [The emergence and the development of insurance in the Ottoman Empire]. Master's thesis, Marmara Üniversitesi, Sosyal Bilimler Enstitüsü, Istanbul, 2007.

Kane, Eileen M. *Russian Hajj: Empire and the Pilgrimage to Mecca.* Ithaca: Cornell University Press, 2015.

Karateke, Hakan T. *An Ottoman Protocol Register: Containing Ceremonies from 1736 to 1808, BEO Sadaret Defterleri 350 in the Prime Ministry Ottoman State Archives, Istanbul.* Istanbul: Ottoman Bank Archive and Research Centre, 2007.

Kawakibi, 'Abd al-Rahman. *Umm al-Qura.* Al-A'mal al-Kamilah lil-Kawakibi. Beirut: Markaz Dirasat al-Wahdah al-Arabiyya, 1995.

Kawakibi, 'Abd al-Raḥman. *Yesodot ha-Islam ha-Salafi: Proṭoḳol ye'idat ha-teḥiyah ha-Islamit bam he-'arim Makah.* 1899. Translated into Hebrew by Shosh Ben-Ari. Tel Aviv: Resling, 2015.

Keeling, Drew. "The Transportation Revolution and Transatlantic Migration, 1850–1914." *Research in Economic History* 19 (1999): 39–74.

Kennedy, Robert E. "The Protestant Ethic and the Parsis." *American Journal of Sociology* 68, no. 1 (1962): 11–20.

Khan, Noor-Aiman. "The Enemy of My Enemy: Indian Influences on Egyptian Nationalism, 1907–1930." PhD diss., University of Chicago, 2006.

Khan, Sayyid Ahmed. *A Voyage to Modernism.* Translated and edited by Mushirul Hasan and Nishat Zaidi. New Delhi: Primus Books, 2011.

Kioumgi, Farid, and Robert Graham. *A Photographer on the Hajj: The Travels of Muhammad 'Ali Effendi Sa'udi (1904/1908).* Cairo: American University in Cairo Press, 2009.

Kipling, Rudyard. "The Deep-Sea Cables." In *The Seven Seas.* London: Methuen, 1896.

Kirby, Richard Shelton, Arthur Burr Darling, Sidney Withington, and Frederick Gridley Kilgour. *Engineering in History.* 1956. Reprint, Mineola, NY: Dover, 1990.

Kirkaldy, Adam. *British Shipping: Its History, Organisation and Importance.* London: Kegan, Paul, Trench, Trübner, 1914.

Kış, Salih. "First Mining School in the Ottoman Empire and Mining Engineer Training Osmanlı Devleti'nde İlk Maden Mektebi ve Maden Mühendisi Eğitimi." *Journal of History Studies* 7, no. 3 (September 2015): 111–21.

Kling,Blair B. *Partner in Empire: Dwarkanath Tagore and the Age of Enterprise in Eastern India.* Berkeley: University of California Press, 1976.

Klorman, Bat-Zion Eraqi. *Yehudey Teyman: Historia, Ḥevra, Tarbut* [Jews of Yemen: History, society, culture]. Vol. 2. Ra'anana: The Open University, 2004.

Kohen, David M. "Insurance in the Ottoman Empire." *Middle East Insurance Review* (May 2009).

Kour, Z. H. *The History of Aden, 1839–72.* London: Frank Cass, 2005.

Ković, Miloš. *Disraeli and the Eastern Question.* New York: Oxford University Press, 2011.

Krishnan, Sanjay. "Seeing the Animal: Colonial Space and Movement in Joseph Conrad's *Lord Jim.*" *Novel: A Forum on Fiction* 37, no. 3 (Summer 2004): 326–51.

Kuehn, Thomas. *Empire, Islam, and Politics of Difference: Ottoman Rule in Yemen, 1849–1919.* Leiden: Brill, 2011.

Kuhn, Thomas S. "Energy Conservation as an Example of Simultaneous Discovery." In *Critical Problems in the History of Science: Proceedings of the Institute of the History of Science, 1957,* edited by Marshall Clagett, 321–56. Madison: University of Wisconsin Press, 1959.

Kuhnke, LaVerne. *Lives at Risk: Public Health in Nineteenth-Century Egypt.* Berkeley: University of California Press, 1990.

Kurpershoek, P. Marcel. *Oral Poetry and Narratives from Central Arabia.* Vol. 5, *Voices from the Desert.* Leiden: Brill, 2005.

Kurt, Burcu. "Buz Temininde Sanayileşme ve Osmanlı İmparatorluğu'nda Kurulan Buz Fabrikaları" [Industrialization in procurement of ice and ice factories established in the Ottoman Empire]. *Ankara Üniversitesi Osmanli Tarihi Arastirma ve Uygulama Merkezi Dergisi* (Ankara University Journal of Ottoman History Research and Application Center) 30 (Autumn 2011): 73–98.

Laffan, Michael Francis. *Islamic Nationhood and Colonial Indonesia: The Umma below the Winds.* London: Routledge, 2003.

Laitin, David, and Said Samatar. *Somalia: Nation in Search of a State.* Boulder, CO: Westview Press, 1987.

Lamplough, A. O. *Egypt and How to See It.* London, 1908.

The Land We Live In: A Pictorial and Literary Sketch-Book of the British Empire. London, 1856.

Latour, Bruno. *An Inquiry into Modes of Existence: An Anthropology of the Moderns.* Cambridge: Harvard University Press, 2018.

———. *On the Modern Cult of the Factish Gods.* Durham, NC: Duke University Press, 2010.

———. *Pandora's Hope: Essays on the Reality of Science Studies.* Cambridge, MA: Harvard University Press, 1999.

———. *Science in Action: How to Follow Scientists and Engineers through Society.* Cambridge, MA: Harvard University Press, 1988.

Lawless, Richard I. "Recruitment and Regulation: Migration for Employment of 'Adenese' Seamen in the Late Nineteenth and Early Twentieth Centuries." In *New Arabian Studies,* vol. 2, edited by R.L. Tidwell, G. Rex Smith, and J.R. Smart, 75–102. Exeter: University of Exeter Press, 1994.

Levy, Jonathan. *Freaks of Fortune: The Emerging World of Capitalism and Risk in America.* Cambridge: Harvard University Press, 2014.

Levy-Aksu, Neomi. "Institutional Cooperation and Substitution: The Ottoman Police and Justice System at the Turn of the 19th and 20th Centuries." In *Order and Compromise: Government Practices in Turkey from the Late Ottoman Empire to the Early 21st Century,* edited by Marc Aymes, Benjamin Gourisse, and Elise Massicard, 146–68. Leiden: Brill, 2014.

Linebaugh, Peter, and Marcus Rediker. *The Many-Headed Hydra: Sailors, Slaves, Commoners, and the Hidden History of the Revolutionary Atlantic.* London: Verso, 2000.

Lockman, Zachary. "'Worker' and 'Working Class' in pre-1914 Egypt." In *Workers and Working Classes in the Middle East: Struggles, Histories, Historiographies,* edited by Zachary Lockman, 71–111. Albany: State University of New York Press, 1994.

Long, J.G. "Foreign Markets for American Coal." In *United States Bureau of Foreign Commerce, Special Consular Reports* 21. Washington, DC: Government Printing Office, 1900.

Low, Michael Christopher. "The Mechanics of Mecca: The Technopolitics of the Late Ottoman Hijaz and the Colonial Hajj." PhD diss., Columbia University, 2015.

———. "Ottoman Infrastructures of the Saudi Hydro-State: The Technopolitics of Pilgrimage and Potable Water in the Hijaz." *Comparative Studies in Society and History* 57, no. 4 (2015): 942–74.

———. "Unfurling the Flag of Extraterritoriality: Autonomy, Foreign Muslims, and the Capitulations in the Ottoman Hijaz." *Journal of the Ottoman and Turkish Studies Association* 3, no. 2 (2016): 299–323.

Lubin, Isador, and Helen Everett. *The British Coal Dilemma.* New York: The Institute of Economics, 1927.

MacArthur, Robert H., and Edward O. Wilson. *The Theory of Island Biogeography.* Princeton, NJ: Princeton University Press, 1967.

MacDuffie, Allen. *Victorian Literature, Energy, and the Ecological Imagination.* Cambridge: Cambridge University Press, 2014.

MacFarlane, Charles. *Constantinople in 1828: A Residence of Sixteen Months in the Turkish Capital and Provinces.* London: Saunders and Otley, 1829.

Macgregor, John. *Commercial Statistics of America: A Digest of Her Productive Resources, Commercial Legislation, Customs, Tariffs.* London: Whittaker and Co., 1847.

Mahan, Alfred Thayer. *The Influence of Sea Power upon History.* London: Sampson Low Marston & Co., 1890.

———. *The Interest of America in Sea Power: Present and Future.* 1897. Reprint, Freeport, NY: Books for Libraries Press, 1970.

———. *The Persian Gulf and International Relations.* London: Robert Theobald, 1902.

Malm, Andreas. *Fossil Capital: The Rise of Steam-Power and the Roots of Global Warming.* London: Verso, 2016.

———. "The Origins of Fossil Capital: From Water to Steam in the British Cotton Industry." *Historical Materialism* 21, no. 1 (2013): 15–68.

———. *The Progress of the Storm: Nature and Society in a Warming World.* London: Verso, 2018.

———. "'This Is the Hell That I Have Heard Of': Some Dialectical Images in Fossil Fuel Fiction." *Forum for Modern Language Studies* 53, no. 2 (2017): 121–41.

———. "Who Lit This Fire? Approaching the History of the Fossil Economy." *Critical Historical Studies* 2, no. 2 (Fall 2016): 215–28.

Manjapra, Kris. "Plantation Dispossessions: The Global Travel of Agricultural Racial Capitalism." In *American Capitalism: New Histories,* edited by Sven Beckert and Christine Desan, 361–87. New York: Columbia University Press, 2018.

Marey, E. J. *La méthode graphique dans les sciences expérimentales et principalement en physiologie et en médecine.* Paris: Libraire de l'Académie de Médecine, 1878.

Marsden, Reginald. *A Digest of Cases Relating to Shipping and Marine Insurance to the End of 1897, with Supplement of Cases to the End of 1910.* London: Sweet & Maxwell, 1911.

Margariti, Roxani Eleni. *Aden and the Indian Ocean Trade: 150 Years in the Life of a Medieval Arabian Port.* Chapel Hill: University of North Carolina Press, 2007.

Marston, Thomas E. *Britain's Imperial Role in the Red Sea Area, 1800–1878.* Hamden, CT: Shoe String Press Inc., 1961.

Martel, Benjamin. "On Water Ballast." *Scientific American: Supplement* 4, no. 97 (November 10, 1877): 1536–38.

Martin, Frederick. *The History of Lloyd's and of Marine Insurance in Great Britain.* Whitefish, MT: Kessinger, 2010.

Marx, Karl. *The Poverty of Philosophy.* 1847. Reprint, New York: International Publishers, 1963.

Masters, Bruce. *Christians and Jews in the Ottoman Arab World: The Roots of Sectarianism.* Cambridge: Cambridge University Press, 2001.

———. "Hajj." In *The Encyclopedia of the Ottoman Empire,* edited by Gábor Ágoston and Bruce Alan Masters. New York: Facts on File, 2009.

———. *The Origins of Western Economic Dominance in the Middle East: Mercantilism and the Islamic Economy in Aleppo, 1600–1750.* New York: NYU Press, 1988.

Mathew, Johan. *Margins of the Market: Trafficking and Capitalism across the Arabian Sea.* Oakland: University of California Press, 2016.

McCarthy, Mike. *Iron and Steamship Archaeology: Success and Failure on the SS Xantho.* New York: Kluwer Academic, 2002.

McNeill, John. *Something New under the Sun: An Environmental History of the Twentieth-Century World.* New York: W. W. Norton, 2000.

Mehta, Uday Singh. *Liberalism and Empire: A Study in Nineteenth-Century British Liberal Thought.* Chicago: University of Chicago Press, 1999.

Melvin-Koushki, Matthew. "Powers of One: The Mathematicalization of the Occult Sciences in the High Persianate Tradition." *Intellectual History of the Islamicate World* 5, no. 1–22017: 127–99.

Mentz, Steve. *Shipwreck Modernity: Ecologies of Globalization, 1550–1719.* Minneapolis: University of Minnesota Press, 2015.

Miesner, Thomas O., and William L. Leffler. *Oil & Gas Pipelines in Nontechnical Language.* Tulsa, OK: PennWell Corp., 2006.

Mikhail, Alan. *The Animal in Ottoman Egypt.* Oxford: Oxford University Press, 2013.

———. *Nature and Empire in Ottoman Egypt: An Environmental History.* Cambridge: Cambridge University Press, 2011.

———. "Unleashing the Beast: Animals, Energy, and the Economy of Labor in Ottoman Egypt." *American Historical Review* 118, no. 2 (2013): 317–48.

Miller, Michael B. "Pilgrims' Progress: The Business of the Hajj." *Past and Present,* no. 191 (May 2006): 189–228.

Minard, Charles Joseph. *Des tableaux graphiques et des cartes figuratives.* Paris: E. Thunot, 1862.

Minawi, Mustafa. *The Ottoman Scramble for Africa: Empire and Diplomacy in the Sahara and the Hijaz.* Stanford, CA: Stanford University Press, 2016.

Mining: Haulage: The Classic 1907 Mining Engineering Text. 1907. Reprint, Periscope Film LLC, 2008. www.periscopefilm.com/mining-haulage.

Miran, Jonathan. *Red Sea Citizens: Cosmopolitan Society and Cultural Change in Massawa.* Bloomington: Indiana University Press, 2009.

Mirowski, Philip. *More Heat than Light: Economics as Social Physics, Physics as Nature's Economics.* Cambridge: Cambridge University Press, 1991.

Mishra, Saurabh. *Pilgrimage, Politics, and Pestilence: The Haj from the Indian Subcontinent, 1860–1920.* Oxford: Oxford University Press, 2011.

Mitchell, Timothy. *Carbon Democracy: Political Power in the Age of Oil.* London: Verso, 2011.

———. *Colonising Egypt.* California: University of California Press, 1991.

Moore, Gene M., comp. "Newspaper Accounts of the *Jeddah* Affair." In *Lord Jim: Centennial Essays,* edited by Allan Simmons and J. H. Stape, 104–39. Amsterdam: Rodopi, 2003.

Moore, Jason W. *Capitalism in the Web of Life: Ecology and the Accumulation of Capital.* London: Verso, 2016.

Morton, Timothy. *Hyperobjects: Philosophy and Ecology after the End of the World.* Minneapolis: University of Minnesota Press, 2013.

Mouhot, Jean-François. "Past Connections and Present Similarities in Slave Ownership and Fossil Fuel Usage." *Climatic Change* 105, no. 1–2 (2011): 329–55.

Moyn, Samuel. "Rights vs. Duties: Reclaiming Civic Balance." *Boston Review,* May 16, 2016.

Muhammad 'Arif ibn Ahmad Munayyir. *The Hejaz Railway and the Muslim Pilgrimage: A Case of Ottoman Political Propaganda.* Translated by Jacob M. Landau. Detroit: Wayne State University Press, 1971.

Munro, J. Forbes. *Maritime Enterprise and Empire: Sir William Mackinnon and His Business Network, 1823–93.* Woodbridge, Suffolk, UK: Boydell Press, 2003.

Nacar, Can. "Labor Activism and the State in the Ottoman Tobacco Industry." *International Journal of Middle East Studies* 46, no. 3 (2014): 533–51

Najder, Zdzisław. *Joseph Conrad: A Life.* Rochester, NY: Camden House, 2007.

Najm, Zayn al-'Ābidīn Shams al-Dīn. *Būr Saʿīd: Tārīkhuha wa-taṭawwuruha mundhu nash'atihā 1859 ḥatā ʿām 1882.* Cairo: al-Hay'a al-ʿĀmma al-Miṣriyya li'l-Kitāb, 1987.

Nanda, B. R. *Mahatma Gandhi: A Biography.* Oxford: Oxford University Press, 1989.

The Nautical Magazine and Naval Chronicle for 1860: A Journal of Papers on Subjects Connected with Maritime Affairs. London: Simpkin, Marshall and Co., 1860.

The Nautical Magazine and Naval Chronicle for 1873: A Journal of Papers on Subjects Connected with Maritime Affairs. London: Simpkin, Marshall and Co., 1873.

Neocleous, Mark. "The Political Economy of the Dead: Marx's Vampires." *History of Political Thought* 24, no. 4 (2003): 668–84.

Netz, Reviel. *Barbed Wire: An Ecology of Modernity.* Middletown, CT: Wesleyan University Press, 2004.

Nikiforuk, Andrew. *The Energy of Slaves: Oil and the New Servitude.* Vancouver, BC: Greystone Books, 2012.

Nuri, Athnasiyus Aghnatiyus. *Rihlat al-Hind: 1899–1900.* Abu Dhabi: Dar al-Suwaydī lil-Nashr wa-al-Tawzīʿ, 2003.

Nye, David. *American Technological Sublime.* Cambridge, MA: MIT Press, 1994.

Ochsenwald, William. *The Hijaz Railroad.* Charlottesville: University Press of Virginia, 1980.

———. *Religion, Society and the State in Arabia: The Hijaz under Ottoman Control, 1840–1908.* Columbus: Ohio State University Press, 1984.

Onley, James. *The Arabian Frontier of the British Raj: Merchants, Rulers, and the British in the Nineteenth Century Gulf.* Oxford: Oxford University Press, 2007.

Önsoy, Rıfat. "Osmanlı İmparatorluğu'nun Katıldığı İlk Uluslararası Sergiler ve Sergi-i Umumi-i Osmânî (1863 İstanbul Sergisi)." *Belleten* 185 (January 1983).

Otter, Christopher. "Planet of Meat: A Biological History." In *Challenging (the) Humanities,* edited by Tony Bennett. Canberra: The Australian Academy of the Humanities, 2013.

———. *The Vital State: Food Systems, Nutrition Transitions, and the Making of Industrial Britain.* Unpublished manuscript, last modified September 20, 2018.

"Ottoman Archives." Turkish Cultural Foundation. www.turkishculture.org /general/museums/ottoman-archives/ottoman-archives-190.htm?type = 1.

Ovington, John. *A Voyage to Surat in the Year 1689*. New Delhi: Asian Educational Services, 1994.

Owen, Roger. *Cotton and the Egyptian Economy, 1820–1914: A Study in Trade and Development*. Oxford: Clarendon Press, 1969.

Özbaran, Salih. "İstanbul'da Kayıkçılık ve Kayık İşletmeciliği." In *Osmanlı İmparatorluğunda Şehircilik ve UlaşımÜzerine Araştırmalar*, edited by Cengiz Orhonlu, 83–103. İzmir: Ticaret Matbaacılık T.A.Ş., 1984.

Ozden, Canay. "The Pontifex Minimus: William Willcocks and Engineering British Colonialism." *Annals of Science* 71, no. 2 (2014): 183–205.

Pachirat, Timothy. *Every Twelve Seconds: Industrialized Slaughter and the Politics of Sight*. New Haven, CT: Yale University Press, 2013.

Palmer, Sarah. "Current Port Trends in an Historical Perspective." *Journal of Maritime Research* 1, no. 1 (1999): 99–111.

Panjwani, Mariam Dossal. "*Godis, Tolis* and *Mathadis*: Dock Workers of Bombay." In *Dock Workers: International Explorations in Comparative Labour History, 1790–1970*, edited by Sam Davies, Colin J. Davis, David de Vries, Lex Heerma van Voss, Lidewijb Hesselink, and Klaus Weinhauer, 1:425–40. Burlington, VT: Ashgate, 2000.

Parker, G.R. *The Commission of H.M.S. Impeccable, Mediterranean Station, 1901–1904*. London: Westminster Press, 1904.

Pearson, Michael. *The Indian Ocean*. London: Taylor & Francis, 2003.

Perkins, Kenneth J. *Port Sudan: The Evolution of a Colonial City*. Boulder: Westview Press, 1993.

Pinto, Karen C. *Medieval Islamic Maps: An Exploration*. Chicago: University of Chicago Press, 2016.

Pirsson, Joseph P. *Additional and Fresh Evidence of the Practical Working of Pirsson's Steam Condenser*. Washington: Gideon, 1851.

Podobnik, Bruce. *Global Energy Shifts: Fostering Sustainability in a Turbulent Age*. Philadelphia: Temple University Press, 2008.

Polanyi, Karl. *The Great Transformation*. Boston: Beacon Press, 1957.

Pomeranz, Kenneth. *The Great Divergence: China, Europe and the Making of the Modern World Economy*. Princeton, NJ: Princeton University Press, 2000.

Porter, Theodor M. *Trust in Numbers: The Pursuit of Objectivity in Science and in Life*. Princeton, NJ: Princeton University Press, 1996.

Port-Saïd, architectures XIXe-XXe siècles, conçu et réalisé par l'Alliance française de Port-Saïd. Le Caire: IFAO, 2005 (in French and Arabic with introduction also translated into English).

Postans, Mrs. "Characteristics of Aden, with the Passage of the Red Sea." *The Illuminated Magazine* (London) 1, 1843.

Povinelli, Elizabeth A. *Economies of Abandonment: Social Belonging and Endurance in Late Liberalism*. Durham, NC: Duke University Press, 2011.

Proceedings of the American Philosophical Society. Vol. 14. Philadelphia: M'Calla and Stavely, 1876.

Quadri, Syed Junaid A. "Transformations of Tradition: Modernity in the Thought of Muḥammad Bakhīt al-Mutī'ī." PhD diss., McGill University, 2013.

Quataert, Donald. "Limited Revolution: The Impact of the Anatolian Railway on Turkish Transportation and the Provisioning of Istanbul, 1890–1908." *Business History Review* 51, no. 2 (1977): 139–60.

———. *Miners and the State in the Ottoman Empire: The Zonguldak Coalfield, 1822–1920*. New York: Berghahn Books, 2006.

———. *The Ottoman Empire, 1700–1922*. New York: Cambridge University Press, 2000.

Rabinbach, Anson. *The Human Motor: Energy, Fatigue, and the Origins of Modernity*. Berkeley: University of California Press, 1990.

Raj, Kapil. "Mapping Knowledge Go-Betweens in Calcutta, 1770–1820." In *The Brokered World: Go-betweens and Global Intelligence, 1770–1820*, edited by Simon Schaffer, Lissa Roberts, Kapil Raj, and James Delbourgo, 105–50. Sagamore Beach, MA: Science History Publications, 2009.

Rediker, Marcus. *The Slave Ship: A Human History*. New York: Penguin Books, 2014.

Reese, Scott S. *Imperial Muslims: Islam, Community and Authority in the Indian Ocean, 1839–1937*. Edinburgh: Edinburgh University Press, 2018.

———. "Patricians of the Benaadir: Islamic Learning, Commerce and Somali Urban Identity in the Nineteenth Century." PhD diss., University of Pennsylvania, 1996.

Reid, Anthony. "An 'Age of Commerce' in Southeast Asia History." *Modern Asian Studies* 24, no. 1 (February 1990): 1–30.

———. "Understanding Melayu (Malay) as a Source of Diverse Modern Identities." *Journal of Southeast Asian Studies* 32, no. 3 (2001): 295–313.

Report from the Select Committee on Steam Communication with India; Together with the Minutes of Evidence, Appendix and Index. London: Ordered by the House of Commons, Parliament, 1837.

Report from the Select Committee on Steam Navigation to India; With the Minutes of Evidence, Appendix and Index. London: Ordered by the House of Commons, Parliament, 1834.

Reports from Committees: Eighteen Volumes; East India Company's Affairs (Parliamentary Papers). Vol. 10, pt. 2. London: H.M. Stationery Office, 1832.

Rida, Muhammad Rashid. *Ta'rikh al-'ustadh al-Imam al-Shaykh Muhammad 'Abduh*. Cairo: al-Manar, 1931.

Rivers, P.J. "Negeri below and above the Wind: Malacca and Cathay." *Journal of the Malaysian Branch of the Royal Branch of the Royal Asiatic Society* 78, no. 2 (2005): 1–32.

Roff, William. "Sanitation and Security: The Imperial Powers and the Nineteenth Century Hajj." *Arabian Studies* 6 (1982): 143–60.

Rudwick, Martin J.S. *Bursting the Limits of Time: The Reconstruction of Geohistory in the Age of Revolution*. Chicago: University of Chicago Press, 2007.

———. "Charles Lyell's Dream of a Statistical Paleontology." In *Lyell and Darwin, Geologists: Studies in the Earth Sciences in the Age of Reform*. Farnham: Ashgate, 2005.

———. *Worlds before Adam: The Reconstruction of Geohistory in the Age of Reform*. Chicago: University of Chicago Press, 2010.

Sadeh, Roy Bar. "Constructing a Pan-Islamic Sphere: Muhammad Rashid Rida's Journey to British India in 1912." Unpublished paper, Tel Aviv University, 2015.

———. "Debating Gandhi in al-Manar during the 1920s and 1930s." *Comparative Studies of South Asia, Africa and the Middle East* 38, no. 3 (2018): 491–507.

Sanderson, G. N. "The Nile Basin and the Eastern Horn, 1870–1908." In *The Cambridge History of Africa*. Vol. 6, *From 1870 to 1905*, edited by Roland Oliver and G. N. Sanderson, 592–679. Cambridge: Cambridge University Press, 1985.

Sariyannis, Marinos. "Ruler and State, State and Society in Ottoman Political Thought." *Turkish Historical Review* 4 (2013): 92–126.

Sarruf, Ya'qub. *Fatat Misr, Riwayah.* Cairo: Matba'at al-muqtataf, 1922.

Sartori, Andrew Stephen. *Liberalism in Empire: An Alternative Story.* Oakland: University of California Press, 2014.

Savage, M. "Discipline, Surveillance and the 'Career': Employment on the Great Western Railway 1833–1914." In *Foucault, Management and Organization Theory,* edited by Alan McKinlay and Ken Starkey, 65–93. London: Sage Publications, 1998.

Schabas, Margaret. *The Natural Origins of Economics.* Chicago: University of Chicago Press, 2007.

Schick, Ernesto. *Railway Flora, or Nature's Revenge on Man.* Berlin: Humboldt Books, 2015.

Schivelbusch, Wolfgang. *The Railway Journey: The Industrialization of Time and Space in the Nineteenth Century.* Berkeley: University of California Press, 2014.

Searight, Sarah. "The Charting of the Red Sea." *History Today* 53, no. 3 (March 2003): 40–46.

Seddon, Mohammad Siddique. *The Last of the Lascars: Yemeni Muslims in Britain, 1836–2012.* Leicestershire: Kube, 2014.

Sekula, Allan. *Fish Story.* London: Mack, 2018.

Şengör, Celal. "Osmanlı'nın İlk Jeoloji Kitabı ve Osmanlı'da Jeolojinin Durumu Hakkında Öğrettikleri." *Osmanlı Bilimi Araştırmaları* 11, nos. 1–2 (2009–10): 119–58.

Şentürk, Ömer Faruk. *Charity in Islam: A Comprehensive Guide to Zakat.* Clifton, NJ: The Light, 2007.

Shapin, Steven. *The Scientific Revolution.* Chicago: University of Chicago Press, 2018.

Shapin, Steven, and Simon Schaffer. *Leviathan and the Air Pump: Hobbes, Boyle, and the Experimental Life.* Princeton, NJ: Princeton University Press, 2011.

Shefer-Mossensohn, Miri. *Science among the Ottomans: The Cultural Creation and Exchange of Knowledge.* Austin: University of Texas Press, 2015.

Sherwood, Marika. "Race, Nationality and Employment among Lascar Seamen, 1660 to 1945." *New Community* 17 (1991): 229–44.

Sibum, Heinz Otto. "Reworking the Mechanical Value of Heat: Instruments of Precision and Gestures of Accuracy in Early Victorian England." *Studies in History and Philosophy of Science* 26, no. 1 (1995): 73–106.

Sieferle, Rolf Peter. *The Subterranean Forest: Energy Systems and the Industrial Revolution*. Translated from the German original by Michael P. Osman. Cambridge: White Horse Press, 2001.

Simmons, Allan, and J.H. Stape, eds. *Lord Jim: Centennial Essays*. Amsterdam: Rodopi, 2003.

Singh, Satyindra. *Blueprint to Bluewater, the Indian Navy, 1951–1965*. New Delhi: Lancer International, 1992.

Singha, Radhika. "Passport, Ticket, and India-Rubber Stamp: 'The Problem of the Pauper Pilgrim' in Colonial India c. 1882–1925." In *The Limits of British Colonial Control in South Asia: Spaces of Disorder in the Indian Ocean Region*, edited by Ashwini Tambe and Harald Fischer-Tiné, 59–93. New York: Routledge, 2008.

Slight, John. "The Hajj and the Raj: From Thomas Cook to Bombay's Protector of Pilgrims." In *The Hajj: Collected Essays,* edited by Venetia Porter and Liana Saif, 115–21. London: British Museum Press, 2013.

Smeaton, John. *A Narrative of the Building and a Description of the Construction of the Edystone Lighthouse with Stone: To Which Is Subjoined, an Appendix, Giving Some Account of the Lighthouse on the Spurn Point, Built upon a Sand*. London: by the author, 1791.

Smil, Vaclav. "Eating Meat: Evolution, Patterns, and Consequences." *Population and Development Review* 28, no. 4 (2002): 599–639.

Smith, Adam. *The Wealth of Nations*. 1776. Reprint, London: Penguin Books, 1979.

Smith, Crosbie. *Coal, Steam and Ships: Engineering, Enterprise and Empire on the Nineteenth-Century Seas*. Cambridge: Cambridge University Press, 2018.

———. "Force, Energy, and Thermodynamics." In *The Modern Physical and Mathematical Sciences*. Vol. 5 of *The Cambridge History of Science*, edited by Mary Jo Nye, 289–310. Cambridge: Cambridge University Press, 2002.

———. *The Science of Energy: A Cultural History of Energy Physics in Victorian Britain*. London: Athlone, 1998.

Smith, Crosbie, and M. Norton Wise. *Energy and Empire: A Biographical Study of Lord Kelvin*. Cambridge: Cambridge University Press, 1989.

Smith, S., ed. *The Red Sea Region: Sovereignty Boundaries and Conflict, 1839–1967*. Cambridge: Cambridge Archive Editions, 2008.

Snouck Hurgronje, Christiaan. *Mekka: Aus dem heutigen Leben*. 2 vols. Den Hague: Martinus Nijhoff, 1888–1889.

———. *Mekka in the Latter Part of the Nineteenth Century, 1885–1889*. Translated by James Henry Monahan. Leiden: Brill, 1931.

Snouck Hurgronje, Christiaan, and E. Gobée. *Ambtelijke adviezen van C. Snouck Hurgronje: 1889–1936, Deel 3*. 's-Gravenhage: Nijhoff, 1965.

Spencer, David. *The Political Economy of Work*. London: Routledge, 2009.

Spooner, Frank C. *Risks at Sea: Amsterdam Insurance and Maritime Europe, 1766–1780*. Cambridge: Cambridge University Press, 1983.

Stausberg, Michael. *Religion and Tourism: Crossroads, Destinations and Encounters*. London: Routledge, 2011.

"Steel Pipes Conquer Desert Waste." *Popular Mechanics,* May 1926, 732–35.

Stolz, Daniel. "'Impossible to Arrive at an Accurate Estimate': Accounting for the Ottoman Debt, 1873–1881." Unpublished manuscript, last modified January 21, 2019.

———. "The Voyage of the Samannud: Pilgrimage, Cholera, and Empire on an Ottoman-Egyptian Steamship Journey in 1865–66." *International Journal of Turkish Studies* 23, nos. 1–2 (2017): 1–18.

Sussman, Herbert. *Victorian Technology: Invention, Innovation, and the Rise of the Machine.* Santa Barbara, CA: ABC-CLIO, 2009.

Tagliacozzo, Eric. "Crossing the Great Water: The Hajj and Commerce from Pre-Modern Southeast Asia." In *Religion and Trade: Cross-Cultural Exchanges in World History, 1000–1900,* edited by Francesca Trivellato, Leor Halevi, and Cátia Antunes, 216–35. New York: Oxford University Press, 2014.

———. *The Longest Journey: Southeast Asians and the Pilgrimage to Mecca.* New York: Oxford University Press, 2013.

Ṭahṭāwī, Rifāʻah Rāfiʻ. *Taʻrīb kitāb al-muʻallim Firād fī al-maʻādin al-nāfiʻah li-tadbīr maʻāyish al-khalāyiq.* Būlāq: Maṭbaʻat Būlāq, 1833. http://nrs.harvard.edu/urn-3:FHCL:24426959.

Taylor, James. "Private Property, Public Interest, and the Role of the State in Nineteenth-Century Britain: The Case of the Lighthouses." *Historical Journal* 44, no. 3 (2001): 749–71.

Taylor, Richard Cowling. *Statistics of Coal: The Geographical and Geological Distribution of Fossil Fuel.* Philadelphia: J.W. Moore, 1848.

The Ottoman Land Laws: With a Commentary on the Ottoman Land Code, of 7th Ramadan 1274. April 21, 1858.

Thobie, Jacques. *L'administration générale des phares de l'Empire ottoman et la société Collas et Michel, 1860–1960: Un siècle de coopération économique et financière entre la France, l'Empire ottoman et les états successeurs.* Paris: L'Harmattan, 2004.

Thomas, David Alfred. *The Growth and Direction of Our Foreign Trade in Coal during the Last Half Century.* London: Harrison and Sons, 1903.

Thomas, Martin. "Managing the Hajj: 1918–1930." In *Railways and International Politics: Paths of Empire, 1848–1945,* edited by Keith Neilson and Thomas K. Otte. London: Routledge, 2006.

Thompson, E.P. *The Making of the English Working Class.* Vintage Books, 1963.

———. "Time, Work-Discipline, and Industrial Capitalism." *Past and Present* no. 38 (December 1967): 56–97.

Tignor, Robert. *Modernization and British Colonial Rule in Egypt, 1882–1914.* Princeton, NJ: Princeton University Press, 1966.

Timbs, John. *The Year-Book of Facts in Science and Art.* London: David Bogue, 1850.

Tok, Alaaddin. "From Wood to Coal: Energy Economy in Ottoman Anatolia and the Balkans (1750–1914)." PhD diss., Boğaziçi University, Istanbul, 2017.

———. "Fuelling the Empire: Energy Economics in the Ottoman Anatolia and the Balkans (1750–1914)." PhD diss., Boğaziçi University, Istanbul, 2018.

———. *The Ottoman Mining Sector in the Age of Capitalism: An Analysis of State-Capital Relations (1850–1908)*. Saarbrücken: Lambert Academic Publishing, 2011.

Tolf, Robert W. *The Russian Rockefellers: The Saga of the Nobel Family and the Russian Oil Industry*. Stanford, CA: Hoover Institution Press, 1976.

Torrens, Hugh. *The Practice of British Geology, 1750–1850*. Farnham: Ashgate, 2002.

Travis, Anthony S. *The Rainbow Makers: The Origins of the Synthetic Dyestuffs Industry in Western Europe*. Bethlehem, PA: Lehigh University Press, 1993.

Trocki, Carl A. "Singapore as a Nineteenth Century Migration Node." In *Connecting Seas and Connected Ocean Rims: Indian, Atlantic, and Pacific Oceans and China Seas Migrations from the 1830s to the 1930s*, edited by Donna R. Gabaccia and Dirk Hoerder, 198–224. Leiden: Brill, 2011.

Tsing, Anna. *The Mushroom at the End of the World: On the Possibility of Life in Capitalist Ruins*. Princeton, NJ: Princeton University Press, 2015.

Tucker, Catherine M. *Coffee Culture: Local Experiences, Global Connections*. London: Routledge, 2011.

UK Parliament, House of Commons. *Report from the Select Committee on Steam Navigation to India: with the Minutes of Evidence, Appendix and Index*. 1834.

United States Office of Naval Intelligence. *Coaling, Docking, and Repairing Facilities of the Ports of the World, with Analyses of Different Kinds of Coal*. 4th ed. Washington: Government Printing Office, 1909.

Urquhart, G.D. *Dues and Charges on Shipping in Foreign Ports: A Manual of Reference for the Use of Shipowners, Shipbrokers, and Shipmasters*. London: George Philip, 1872.

Vandal, Albert. *Une ambassade française en Orient sous Louis XV: La mission du Marquis de Villeneuve*. Paris: E. Plon, Nourrit, 1887.

Wagner, Mark S. *Jews and Islamic Law in Early 20th-Century Yemen*. Bloomington: Indiana University Press, 2015.

Weeks, Kathi. *The Problem with Work: Feminism, Marxism, Antiwork Politics, and Postwork Imaginaries*. Durham, NC: Duke University Press, 2001.

Wegerich, Alexis. "Digging Deeper: Global Coal Prices and Industrial Growth, 1840–1960." PhD diss., University of Oxford, 2016.

Weir, William Viscount. *The Weir Group: The History of a Scottish Engineering Legend*. London: Profile Books, 2013.

Whately, Mary L. *Letters from Egypt to Plain Folks at Home*. London: Seeley, Jackson & Halliday, 1879.

Wheeler, Brannon. *Mecca and Eden: Ritual, Relics, and Territory in Islam*. Chicago: University of Chicago Press, 2006.

White, David L. "Parsis in the Commercial World of Western India, 1700–1750." *Indian Economic and Social History Review* 24, no. 2 (1987): 183–203.

White, John H., Jr. *Wet Britches and Muddy Boots: A History of Travel in Victorian America*. Bloomington: Indiana University Press, 2012.

White, Sam. *The Climate of Rebellion in the Early Modern Ottoman Empire*. Cambridge: Cambridge University Press, 2011.

Wick, Alexis. *The Red Sea: In Search of Lost Space*. Oakland: University of California Press, 2016.

Willcocks, William. *Egyptian Irrigation*. London: E. & F.N. Spon, 1899.

Willis, John M. *Unmaking North and South: Cartographies of the Yemeni Past, 1857–1934*. London: Hurst, 2012.

Wilson, Frances. *How to Survive the Titanic: The Sinking of J. Bruce Ismay*. New York: Bloomsbury, 2011.

Wilson, Gordon. *Alexander McDonald: Leader of the Miners*. Aberdeen: Aberdeen University Press, 1982.

Wilson, J.H. *On Steam Communication between Bombay and Suez, with an Account of the Hugh Lindsay's Four Voyages*. Bombay: American Mission Press, 1833.

Wink, Andre. *Al-Hind: The Making of the Indo-Islamic World*. Leiden: Brill, 2002.

Wise, Henry. *An Analysis of One Hundred Voyages to and from India, China &c*. London: J.W. Norie and Co., 1839.

Wishnitzer, Avner. "Into the Dark: Power, Light and Nocturnal Life in 18th Century Istanbul." *IJMES* 46 (2014): 513–31.

———. *Reading Clocks Alla Turca: Time and Society in the Late Ottoman Empire*. Chicago: University of Chicago Press, 2016.

———. "Shedding New Light: Outdoor Illumination in Late Ottoman Istanbul." In *Urban Lighting, Light Pollution and Society*, edited by Josiane Meier, Ute Hasenöhrl, Katharina Krause and Merle Pottharst, 66–88. New York; London: Routledge, 2015.

Witkam, Jan Just. "The Battle of the Images: Mekka vs. Medina in the Iconography of the Manuscripts of al-Jazūlī's Dala'il al-Khayrat." *Beiruter Texte und Studien* 111 (2007): 67–82.

———. "The Islamic Pilgrimage in the Manuscript Literatures of Southeast Asia." In *The Hajj: Collected Essays*, edited by Venetia Porter and Liana Saif, 214–23. London: British Museum Press, 2013.

Wolfe, Michael. *One Thousand Roads to Mecca: Ten Centuries of Travelers Writing about the Muslim Pilgrimage*. New York: Grove Press, 1998.

Wolmar, Christian. *Railways and the Raj: How the Age of Steam Transformed India*. London: Atlantic Books, 2017.

Woodhouse, C.M. *The Battle of Navarino*. London: Hodder and Stoughton, 1965.

Woods, Rebecca J.H. "Breed, Culture, and Economy: The New Zealand Frozen Meat Trade, 1880–1914." *Agricultural History Review* 60, no. 2 (2012): 288–302.

———. "Nature and the Refrigerating Machine: The Politics and Production of Cold in the Nineteenth Century." In *Cryopolitics: Frozen Life in a Melting World*, edited by Joanna Radin and Emma Kowal, 89–116. Cambridge, MA: MIT Press, 2017.

Wright, Tim. *Coal Mining in China's Economy and Society, 1895–1937*. New York: Cambridge University Press, 1984.

Wu, Shellen Xiao. *Empires of Coal: Fueling China's Entry into the Modern World Order, 1860–1920*. Stanford, CA: Stanford University Press, 2015.

Wyse, Sir Thomas. *Education Reform; or, The Necessity of a National System of Education.* London: Longman, Rees, Orme, Brown, Green & Longman, 1836.

Yalçınkaya, M. Alper. *Learned Patriots: Debating Science, State and Society in the Nineteenth-Century Ottoman Empire.* Chicago: University of Chicago Press, 2015.

Yener, Emir. *From the Sail to the Steam: Naval Modernization in the Ottoman, Russian, Chinese and Japanese Empires, 1830–1905.* Saarbrücken: Lambert Academic Publishing, 2010.

Yıldırım, Kadir. *Osmanlılar'da İşçiler (1870–1922).* Istanbul: İletişim, 2013.

Ziad, Homayra. "The Return of Gog: Politics and Pan-Islamism in the Hajj Travelogue of 'Abd al-Majid Daryabadi." In *Global Muslims in the Age of Steam and Print,* edited by James L. Gelvin and Nile Green, 227–49. California: University of California Press, 2014.

Zilfi, Madeline C. *Women and Slavery in the Late Ottoman Empire: The Design of Difference.* Cambridge: Cambridge University Press, 2010.

Zimmerman, Erich W. "Why the Export Coal Business of America Should Be Built Up—III." *Coal Age* 19 (1921).

Index

Founded in 1893,
UNIVERSITY OF CALIFORNIA PRESS
publishes bold, progressive books and journals
on topics in the arts, humanities, social sciences,
and natural sciences—with a focus on social
justice issues—that inspire thought and action
among readers worldwide.

The UC PRESS FOUNDATION
raises funds to uphold the press's vital role
as an independent, nonprofit publisher, and
receives philanthropic support from a wide
range of individuals and institutions—and from
committed readers like you. To learn more, visit
ucpress.edu/supportus.